Praise for *Product Realization: Going from One to a Million*

Prof. Thornton does an excellent job demystifying the process required to take a working prototype all the way through high-volume manufacturing. In her book, she sheds light on the "unknown unknowns" of manufacturing and clearly explains each step of the journey based on her extensive first-hand experience. All companies building high-volume products should read this book.

Scott Miller
CEO, Dragon Innovation

Dr. Thornton has put together nothing short of a masterpiece in *Product Realization*. Nearly every hardware startup founding team I've worked with is lost when moving from a prototype to full scale production. Thornton provides a tactical roadmap for how to get from here to there, profitably. Do yourself a favor and keep this book within an arm's reach as you scale!

Erica Iannotti
Growth Strategist, Tech to Market Advisor,
Serial Entrepreneur Founder, Manufacturing Corps
& Hardware Scaleup

Product Realization picks up where so many product design books end. Prof. Thornton clearly explains what it takes to bring a product from completion of the concept development stage through the necessary steps of pilot testing, manufacturing planning, tooling design, quality management, supply chain ramp-up, and full-scale production. Here is the book that explains it all — chock full of shop-floor wisdom, fascinating stories and compelling examples.

Steven Eppinger
Professor of Management Science
and Engineering Systems, Massachusetts
Institute of Technology

Anna Thornton's book is the book we have always wanted for our hardware entrepreneurs. Most books about launching new products focus on the software and services side. A lot of the advice in those books does not translate directly to hardware entrepreneurship. Hardware is hard and teams need to learn how to work around the unique challenges of managing the transition from prototyping to mass production, and managing hardware operations with its unique cash flow challenges and impact on unit and overall economics. This book is a go-to resource for new and experienced hardware teams to help them plan for and execute a new hardware startup successfully and avoid common pitfalls. Highly recommended.

Bill Aulet
Managing Director, The Martin Trust Center for MIT
Entrepreneurship & Professor of the Practice, MIT Sloan School
Author of *Disciplined Entrepreneurship*

Taking a new product from idea to full scale production when it involves manufacturing a piece of hardware or tangible product is a complex undertaking. The advent of 3D printing and online fundraising sites, like Kickstarter, have made it easier for folks to build realistic prototypes and raise significant capital for interesting product ideas. However, as many failed startups can attest, taking those prototypes to scale and delivering it to customers are a greater challenge than most imagine. Anna Thornton's book "Product Realization: going from One to a Million," is exactly what those startups needed. She covers the process from concept to production, breaking it into key steps and provides detailed checklists, practical examples, and documents that illuminate the way to successful implementation. In reality, a product needs to not only address the user's needs but also be easy to manufacture, assemble, ship and support. This book outlines the planning processes from product design and planning to manufacturing and production planning and eventual distribution and support. It should be a must-read reference for anyone who intends to successfully build a product and bring it to market.

Desh Deshpande
Entrepreneur & Life Member of MIT Corporation

Design courses help students design and develop a great prototype. Business classes help the student to consider how the design can be profit making. And manufacturing classes help the student understand the many ways a product can be mass produced. However, there is a critical gap that an designer/entrepreneur must know that to make the leap from a single prototype to something that is actually mass producible using modern global strategies. This book fills this gap in an engaging and concrete way based on Prof Thornton's deep experience in industry.

Maria Yang
Professor of Mechanical Engineering
and Engineering Systems, MIT

Anna Thornton's Product Realization: Going from One to a Million contains the critical information and roadmap hardware entrepreneurs need as they take their concepts from prototype to production. The emergence of new prototyping tools such as 3D printers, desktop mills, and laser cutters as well as electronics such as the Arduino and Raspberry Pi has spurred on a renaissance in product entrepreneurship. Prototyping product concepts and gathering customer feedback has never been easier. Yet the knowledge required for transitioning an idea from prototype to production is scattered and rarely covered in sufficient depth. This book finally addresses that gap and is the resource product developers have been desperately lacking. I look forward to recommending Anna's book for my Product Entrepreneurship courses at Cornell and Hardware Programs at Rev: Ithaca Startup Works.

Ken Rother
Managing Director eLab and Visiting Lecturer
of Management, Johnson Graduate School of Management

An excellent, practical guide for first time entrepreneurs building physical world products. Anna draws from a variety of real world situations to provide advice in a format that is easily digested and applied to a wide range of industries.

Laila Partridge
Managing Director, STANLEY+Techstars Accelerator

Product Realization

Product Realization

GOING FROM ONE TO A MILLION

Anna C. Thornton

Illustrations by Karyn Knight Detering

Registered Office
John Wiley & Sons, Inc., 111 River Street, Hoboken, NJ 07030, USA

Editorial Office
The Atrium, Southern Gate, Chichester, West Sussex, PO19 8SQ, UK

For details of our global editorial offices, customer services, and more information about Wiley products visit us at www.wiley.com.

Wiley also publishes its books in a variety of electronic formats and by print-on-demand. Some content that appears in standard print versions of this book may not be available in other formats.

Limit of Liability/Disclaimer of Warranty

Library of Congress Cataloging-in-Publication Data
Names: Thornton, Anna C., 1968- author. | Detering, Karyn Knight,
    illustrator.
Title: Product realization : going from one to a million / Anna C. Thornton;
    illustrations by Karyn Knight Detering.
Description: Hoboken, NJ, USA : Wiley, 2021. | Includes bibliographical
    references and index.
Identifiers: LCCN 2020030613 (print) | LCCN 2020030614 (ebook) | ISBN
    9781119649533 (hardback) | ISBN 9781119649663 (adobe pdf) | ISBN
    9781119649656 (epub)
Subjects: LCSH: New products.
Classification: LCC TS170 .T485 2021 (print) | LCC TS170 (ebook) | DDC
    658.5/75--dc23
LC record available at https://lccn.loc.gov/2020030613
LC ebook record available at https://lccn.loc.gov/2020030614

Cover Design: Wiley
Cover Image: Karyn Knight Detering

Set in 10/12pt and ITC Garamond Std by SPi Global, Chennai, India

10 9 8 7 6 5 4 3 2 1

This book is dedicated to my husband, Robbie, and my daughters Alexandra and Chloe.

CONTENTS

ACKNOWLEDGEMENTS

I want to acknowledge the numerous people who helped make this book happen. Most importantly, my daughters and husband who have put up with me writing, editing, and taking over the kitchen table. Thank you to Karyn Knight Detering whose illustrations helped me organize my thoughts and keep my sense of humor. To my father Roy Thornton and editor Céilidh Erickson, who read every word made this a much better book. This book wouldn't have been written without the encouragement and prodding of Elaine Chen and Steven Eppinger. BU College of Engineering gave me the space and resources to develop the course on Product Realization. My collegues in 730 – Gerry Fine, Bill Hauser, and Greg Blonder – have been great sounding boards and helped me become a better teacher. Thanks also to Ken Rother, Clive Bolton, Ben Flaumenhaft, and Steve Hodges, who all contributed their expertise and experiences. A shout out to the team at Dragon including Scott Miller and Bill Drislane from whom I learned a lot. Thanks to all of the clients I have worked with, especially the people at Boeing, SRAM, and Fresenius Kabi. A special thanks to my friends Sarah and Rita, my mother, and coaches (Rita Allen, Cecile J. Klavens, and Susan Farina) who supported and encouraged me. Thanks to all those who wrote vignettes and supplied pictures. And finally, to all of my students, who read my first very rough drafts and asked hard questions, you are why I love to teach.

Please visit the website www.productrealizationbook.com for additional references and resources.

Chapter 1
Introduction

'THE PATH AHEAD'

1. INTRODUCTION

2. ARE YOU READY TO START?

3. PRODUCT REALIZATION PROCESS

4. PRODUCT MGMT

Product Realization: Going from one to a million, First Edition. Anna C. Thornton.
© 2021 John Wiley & Sons, Inc. Published 2021 by John Wiley & Sons, Inc.

New technology and new products have the potential to transform our lives and our society. Much is written about how to get a spark of an idea and translate that into a prototype and an initial business plan. However, surprisingly little is written about the thousands of complicated steps required to get from that prototype to a finished product in the hands of the customer. Unfortunately, teams almost always underestimate the pain, work, time, and resources involved. As a result, many companies launch new products late, over budget, and with substandard quality.

Product realization (also called launch, transition to production, piloting, or production ramp) starts when the product development team has a looks-like/works-like prototype, has defined the product geometry and material, has specified manufacturing methods, and is ready to produce at volume. Most groups believe that if the prototype works and there is a market, it will only take a few months to manufacture the product and start selling it. Whether the new product is a small widget or a complicated aircraft, many products arrive in the market later than anticipated. Many products also arrive with fewer features than planned or are over budget. There are invariably more complications and costs than the team initially predicted. By its very nature, product realization is an iterative, painful, but ultimately rewarding process.

There are only two near-certainties in product realization: there will be more work than teams plan for, and almost nothing will be done perfectly the first time. Parts will not come out of the mold as expected, packaging will fail to protect products from breaking, and a supplier will not ship a critical part on time.

This book is designed to help students, engineers, start-ups, and organizations navigate the complex and highly interrelated activities of getting a product into production. This book is not intended to help you come up with a brilliant product idea or market it – there are enough of those books. By understanding the road ahead with all its potholes and detours, teams will better anticipate potential problems before they significantly compromise their business plan. The lessons in this book were gleaned from experiences with over a hundred companies ranging from zero revenue start-ups to multi-billion-dollar companies. While on the surface, the product realization process looks very different for an aircraft vs. a medical device vs. a new drone, most industries use similar methods, principles, and documents. Independent of size, every company must define the product, design their production system, and get everything to work while balancing the competing goals of cost, quality, and schedule.

1.1 Examples

The launch of the Tesla Model 3 appeared in over 500 *New York Times* articles from January 2017 to May 2018. Since Tesla announced the Model 3, it has become painfully apparent that Elon Musk and his team significantly underestimated the time it would take to bring a high-volume car with dramatically

new technology to the market, while at the same time building a highly automated manufacturing plant. In April 2017, Tesla's market valuation of 50.9 billion USD was higher than that of General Motors, and Tesla promised production of over 500,000 cars in 2018. However, by the final week of March 2018, Tesla had only produced 2,000 Model 3 vehicles. By mid-May of 2018, Tesla had shut down production to address critical production issues. Tesla increased production dramatically but by Q3 2019, they were still only producing at an annualized rate of around 319,000 [1, 2]. While Tesla has not been forthcoming about the exact reasons for the delays in the production ramp, Tesla has hinted at bottlenecks, supplier delays, delivery challenges, quality issues, and over-automation. Tesla is not unique in its struggles. Other highly publicized delays include:

- The Joint Strike Fighter (JSF) contract (now F-35)[1] award to Lockheed Martin was announced in 2001 with a plan for combat-ready aircraft by 2010. The F-35 has been significantly over budget and behind schedule. It is likely to cost over one trillion USD over the life of the program. As of the publication of this book, Lockheed Martin was still struggling with critical technical deficiencies as they got ready to increase production rates significantly [3].
- The Boeing 787 was plagued with delays due to documentation errors, supplier delays, assembly errors, supply chain issues, and battery quality issues. The initial cost was budgeted at $6 billion, but it has been estimated that the total cost was probably closer to $32 billion [4].
- GTAT's attempt to mass-produce sapphire screens for Apple was plagued with production and yield issues. The yield issues were likely a contributor to the bankruptcy of the company in 2014 [5].

Problems with product realization are not unique to large companies. Companies such as GlowForge (a laser cutter) and Coolest Cooler (a cooler with a battery powered blender), that launched their products on crowdsourcing platforms, have been years late in delivery or never met all of their promised deliveries. Figure 1.1 shows an example of the accumulated delays in one crowdsourced product that raised close to 3 million USD. The original launch promised delivery in 7 months, but the first units did not ship until 27 months after the crowd-sourced campaign. The timing of each announcement is shown by a horizontal bar with an arrow indicating the delay in the promised product delivery date. The company sent out an update 33 months after their launch, closing the company and apologizing to their customers with excuses about tight finances and increased prices. At the time this book was written, not all of the backers had their products, and the company had not sent an update for six months.

---

[1]Image source: Defense Visual Information Distribution Service. Public Domain Photo by A1C Brooke Moeder, *Luke F-35 surpasses 35K sortie milestone*. The appearance of US Department of Defense (DoD) visual information does not imply or constitute DoD endorsement.

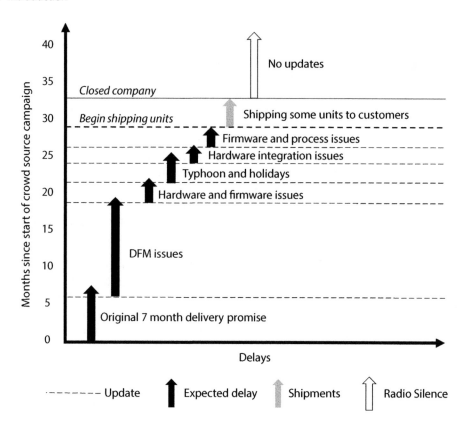

FIGURE 1.1
Example of delays in promised delivery dates

An article by Jensen and Özkil [6] found that over 40% of the Kickstarter products that they studied were more than a year late, and of those, half never delivered products. Jensen and Özkil found various reasons for these delays, but the most common reasons were issues in product delivery, issues found in quality testing late in production, and failure to ensure that the product was designed for manufacturing (DFM).

Each of the companies in Table 1.1 failed for its own reasons [7]. However, several themes emerge when looking at them as a group:

- **The technology was not ready for mass production.** Many product introductions failed because companies started product realization without ensuring the technology was mature enough. The prototype worked, but they could not produce in volume. If they were able to ship products, they delivered the product either at a lower quality than promised or with missing features.
- **The production system was not mature enough to support mass production.** While the product technology was ready, the methods to produce at volume were not. Some companies failed to develop a reliable supply chain, other companies

Product	History
CST-01 (Central Standard Timing)	Raised over $1 million USD in crowdfunding pledges in 2013 to build the "world's thinnest watch." Failed to deliver because (i) the technology was not stable enough, and (ii) they did not have the skills to make the product.
Elio P4 Scooter (Elio Motors)	Raised $17 million USD, promising to launch a three-wheeled electric autocycle by 2014. As of 2019, the company had yet to deliver any products.
Zano drone (Toquing Group)	The promise of a tiny camera drone helped the company raise $4 million USD on Kickstarter in 2015, but the company soon filed for bankruptcy. Investigation revealed that the drone was never operational.
Coolest Cooler	This was the most-funded Kickstarter campaign of 2014. Coolest was able to ship a small number of products but ran into funding issues before it was able to fulfill all orders.

Table 1.1
Examples of start-up products that have failed
Source: From [7]

were unable to find reliable manufacturing partners, while others were not able to produce with consistent quality.

- **It took too much time and too much money to get through the launch process.** Many times failed companies ran out of cash before mass production, and at other times, the lack of funds caused the product to be launched with sub-par quality, leading to poor customer satisfaction and low follow-on sales.

Inset 1.1: Icons

 Throughout the text, you will see versions of these characters that were inspired by two sources: Mingyur Rinpoche's guides to mindful meditation and Ted Urban's TED talk on "Inside the mind of a master procrastinator" [8, 9]. The image is intended to highlight problems to avoid because of their significant downstream consequences. In addition, a number of other icons will be used to help you navigate the text.

	Terms and jargon are used during product realization		Chapter summary and takeaways
	Checklists to make sure you have completed everything		Documents you will need to create to support the product realization process

1.2 Building Ten Thousand is Very Different from Building One

 Initial product design is about getting one item to work once under one set of conditions. Product realization is about getting a million to work every time under a million different conditions. A single prototype can be hand-built by expert makers in a few weeks. To impress investors or faculty, the prototype only has to work once for the camera. The single prototype is typically operated by an expert who knows the quirks or bugs of the device and can avoid them.

By contrast, producing at scale and selling into the market involves building thousands of identical products that can be used by any customer in a wide range of environments. Manufacturing at high volumes is challenging because manufacturing processes introduce many sources of variation, all of which conspire to derail production, testing, shipping, and the customer's experience. Additionally, customers often use the products in unpredictable ways with unpredictable results. Finally, as products go from prototype to mass-produced product, the number of people and organizations involved grows exponentially, further complicating the process and increasing the opportunities for defects and miscommunication.

A successful product must solve an actual customer problem, be of high quality, and be delivered to the consumer on time and at the right price. Launching products requires a complex interplay of activities, including:

- Making sure the product is ready for production.
- Assessing product and process safety and liability issues.
- Addressing environmental issues such as hazardous material handling and waste disposal.
- Designing the production system to build the product consistently at low cost.
- Running several pilot plant builds to test the engineering, design, and manufacturing system.
- Building a logistics system to get the materials to the factory on time and deliver the product to the customer.
- Ensuring the product conforms to legal requirements.
- Defining a quality control system to test for issues before products get into the hands of the customer.
- Creating a customer support system.

Successful product realization requires coordinating a large number of activities to define, test, and deploy all of the systems to mass produce at high rates. It is not a matter of merely flipping a switch at the factory. Instead, each aspect of the production system has to be tested and debugged. No

Inset 1.2: Why is Understanding Jargon Important

 The product realization process is full of jargon, the meaning of which may not be obvious. Part of learning about product realization is learning how to talk to people in their language. For example, the box used in sand casting is called a flask. The term "wall wart" refers to the small power adapter that you plug a charger into (it only makes sense when you think about how ugly power adapters are). First test shots are the initial parts produced on a new injection molding tool. Throughout this book, we will use common product realization and manufacturing jargon. Text boxes with the dictionary symbol will be used to highlight terminology that readers may not be familiar with. There is also a comprehensive glossary at the back of the book.

company can go from first test shot (Inset 1.2) to full production rates instantaneously, because no company can get everything right the first time. During piloting, teams will scramble to redesign parts, re-cut tools, work with suppliers to improve quality, change assembly sequences, and undertake a myriad of other efforts to hit their launch goals. Here are some unexpected problems that teams often run into:

- **As the speed and volume of production increase, more unexpected issues will arise with less time to fix them**. Many quality problems only become apparent when teams start building many copies of the same product at a high rate. First, producing more product gives more opportunities for low-chance failures to  happen. Second, processes do not produce the same quality at high volumes as those that operate at lower rates.
- **In-line testing does not capture defects**. Teams can carefully test their products both during and after assembly, but the tests often fail to replicate many of the conditions that the product will encounter in the field. As a result, the defects aren't found until after a customer finds them.
- **The product fails during the warranty period due to unexpected stresses**. The product will be dropped, shaken, and generally abused by the users. It is easy to overlook usage scenarios that damage the product, especially for new technology the customer has never used before.
- **Companies are surprised by legal regulations late in the process**. Each product will need certifications to comply with EMF (electromagnetic field) exposure, EMI (electromagnetic interference), safety, and environmental laws. You do not want to be adding EMF shielding or redesigning charging circuitry late in piloting because you did not plan for passing key certification tests.
- **Customers find things they do not like just before launch**. Often, customers cannot give meaningful feedback on the product until they have a working unit using the final production-intent materials.

Unfortunately, this feedback usually occurs after building expensive tooling, and when design changes are costly and time-consuming.

- **The forecast may not be accurate**. The procurement team must order all of the right materials well ahead of production. However, at the time orders need to be placed, the forecast is often very uncertain. Companies may need to lay out significant cash for excess inventory early or face potential material shortages later on.

To get through product realization, product development and manufacturing companies have developed highly interconnected tools, documentation, methods, and processes to manage these complexities. These product realization processes are put in place to reduce the chance of error, reduce variability, and improve the consistency and reliability of the product and production systems.

1.3 Product Realization is a Marathon

 Product realization is like running a marathon. Everyone wants to focus on the promise of the first mile and the nail-biting last mile, but the races are won or lost in the middle 24.2 miles (and most often in the Boston Marathon on Heartbreak Hill).

The first "mile" of any venture is the exciting one. The team gets to talk to customers, come up with ideas, create branding, hack hardware, and design cool t-shirts. The first mile is the phase most books, innovation programs, and undergraduate and graduate business curricula focus on. Business students come out of business school classrooms armed with business plans. Engineering students spend most of their design classes building a single prototype and thinking they have learned how to develop a product. Building only one often provides a false sense of security to new graduates who underestimate all the subsequent steps it takes to get a product into the hands of a customer. Both business school and engineering students leave university assuming that that first mile is most of the marathon.

The last mile, getting the product across the finish line, also gets attention: business magazine articles are usually focused on what happens when the product arrives in the customer's hands. On the drier academic side, the post-finish-line period is also the focus of significant literature on optimizing operations once production is running at full rates.

The middle 24.2 miles – from a great idea to the first product on the shelf – is long, painful, and full of challenges. New teams are excited about their new product and have incredible energy going into the process, often naïvely not understanding the length and difficulty of the marathon ahead of them.

With so little discussion of what happens in between the beginning and the end of this race, it's no wonder that companies or teams often underestimate the length and difficulty of the journey. Just as in a marathon, those who train and prepare have a higher chance of success. Naïve, untrained people get removed from the race on a stretcher.

Product developers who train for the product realization marathon can anticipate many of the challenges, make sure issues are proactively addressed, and have the right resources on board. By knowing what is ahead, teams reduce the pains and unknowns of the process and increase the odds of a successful outcome.

1.4 The Factory is Not a Giant 3D Printer

Most inexperienced (and even some experienced) teams think of the process of sending a design to the factory is similar to sending a file to a 3D printer and hitting print. They assume that after selecting a factory or contract manufacturer (CM), they can send the factory the drawings and a check, and ta-da! – a product shows up via DHL in full working order in three months. Some manufacturers may even promise you that service, but you would be foolish to believe it.

Even if the product development team wholly outsources manufacturing, organizations still need to understand the processes that the product will go through to get produced. The product development team will need to test and provide feedback on production samples and continue to refine the design. Teams need to actively partner with the contract manufacturer or factory; they should understand enough about the process to respond quickly to any questions and to avoid potential errors. In short, teams can outsource the manufacturing itself, but not the *responsibility* for the manufacturing.

1.5 Three Rules

Most failures of product realization result from violating one of the three fundamental rules of product design:

1. Understand and design within the fundamental laws of physics
2. Ignore manufacturing at your peril
3. Know your costs and cash flow

You cannot break the fundamental laws of physics. Ultimately, hardware is governed by the laws of physics, mechanics, and electronics; a product that violates fundamental first principles will never work. This principle probably seems patronizingly apparent, but developers get into trouble when they promise something that is not feasible. In a notorious case, Elizabeth Holmes raised over $700 million USD in 2013 on the promise that Theranos would deliver at-home medical testing from a single droplet of blood, based on little more scientific foundation than her desire to make it a reality [10].

Teams need to use sound engineering and analysis. In the author's experience, most quality issues, recalls, and product failures happen because the physics of either the design, materials, or manufacturing was overlooked in favor of visual appeal or unachievable functionality. As much as teams want to get eight hours of power out of a minuscule battery or pack a Wi-Fi chip, Bluetooth, speaker, and headphone jack into a device the size of a quarter, the fundamental laws of physics limit what they can design using current technologies.

You cannot get white plastic to match white paint. Many of the practical realities of what happens when you actually have to build a product only become apparent through experience. For example, teams who have not designed products before often try to make all-white products including a combination of molded plastic and painted parts. When they get the samples back, even when the Pantone colors are ostensibly identical, the two whites will never appear to match. In different lighting, they will look vastly different, and one will always look dirty. This rule, and thousands of others, is knowledge designers acquire through experience and critical evaluation of competitor products. Teams – even seasoned ones – should consult experts about the feasibility of producing their product. Caution: if your organization has no experience in a field and the first response of the team is "this should be easy," it most likely will be a problem that will crop up 10 days before shipping to the customer.

 **Insufficient cash flow is the number one reason that start-ups fail**. Engineers should not live in a technical bubble. Product development requires an appreciation of the broader realities of cost, quality, resource limitations, and schedule. Too many business plans focus solely on the cost of the physical product and return on investment (ROI) without understanding the cash required to get the product built. Unfortunately, building hardware is a cash-intensive process: teams need to outlay money for development, tooling, and samples long before any returns are actualized. By avoiding the pitfalls that we talk about in this book, teams can reduce the chance of unexpectedly running out of money before the product goes to market by more accurately predicting how much funding they should raise in the first place.

1.6 Why Learn about Product Realization?

No orchestra would perform without practice, or without at least reviewing the music first. Unfortunately, during most product launches (especially with start-up teams), the players are learning to read music, learning a new instrument, and performing at the same time. This book is intended to educate the reader on all of the activities involved in product realization so that teams can be better prepared.

The decisions that your team makes in the early phases of design will have a significant downstream impact on your ability to get product smoothly into mass production. For example, a choice to solder wires rather than using an edge connector can reduce the COGS (cost of goods sold) (Section 8.3) but can also increase the risk of cold-soldering failures that crack and drive defects. In another example, the decision to sell your product in multiple countries may increase sales but end up costing more in certifications and in-country support.

It is very easy to get distracted by interesting technical challenges or trying to do too much. The following are several examples of issues that can distract or take focus away from what is most important.

Trying to make too much yourself. Many teams get caught up in the fervor of innovation and want to produce the whole product themselves when outsourcing makes more sense. Teams end up doing jobs badly that a supplier could do cheaper, faster, and with better quality. For example, teams may try to manage in-country distribution only to discover they cannot hire enough people at peak times; it can be much cheaper to outsource incoming deliveries and outgoing orders to a logistics company.

Selecting the cheapest manufacturer and ending up with quality issues. It is often tempting to pick the lowest-cost provider, but that often comes at the expense of poor quality or working with an uncooperative and unresponsive CM.

Promising aggressive timelines before understanding how long product realization takes. It is tempting to promise customers that you will deliver product quickly. However, if you promise aggressive delivery timelines, you will need to rush through critical product testing or make your customers unhappy by being late. It is better to under-promise and over-deliver than the opposite.

Waiting to engage with service partners. It is tempting to postpone engaging with downstream partners because "it should be easy" or "we're not at that stage yet." For example, packaging design will always take longer than expected, but companies often delay the packaging because they think that "Packaging design won't take very long, it is just a box." Companies often delay contracting with third-party logistics providers (3PL) until the last minute and end up scrambling and paying a lot to get their products delivered to customers. By understanding partnership needs early, the company has a longer window to assess potential suppliers and deliberately choose the right ones.

Waiting to estimate costs until late in the process. Teams do not want to advertise a target MSRP (manufacturer's suggested retail price) only to determine late that their landed costs (the cost of the product including delivery) are too high. Early estimates can indicate whether the teams have a chance of meeting target costs.

Assuming that it will be easy to minimize cost through volume. Many products make it to market with a higher than expected COGS, but teams believe that they can reduce those costs when they reach peak volume. Unfortunately, many engineers overestimate those cost reductions. The unit price difference between one unit vs. 100 can be quite dramatic, but the marginal savings drop dramatically as the volumes go from 1,000 to 10,000.

Not planning carefully for cash flow. When planning budgets or fundraising, understanding the real cost to get a product to the finish line can mean the difference between failure and success. It is horrible to run out of cash just at the time when you are ready to start selling to customers.

Not finding manufacturing issues until late in the pilot process. Teams need to evaluate their designs for both manufacturability and ease of assembly as early as possible. You do not want to find out a part is not manufacturable after you have cut and paid for your tools.

Adding features late in the process. The product management team will always be tempted to add features or product variety very late in the process. Because you do not leave yourself enough time to fully test the new features while increasing the complexity of the production system, quality failures inevitably arise. Having good discipline around sticking as close to the minimum viable product as possibly will reduce the chance of failure.

Not understanding the usage of the product early and designing for it. Too often, reliability and durability requirements are specified very late in the design process leading to expensive redesigns. Late design changes – for example, to increase reinforcements or reduce thermal loading – can delay product launch and drive up costs.

1.7 Book Structure

This book will walk readers through the process of going from a prototype through piloting to production ramp, and will introduce teams to the concepts, tools, and challenges of each step.

It is important to note that although the chapters in this textbook have to be laid out in a sequence, this does not mean that your team will be going through these processes one at a time or necessarily in order. Many processes will need to be executed simultaneously, and teams will iterate between them many times. For example, organizations need to plan the pilot process before they can appreciate the context of the quality planning process, but they also need to grasp the quality planning process to design the right set of pilots.

This book is written for a range of audiences including graduate and advanced undergraduate courses, start-up teams, and larger companies. It is written from the point of view of a mechanical engineer with 20+ years directly involved in product design, and who still spends significant time on factory floors. As a result, the book focuses mainly on the engineering, cost, and scheduling issues related to getting a product through the piloting phases and into mass production, and less on marketing and financing. Because it is not possible to list all of the references in this text (and it would quickly become out of date), the website *productrealizationbook.com* contains additional references and resources for the reader.

The textbook is broken roughly into five sections shown on the map in Figure 1.2.

- The first section (**The path ahead**) ensures the team is ready to start product realization (Chapter 2), gives the reader background on the product realization process (Chapter 3), and introduces several important product realization project management tools (Chapter 4).
- The second section (**Product planning**) describes getting the product design ready, including ensuring a comprehensive specification document (Chapter 5), defining all aspects of the product design (Chapter 6), defining how quality will be verified and validated in the pilot runs (Chapter 7), and predicting product costs and managing cash flow (Chapter 8).
- The third section (**Manufacturing planning**) focuses on getting the manufacturing system ready, including background on manufacturing systems (Chapter 9), ensuring the product is manufacturable (Chapter 10), defining the process so it can be executed (Chapter 11), designing and producing tooling (Chapter 12), and managing quality during production (Chapter 13).
- The fourth section (**Production planning**) focuses on the management of the supply chain. It includes how to design your supply chain (Chapter 14), how to plan for production to ensure sufficient material (Chapter 15), and how to get your product to your consumer (Chapter 16).
- The fifth and final set of chapters (**Selling your product**) covers the certifications your product will need (Chapter 17), how to set up ongoing customer support (Chapter 18), and what happens once you are at full production rates (Chapter 19).

'THE PATH AHEAD'

1. INTRODUCTION

2. ARE YOU READY TO START?

3. PRODUCT REALIZATION PROCESS

4. PROJECT MGMT

PRODUCT PLANNING   PRODUCTION PLANNING

MANUFACTURING PLANNING

14. SUPPLY CHAIN

9. MANUFACTURING SYSTEMS

5. SPECIFICATIONS

10. DESIGN for MANUFACTURING & DESIGN for X

6. PRODUCT DEFINITION

15. PRODUCTION PLANNING

11. PROCESS DESIGN

7. PILOT-PHASE QUALITY TESTING

12. TOOLING

16. DISTRIBUTION

13. PRODUCTION QUALITY

8. COSTS & CASH FLOW

SELLING YOUR PRODUCT

17. CERTIFICATION/LABELING

18. CUSTOMER SUPPORT

19 MASS PRODUCTION

FIGURE 1.2
Chapter map

When learning about the product realization process, you may find it easy to get lost in the weeds and lose sight of the overall production realization process. Figure 1.2 shows the relationships between the chapters and will be used as a map and guide throughout the book.

Summary and Key Takeaways

❏ Building ten thousand is very different from building one.
❏ Products fail during product realization for many reasons, including failures in technology readiness, production system maturity, and cash flow.
❏ Product realization is a complex, multifunctional process that involves a large number of people within and across organizations.
❏ During product realization, teams will need to balance cost, quality, and schedule. Ultimately, teams need to achieve all three to create a successful product.
❏ Understanding the road ahead will help teams better avoid problems and help them plan for the needed resources to accomplish their goals.

Chapter 2

Are You Ready to Start?

Entering product realization before the product is ready increases cost and reduces the chance of product success. This chapter defines what "readiness" means: namely, the product meets customer needs, the technology has been thoroughly tested, the manufacturing processes are mature enough, and all documentation has been prepared.

Product Realization: Going from one to a million, First Edition. Anna C. Thornton.
© 2021 John Wiley & Sons, Inc. Published 2021 by John Wiley & Sons, Inc.

Checklist 2.1: Product Definition Maturity Checklist

❑ **Is your design concept ready?** Does it meet the customer needs at a reasonable price point? Does the product have a viable business model?

❑ **Is the technology used in the product mature enough?** Has the team ensured new untested technology will perform reliably? Does any of the technology have any fundamental reliability or quality flaws?

❑ **Is the prototype mature enough?** Is the prototype a true engineering prototype that represents all of the production-intent details?

❑ **Is the product definition mature enough?** Has the team documented the product so it can be transferred to the factory and be built to specifications?

❑ **Are the manufacturing processes mature enough?** Have any new manufacturing process technologies been matured and shown that they can operate at full rates?

❑ **Do you have enough time?** Have you accurately assessed the time required to get the product ready for production or are you going to be late?

❑ **Do you have enough cash on hand?** Have you thought through the cost to do the necessary pilot runs and the actual cost to launch the product?

❑ **Have you addressed all of the readiness risks?** If you are going into piloting with a less than mature product, have you created a risk management plan?

Companies often rush into production with an immature concept, trying to get ahead of the competition or meet unrealistic customer promises. Rushing into product realization too soon invariably leads to downstream failures that ultimately cost the team time and money. When teams move into product realization, the cash flow required increases dramatically: more people have to be hired, non-recurring engineering costs accumulate, and materials have to be purchased. It is better to spend an extra month getting ready when your burn rate is $10,000 per month than to spend an extra month in piloting when your burn rate is 10 times that amount.

While it is tempting to "just get started and figure it out later," teams need to ask themselves the hard questions about their readiness before committing the resources and capital to start production. Figure 2.1 and Checklist 2.1 summarize the questions and actions to take if the answer to any of the readiness questions is no. The following sections go into more detail on each of the measures of readiness.

2.1 Is Your Concept Ready?

Many products fail because they don't meet the needs of the customer at the right cost. Before starting product realization, it is crucial to ensure that you have a design concept that meets an important customer need and that you have a sustainable business model. Here are a few examples of products that went through product realization – or most of it – and failed because they failed to satisfy this fundamental readiness measure. In each of these cases, the design of the product didn't meet a customer need at the right price.

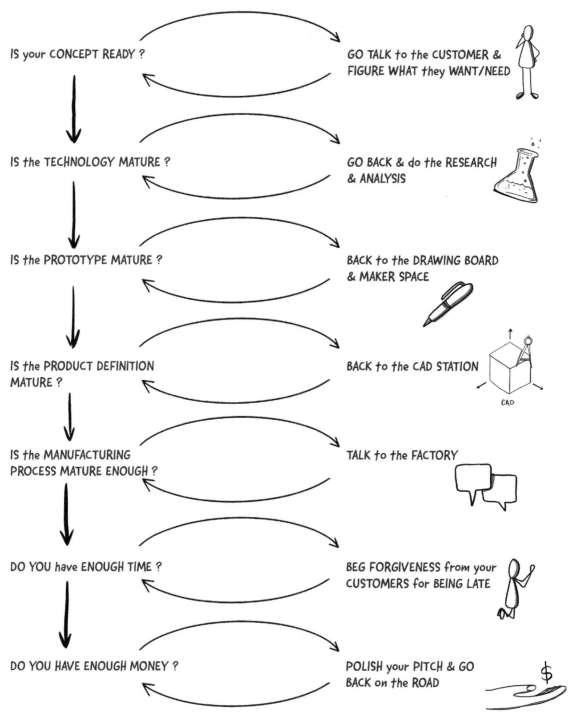

IS your CONCEPT READY ?

GO TALK to the CUSTOMER & FIGURE WHAT they WANT/NEED

IS the TECHNOLOGY MATURE ?

GO BACK & do the RESEARCH & ANALYSIS

IS the PROTOTYPE MATURE ?

BACK to the DRAWING BOARD & MAKER SPACE

IS the PRODUCT DEFINITION MATURE ?

BACK to the CAD STATION

IS the MANUFACTURING PROCESS MATURE ENOUGH ?

TALK to the FACTORY

DO YOU have ENOUGH TIME ?

BEG FORGIVENESS from your CUSTOMERS for BEING LATE

DO YOU HAVE ENOUGH MONEY ?

POLISH your PITCH & GO BACK on the ROAD

FIGURE 2.1 Are you ready?

Juicero was the darling of the investor community around 2016. It raised over $120 million to produce an $800 machine that squeezed juice from a bag of pre-cut cleaned fruits and vegetables. An exposé in *Bloomberg News* [11] revealed that the fruit and vegetable packs the company supplied could just as easily be squeezed by hand. After the publication of this video, Juicero became a widely mocked failure [12]. A teardown by the blog Bolt highlighted that over-engineering the machine had driven up complexity and therefore cost:

> Our [Bolt's] usual advice to hardware founders is to focus on getting a product to market to test the core assumptions on actual target customers, and then iterate. Instead, Juicero spent $120 M over two years to build a complex supply chain and perfectly engineered product that is too expensive for their target demographic. [13]

The author recently had a conversation with a young engineer who was enamored of what he considered high-quality engineering that went into the Juicero. Although he had read all of the articles referenced here, he kept insisting that the engineering was excellent. It took a while to convince him that it was in fact poorly engineered. While the complicated technical details were impressive in their own right, the machine did not meet the needs of the customer: it was too expensive and did not do the job as well as someone kneading the bag with their hands. Sound engineering is measured by meeting the customer need at the right price point and quality, not by part count or complex mechanisms.

Another new beverage product that failed around the same time was the **Keurig Kold**, an attempt to expand Keurig's already popular hot coffee and tea product line into the realm of cold beverages using individual serving pods. They invested 100 million USD in fiscal year 2015 with a plan to spend a similar amount in subsequent years. They had agreements with both Coca-Cola and the Dr. Pepper Snapple Group to distribute their proprietary drinks. Within one year, though, Keurig pulled the product from the market. Several factors likely contributed to the failure. The unit was expensive and bulky, and the price point for each pod was higher than the cost of a single can of Coke [14]. Customers were just not willing to pay the cost of the device and the pods to get something they could already conveniently purchase ready-made and at lower cost.

Google Glass, which launched in 2013, was an attempt at bringing the technology and capabilities of a smartphone to a wearable device through the application of augmented reality (AR). When Google released Google Glass the product had several fatal flaws, including clunky aesthetics, an unwieldy user interface, and short battery life. The ultimate reason for the failure was that while the technology was impressive, it didn't solve any discernible customer problem and thus never found a user base [15].

2.2 Is the Technology Mature Enough?

"We have a cool new technology" is often interpreted as "we are ready to sell to customers," but that is rarely a useful metric for readiness. A working bench test can demonstrate the feasibility of the technology, but it cannot demonstrate safety, reliability, scalability, performance, or manufacturability. For technology to be truly mature, it must meet performance targets, be producible at volume, and work reliably for all customers in all environments.

The **EV1 from GM**, launched in 1996, was one of the first mass-produced fully electric vehicles. There was significant media buzz about the product when it was launched, and many potential customers expressed interest. However, the car was pulled from the market just a few years later because it could not provide the basic functionality that most car buyers needed. Its heavy battery could only hold enough charge for short-distance driving, and the interior could only seat two passengers. Electric vehicles eventually came to market (e.g. GM's Bolt, the Tesla, and Nissan's Leaf) only when battery technology became sufficiently advanced to allow drivers a driving experience comparable to that of traditional vehicles [16].

VW had to recall eleven million cars and spent 18 billion USD in expenses related to the excessive diesel engine emissions from their cars. VW couldn't develop an engine that would have the excellent fuel mileage they were targeting and still pass the fuel emission standards in most countries. They covered this technology failure by putting software in their product that changed the performance during the emissions tests so the product would pass regulatory requirements [17]. This action resulted in the loss of customer confidence, was unethical and illegal, resulted in large fines, and more importantly, damaged the environment.

In 2019, the **Boeing 737 MAX**, a retooled version of the classic 737, was grounded after two plane crashes. Its new flight control system had a single point of failure when the angle-of-attack sensor failed. Pilots were not informed of the difference between this control system and that of the 737s they were trained on. A number of organizational failures allowed the single point of failure to be designed into the system, and pilots were not trained to respond correctly to that scenario quickly enough to avoid a crash [18].

These kinds of technological failures are avoidable if teams carefully test for technology readiness. In the 1970s, NASA developed a Technology Readiness Level (TRL) standard (Table 2.1) [19] to assess the technology to be used in the Jupiter orbital mission [20]. Manufacturing readiness levels (MRLs) [21] were later created to assess whether military products could be produced at volume. The TRL and MRL levels were designed to ensure that all team members understood when the technology and manufacturing process were mature enough to be included on a mission. The TRL and

TRL	Definition
1	Basic principles observed and reported.
2	Technology concept and/or application formulated.
3	Analytical and experimental critical function and/or characteristic proof of concept.
4	Component and/or breadboard validation in laboratory environment.
5	Component and/or breadboard validation in a relevant environment.
6	System/subsystem model or prototype demonstration in a relevant environment.
7	System prototype demonstration in an operational environment.
8	Actual system completed and qualified through test and demonstration.
9	Actual system proven through successful mission operations.

Table 2.1
Technology
Readiness Levels
Source: From [19]

MRL frameworks have been used across many industries to ensure that sufficient resources are allocated to less-mature technologies and to manage the likely risks in a project containing multiple new technologies. Typically, product realization can start when the technology reaches at least a level 8 on the scale of 1 to 9.

2.3 Is the Prototype Mature Enough?

A technology may be mature and ready for use, but that still doesn't mean that the product concept is ready to enter product realization. Prototypes allow teams to test concepts in the real world to see how well the current state of the design works. Issues that aren't obvious during concept formulation and initial design become very apparent once the prototype is physically built and used by real users.

Prototypes fall into one of several categories: works-like, looks-like, works-like/looks-like, or engineering. Works-like models demonstrate the function of critical technologies. Typically, a works-like prototype of an electromechanical product is composed of a breadboard style PCBA (printed circuit board assembly) with the most important mechanical systems built using rapid prototyping methods. It often looks like a tangled bunch of wires, but it can be used to demonstrate technical feasibility. Works-like prototypes are used to validate the TRL through level 5 as described in the preceding section. The looks-like prototype is generally a non-functioning aesthetic prototype made using low-volume prototyping methods. As the name suggests, the looks-like is used to demonstrate the final product user interface, and color, material and finish (CMF).

It is only when the technology and the product form and finish are combined that the product is ready to pass to the next step. The looks-like/works-like prototypes prove that the functional elements can fit within the envelope of the final product. Once this is completed, TRL level 7 can be achieved. Finally, teams should produce an engineering prototype that is as close to final production as possible without using high-volume manufacturing techniques. Engineering prototypes test not only the function but the aesthetics, assembly, and user interactions.

Table 2.2 lists the types of prototypes and their relevant characteristics. Figure 2.2 shows photos of various stages of prototyping for the Embr Wave designed and built by Embr Labs. Version 1.0 and 1.5 were used to test the technology and get user feedback; but only when the aesthetics and technology were married was the product ready for design verification testing (DVT) (Chapter 3). These units were built by the factory using part manufacturing methods close to those of final production, but were assembled by hand.

2.4 Is the Product Definition Mature Enough?

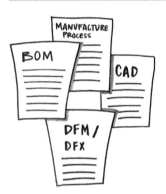

In addition to having the prototype working and tested, all of the product definition and documents should be ready for hand-off to the factory. Checklist 2.2 provides a list of the information that should be ready. Chapter 6 describes these documents in more detail.

In addition to these considerations, teams need to understand the readiness of software, firmware, and any additional services, apps, or capabilities. For example, in the start-up example in Figure 1.1, the hardware maturity and expected manufacturing rate were responsible for most of the delay, but team's inability to deliver the software was also a major contributor.

Checklist 2.2: Product Realization Readiness Checklist

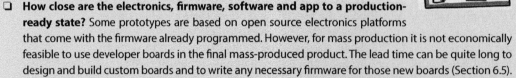

❑ **Do you have a complete bill of materials (BOM)?** It is not possible to estimate the cost of the product or plan the supply chain unless there is a comprehensive bill of materials (Section 6.2).

❑ **Are all of the parts designed and engineering drawings created?** Drawings are different from CAD and include all of the relevant information needed by manufacturing to produce the parts and assemblies to specifications (Section 6.4).

❑ **How close are the electronics, firmware, software and app to a production-ready state?** Some prototypes are based on open source electronics platforms that come with the firmware already programmed. However, for mass production it is not economically feasible to use developer boards in the final mass-produced product. The lead time can be quite long to design and build custom boards and to write any necessary firmware for those new boards (Section 6.5).

❑ **Are the manufacturing processes for each part determined and the part reviewed for manufacturability?** Each part should have a manufacturing process selected. If processes are selected late, the parts may need to be redesigned to accommodate the selected process (Section 10.1).

❑ **Do you have suppliers identified for custom parts and can they make them?** It is necessary to both select a vendor and verify they can meet your design requirements. For example, if teams need odd-shaped batteries or motors, it is critical to determine whether that geometry is possible before locking in the rest of the design (Chapter 14).

Table 2.2 Prototype maturity

Type of prototype	Demonstrates that . . .	Maturity of works-like prototype	Maturity of looks-like prototype	Maturity of manufacturing processes	Readiness for product realization
Proof of principle/ concept	Technology can work in limited conditions.	Parts are not finalized. Space and layout are not optimized. Not all functions are designed.	Typically, the team has just a sketch of the housing or a 3D printed structure.	No thought has been given to manufacturing strategy yet.	Awful. There are too many unanswered questions to start piloting. Significant design effort is needed.
Functional (also called works-like)	Almost all functions will work.	The prototype demonstrates most functions. Detailed part design is not complete and final components are not selected. Layout is not optimized for the final product. The prototype is often built with off-the-shelf tech such as Arduino.	Typically, the team has just a sketch of the housing or a 3D printed structure.	No thought has been given to manufacturing strategy yet.	Poor. The parts, components, and layout need to be detailed, custom circuits need to be designed, and it need to be married to the looks-like prototype.
Looks-like	The 3-D structure and finishes of the final product are complete and parts of the user interface are modeled.	The prototype demonstrates some of the passive user interface functions.	Materials, internal structures, assembly sequences, and marriage to the functional parts are not complete.	No thought has been given to manufacturing strategy yet.	Poor. There is much engineering to be done to marry the works-like and looks-like prototypes.
Works-like/ looks-like	The functionality can be incorporated into the industrial design.	The layout is almost complete, but not all design decisions are made about components and subsystems. Prototypes are built using prototype/hand-built methods.	Rapid prototyping is used to generate mechanical structures. The design is not optimized for manufacturing.	Manufacturing processes are selected for a couple of critical parts. Prototypes are not usually made with production materials.	Not great. There is significant detailing and thought to get the design completed and the manufacturing strategy ready.
Engineer-ing prototype	The product can be built and is functional using production-intent parts.	The layout is finalized. Manufacturing is simulated using low-volume production rates and rapid prototyping.	All features are defined. Rapid prototyping techniques are used to create the parts.	Design for manufacturing has been applied to all parts, the bill of materials (BOM) is complete, and parts are designed to be made in mass production.	Good. The only outstanding issues are the DFM changes and other adjust-ments based on feedback from parts suppliers.

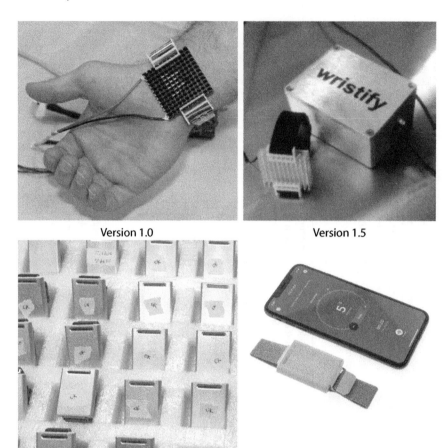

Version 1.0 Version 1.5

FIGURE 2.2
Prototyping phases
for Embr Wave
Bracelet
Source: Reproduced
with permission
from Embr Labs

Design verification units Final product

2.5 Is Manufacturing Mature Enough?

The second-to-last step in checking for readiness is an assessment of the manufacturing process readiness. Some products may require not only the addition of unique technology but also the development and refinement of new manufacturing techniques. The Department of Defense's (DOD) MRLs give a comprehensive description of the maturity of the manufacturing processes (Table 2.3) [21]. Ideally, product realization should not start until the manufacturing technology is at level 8.

GTAT Corporation had been producing single crystal sapphire in smaller sizes and relatively low production rates. Their inability to meet their contractual obligations to build a larger crystal for Apple contributed to

MRL	Definition
1	Basic manufacturing implications identified.
2	Manufacturing concepts identified.
3	Manufacturing proof of concept developed.
4	Capability to produce the technology in a laboratory environment.
5	Capability to produce prototype components in a production-relevant environment.
6	Capability to produce a prototype system or subsystem in a production-relevant environment.
7	Capability to produce systems, subsystems or components in a production-representative environment.
8	Pilot line capability demonstrated. Ready to begin low-rate production.
9	Low-rate production demonstrated. Capability in place to begin full-rate production.
10	Full-rate production demonstrated and lean production practices in place.

Table 2.3
Manufacturing
readiness levels
Source: From [21]

bankrupting the company. They probably overestimated the maturity of their manufacturing technology and were unable to scale their existing rates and sizes quickly enough [5].

Tesla struggled for the first few years with getting their automation and manufacturing systems to produce all-electric cars reliably at the desired rate. They assumed they could implement extensive automation without planning for a long period of debugging the product and process. The delays significantly limited their ability to meet demand.

2.6 Is there Enough Cash and Is there Enough Time?

It's a cliché in manufacturing and construction that teams always over-promise on cost and schedule but rarely deliver as promised. However, this is not a good way to run a business. Even in large companies, teams need to know in advance how much time and money it will take to produce and launch a product. If you have missed deadlines and run low on cash, it will become even harder to raise the funding necessary to complete production. Section 4.2 describes how to plan the project schedule, and Chapter 8 describes how to estimate costs and cash flow early in the process.

Coolest Cooler, a wheeled cooler with a battery powered blender, which raised over $14 million on Kickstarter, delivered to about two-thirds of

Vignette 1: Are You Ready for Production?

Scott N. Miller, CEO and founder of Dragon Innovation

Building high-volume hardware products is hard because it involves so many different disciplines, all of which have to come together seamlessly for the company to be a success. Over the last 25 years, I've had the opportunity to be part of teams that have launched new products at Disney Imagineering (Walking Robotic Dinosaurs) and iRobot (the first four versions of Roomba), as well as helping hundreds of other companies manufacture their own products via Dragon Innovation.

One of the biggest stumbling blocks is accurately assessing design maturity for manufacturing and assembly. I often use the analogy of an alphabet, where "A" represents the idea, and "Z" represents the product in the customer's hands. Many times companies complete what they consider to be the final working prototype and assume that they are around the letter "M" when, in practice, they are closer to the letter "C."

The advent of affordable desktop 3D printing was a game-changer for prototyping. However, the technology is so good these days, it's often difficult to tell the difference at first glance between an injection molded part and a 3D printed one that has been filled (to hide the layers) and painted. This aesthetic realism has an unintended downside in that it can lure teams into thinking they are ready to hand over the files to a factory for tool-making when in fact there is still much more work to be done. As a result, the teams do not fully account for how long the process actually takes or how much engineering is required to convert their CAD into parts that can actually be tooled.

To convert the 3D printed model to production parts, we like to start with Design for Assembly (DFA). There are many methods, but we focus on two: reducing part count and making the assembly simple. Reducing part count means you can't forget to order a part you don't need. Second, it is critical to make the assembly as simple as possible; there is a 20% year-on-year turnover in China, and you don't want to have to spend weeks training a new worker just to have them leave. The next step is Design for Manufacturing (DFM), which happens on the part level, and including fabrication and material selection, drafts and rounds, parting planes, etc. The challenge is that these DFA/M changes to the geometry can impact the function and quality of the product, so iteration is often required.

To illustrate some of the latent challenges, I'll share the story of a customer who came to us for help with designing a very clever game system. The product was made up of stackable tiles, each with a small LCD screen (imagine playing scrabble, where the blocks could each be flashed with a different letter and would be able to detect the letter shown on their neighbors when lined up in a row). Each block was 3D-printed with perfectly straight sides so that they could also be stacked vertically. However, for injection molding, it's necessary to add a small angle (called draft) to accommodate shrinkage and make sure you can get the part off the tool without scratching it. Adding even a small angle (0.25°) to the wall resulted in the blocks looking like the Leaning Tower of Pisa. To get a straight wall required a much more complex tool (imagine slides on four sides), which in turn extended the schedule and put shipping for the holiday season at risk.

To make sure you are ready to move into manufacturing, keep in mind the alphabet analogy and factor in the DFA/M process. Even the smallest details can have a disproportionate effect and create unintended consequences. The last 20% of the work does often require 80% of the time.

Text source: Reproduced with permission from Scott N. Miller
Logo source: Reproduced with permission from Dragon Innovation

the roughly 60 000 backers, but went bankrupt five years after its crowd-source campaign [22]. Despite raising significant money, they didn't have enough cash on hand at the end to fund the production rates needed to deliver on their promises. The cooler worked and the customers who received them were happy with the product. They were able to deliver some products to their backers. To generate the cash to deliver to their backers, they had to start selling on Amazon, further irritating their initial backers [23].

Glowforge, a laser cutter for home use, was unable to deliver to several international customers because Glowforge didn't understand early enough the cost, time, and legal implications involved in selling internationally. Most companies start by supplying only the US, Canada, and Europe. Even if you are delivering a handful of units to a country, the product requires a full certification (Chapter 17), which is time-consuming and expensive. In addition, distribution internationally is expensive. Glowforge apparently didn't appreciate the cost and complexity of delivering outside the US and European Union (EU) regions and had to disappoint customers [24].

2.7 How Ready is Ready?

Oftentimes, budget and promises to customers mean that you have to start production before you are fully ready. While risky, moving toward production with a less-than-ready product is possible if:

- **Changes are unlikely to have ripple effects on the rest of the design**. For example, if the LCD (liquid crystal display) design is not completed, it may be okay going forward with tooling the surrounding parts as long as the changes are unlikely to impact the fit and function of the rest of the product. As another example, you might start ordering parts for the charging circuitry before the battery supplier is selected. If the volume and power requirements are well within the available COTS (commercial off-the-shelf) parts available, you can probably take that risk.
- **There is a more expensive backup plan**. Often, technological risk is caused by assuming that a low-rate production process can be sped up easily or by selecting components for cost rather than quality. Teams can keep a backup plan ready just in case the quality or speed can't be achieved. For example, an inexpensive motor might be selected initially to keep costs down, but if it does not meet specifications, a more expensive motor can be swapped in at short notice.

However, no matter how much you plan for contingencies, late changes always come with risk. Unintended consequences will arise and derail what seemed like perfect plans. The more risks that can be identified and managed proactively, the less likely those risks are to result in a delay or a cost increase.

Summary and Key Takeaways

- ❏ Just because you have a prototype and a single customer doesn't mean you should start product realization.
- ❏ Product teams should ask themselves several questions:
    - ❏ Is your concept ready?
    - ❏ Is your technology mature enough?
    - ❏ Is the prototype mature enough?
    - ❏ Have you wholly defined your product?
    - ❏ Is your manufacturing plan ready?
    - ❏ Do you have enough cash and time?
- ❏ It is possible to start product realization with a design that isn't theoretically mature enough, but teams need to actively manage those risks and put in backup plans to keep an immature design from impacting cost, schedule or quality.

Chapter 3
Product Realization Process

'THE PATH AHEAD'

1. INTRODUCTION

2. ARE YOU READY TO START ?

3. PRODUCT REALIZATION PROCESS

4. PRODUCT MGMT

Product realization is a complex, highly iterative, and often unpredictable process. Before starting, teams need to know the road ahead and plan for all of the work that needs to be done between starting discussions with the factory and getting the first saleable units off the line. Teams need a significant volume of documentation to ensure that the product and process designs can meet the demand for a product at the right quality and price.

Product Realization: Going from one to a million, First Edition. Anna C. Thornton.
© 2021 John Wiley & Sons, Inc. Published 2021 by John Wiley & Sons, Inc.

Product realization is the last step in the overall process of getting a product from an idea to mass production. Unfortunately, there is very little in the literature on product realization, and most engineers only learn about it on the job. Before diving into the details, it is important to understand how product realization fits within the product development frameworks taught in academia and those used in industry. The first section of this chapter describes how product realization relates to the overarching product development process. The second section describes how the industry you are in, the technology you are using, and the customers you are selling to, may influence the process. The third section describes the piloting process: how production transitions from the initial prototype through to mass production.

3.1 Product Development Processes

There are many product development frameworks, but most companies follow a structure similar to the one described by Ulrich and Eppinger [25]. The specific activities in each step, the formality of the process, and the number of iterations depend on the company-specific processes and technology. In addition, the product development process can take a waterfall approach (sequentially executing each step) or an Agile or Lean approach (rapid multiple iterations of each stage).

The left-hand part of Figure 3.1 shows the basic stages of product development as defined by Ulrich and Eppinger. Teams start by understanding the customers and their needs (*product planning stage* or *ideation stage)*. Recently, industry and academia have put increased attention on product planning. Ensuring that you have the right concept for the right need is critical to creating a successful business. As pointed out in Chapter 2, a perfectly manufactured product that does not meet a customer need is still a failure. Design Thinking, the development model described in *Disciplined Entrepreneurship* [26], or *Lean Start-up* [27] provide structured methods to define the market, find unmet needs, and iterate and test concepts with customers. The top right list of Figure 3.1 shows one example of a product planning framework defined by Aulet.

Once the product idea is developed enough to start designing the actual product, the engineering team, the brand management team, and the industrial design team (collectively referred to as the development team) work together to generate several potential design concepts and solutions (*concept design stage)*. The development team then evaluates these potential designs to determine which approach is most likely to meet customer needs. Engineering then defines the product architecture and the major subsystems of the product (*system-level design stage)*. The detailed design comes next, resulting in a product that is complete enough to enable production (*detail design stage)*. The product design is tested and refined, and then the product is mass-produced (*production ramp)*.

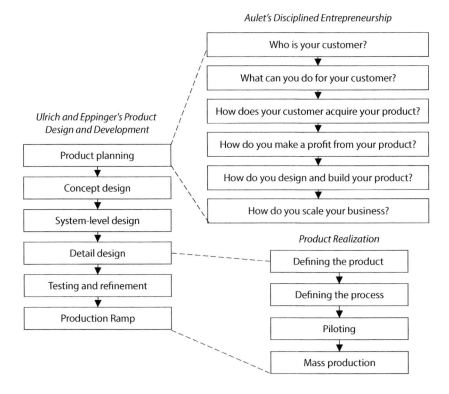

Aulet's Disciplined Entrepreneurship

Who is your customer?

What can you do for your customer?

How does your customer acquire your product?

How do you make a profit from your product?

How do you design and build your product?

How do you scale your business?

Ulrich and Eppinger's Product Design and Development

Product planning

Concept design

System-level design

Detail design

Testing and refinement

Production Ramp

Product Realization

Defining the product

Defining the process

Piloting

Mass production

FIGURE 3.1 Product realization process and its relationship to other product development frameworks [25, 26]

Product realization comes at the end of the traditional product development process. It starts at the tail end of the *detail design stage* and continues through *production ramp* including mass production. Roughly, the product realization process occurs in four stages.

- **Defining the product**. The factory needs a complete definition of the product to be able to start mass production. The definition is more than just the detailed models of the parts. Before transitioning the design to the factory, the development team needs to provide a comprehensive and accurate set of documents including the specification document, drawing package, bill of materials, CMF (color, material, and finish) files, and the electronics package (Chapters 5 and 6).
- **Defining the process.** To produce the product, not only must the product be designed, but the manufacturing team must select (or develop) and fully specify the processes needed to make the product (Chapter 10). The process definition includes defining the manufacturing processes (Chapter 9), defining part fabrication and assembly processes (Chapter 11), designing and building tooling (Chapter 12), and establishing the supply chain (Chapter 14).
- **Piloting**. Even after the team designs the production system, they cannot simply turn the production line on like a light switch and run at full production rates. In the same way that the product development team builds a series of prototypes during design and refinement,

the factory will likewise run a series of pilots (also called pilot runs, builds, or pilot builds). With each subsequent pilot, the manufacturer increases the number of units built. In addition, with each pilot run, rates increase and the line gets closer to the final mass production configuration. The purpose of the pilot phase is to identify any problems with the design, parts, and production system before going to full production rates. This pilot process is the focus of the rest of this chapter.

- **Production ramp**. After the operations organization is assured that the process can reliably build quality products at the full rate, production rates are increased. The operations organization plans material purchases and production schedules to meet the demand (Chapter 15). Lastly, the organization must design the distribution system needed to deliver the product to the customer (Chapter 16). Even after the product realization process is complete, the team will continue to manage production, continually improve the product, and support the product in the field (Chapter 19).

Throughout the product realization process, the team must continue to ensure that the product meets cost, quality, and schedule targets. Unfortunately, you always have to make tradeoffs. The general rule of thumb is that you can get two out of the three. For example, you can get low cost and great quality but that takes time. The team has to constantly balance between the three metrics to find the best solution.

- **Cost** (Chapter 8). The process needs to ensure the COGs (cost of goods sold) and landed cost meet targets. In addition, the teams need to ensure they do not run out of cash. During the pilots, significant money is spent on services, parts, tools, and samples, and it is easy to spend more than you have.
- **Quality** (Chapters 7, 13, and 17). During the pilots, the teams will constantly check to see whether the products meet the quality targets. The testing checks if the product, as built, is both durable and reliable, and whether it meets all of the legal standards of the countries where it will be sold.

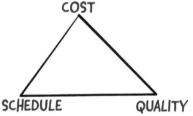

- **Schedule** (Chapters 3 and 15). The process must meet delivery promises. This involves three factors. First, the product has to be ready for mass production. The pilot process defines the overall delivery schedule, and teams need to plan the number of pilot runs and when they need to be complete to launch the product on time. Secondly, you need to ensure you have enough capacity in the factory to produce at a sufficient rate to meet customer demand. Finally, you have to have all of the materials from your suppliers in a timely fashion.

As teams transition from ideation and product design into production, several organizational and cultural changes will be needed (Table 3.1). The size of the core team and of the extended virtual teams (contract manufacturers,

Characteristics	Early design	Product realization
Number of products made	Single units	>10 000
Team size	4–5	100+
Management style	Flexible	Standardized and disciplined
Reliability	Focus on getting it to work once	Needs to work every time
Product testers	Designers	User
Legal	No approvals	Fully certified
$$	Get it working at any cost	Focus on COGS and efficient use of limited cash flow
Schedule	Flexible	Need to meet delivery promises

Table 3.1
Organizational changes as teams transition to product realization

part suppliers, and service suppliers) grows dramatically as experts join the core team, and suppliers integrate into the process. The organizational style will change from informal to formal. Shifting into product realization mode requires additional management structures, discipline around document management, and structured processes to approve changes.

3.2 Industry Standards

Depending on the industry or product, industrial or government standards groups may impose regulations and requirements not only on products but also on manufacturing methods, work environments, hazardous materials use, and waste disposal. Industry standards serve several purposes: some are put in place to standardize processes, and others to ensure consumer safety. For example, the APQP (Advanced Product Quality Planning) [28] methods were developed to reduce the paperwork for suppliers while ensuring consistent quality from the supply base. A single automotive supplier will typically supply to more than one automotive client. Before the development of these standards, suppliers had to maintain multiple versions of the same documents to satisfy the requirements of their varied customers. The automotive standards meant that suppliers only had to follow and document one product realization process for all of their customers, thus improving outcomes and reducing overhead.

In the case of highly regulated industries, such as aerospace or medical products, the standards ensure that companies use best practices that are known to improve safety. For example, the CGMPs (Current Good Manufacturing Processes) [29] defined by the US Food and Drug Administration (FDA) ensures that manufacturing processes do not go out of control unexpectedly and that any defective products are caught before they get to the consumer. Table 3.2 contains a list of industries and customer types that have unique product realization requirements. For each of these, experts should be consulted (either as employees or outside consultants) to understand the constraints imposed by the market or product niche.

Vignette 2: De-risking During Pilots

Santiago Alegria, Founder, Íko Systems

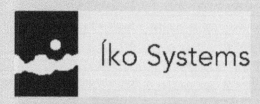

Íko Systems is an agri-tech company developing a smart micro-greenhouse for the modern home. We like to tell people that our product is like a Keurig machine but for plants. All you do is replace seed pods every two months, add water, and watch your herbs grow. The product lies at the intersection of hardware, software, and botany, and that causes development to be pretty complex. During our alpha program (engineering validation testing, or EVT), we decided to de-risk testing by sending customers pre-germinated plants. We started our beta program with plans to send customers seed pods to germinate in the product, but we quickly found this to be much more difficult than we thought. Most of the seeds fell too far into the soil or were crushed during shipping. Our handmade ceramics – which contain the seeds and media – had too many pore inconsistencies and quickly became clogged with nutrients, preventing water from being absorbed. Most importantly, the roots and seed were not getting enough oxygen, so they were growing slowly or not at all.

The problems were compounded by a number of factors. First, we decided to go with an outside vendor for pods for more consistent pores, but they were unable to meet our small orders in a timely fashion . . . we were just too small. Second, the cycle time to learn if a new pod would work reliably was several weeks – the time to germinate a plant.

As the project stands today, we are producing reliable pods and are able to get them into the hands of the customer with no shipping issues. We also upgraded our systems with airstones to oxygenate the roots. Had we started the de-risking process earlier and in parallel with the EVT phase, we could have saved several months of struggling and testing the Betas.

Text source: Reproduced with permission from Santiago Alegria.
Photo source: Reproduced with permission from Íko Systems.
Logo source: Reproduced with permission from Íko Systems.

Product type/customer	Description of specific requirements
Medical devices, materials, and implants	Depending on the class of device or material, different requirements are imposed on the product, packaging, sterility, design changes, and manufacturing processes by the US Food and Drug Administration and other countries' medical regulatory bodies.
Safety-critical products (e.g., fire alarms)	There are unique regulations around safety-critical items to ensure sufficient fail-safes, and to minimize false positive and false negative alarms.
Products for children	ASTM (American Society for Testing and Materials) has a standard that defines the safety requirements for any product used by children in the US [30]. A quick scan of the Consumer Product Safety Commission (CPSC) website will highlight just how many products are recalled for not following those regulations [31].
Data gathering devices and software	If critical data (personal or other) is being transmitted back to a central database, the use and security of that data may be regulated, especially in Europe.
Non-medical biocompatible products	Some products are not medical products but can cause medical problems. For example, watches, which are in contact with the skin, can cause irritation. Depending on the usage and applications, it may be necessary to do bio-compatibility testing (Inset 7.9).
Defense-related products	Some technologies need approval before they are exported if they are considered critical military and defense. ITAR (International Traffic in Arms Regulations) and EAR (Export Administration Regulations) are the US regulations that oversee export of defense products [32, 33]. These products fall into two classes: the US Munitions List (USML), which is primarily for defense-related applications, and the Commerce Control List (CCL), which addresses items that are commercial but could have military applications, such as specialized materials, electronics design, and information security.
B2B (business-to-business) products	B2B organizations sell products to other corporations. Some industry groups impose common requirements on their suppliers, such as the automotive and aerospace industries.
Cannabis-related products	At the time this book was written, cannabis-related industries were rapidly growing and the regulations varied state-to-state. Because of the gray area between state and federal laws, oversight and rules are in flux. When developing products for this industry, businesses must not only look at today's state and federal regulations but look forward 3–5 yr to try to predict what new regulations may emerge.
Construction materials	Each state (and sometimes each city) has its own regulations that need to be satisfied to allow products to be installed in new or renovated construction projects.
Products sold outside major markets	Each country has regulations and requirements for local representation, certification, and shipping. These may be subject to change with little warning. Graft and kickbacks can be the norm in some countries, but engaging in those practices is typically illegal.
Lasers or toxic chemicals	Some technologies are regulated and require additional certifications.

Table 3.2 Unique product realization process factors

Industry and regulatory standards are often publicly available. While they may not directly apply to your product, they are good resources to learn about best practices which can be applied to a broad range of products. Here are a few that may be relevant to readers:

- **ISO 9001 Quality Management Systems** [34] is a general standard used to define quality systems across a wide variety of industries. Many of the following standards are derived from these practices.
- **Current Good Manufacturing Practices** are regulations that govern how medical devices are produced [29]. CGMPs outline in detail the following: design control, design verification, process validation, and incoming inspection requirements. These requirements are not very specific (i.e. exact templates or software are not specified by the regulations), and thus give companies some leeway with implementation. Different countries or economic groups may have their own standards; for example, medical devices in the EU are governed by Regulation 2017/745 [35]. There are also some international standards used to certify manufacturing facilities but are not necessarily legally required (e.g., ISO 13485:2016 [36] and ISO 14971:2012 [37] for risk management of medical devices).
- **ISO/TS 16949 (replaced QS9000) and Advanced Product Quality Planning (APQP)** [28, 38] are a set of processes originally developed in the late 1980s by the Automotive Industry Action Group (AIAG), a consortium initially formed by Ford, Chrysler, and GM. APQP describes several required processes, including managing failure risk through the application of FMEA (failure modes and effects analysis), using SPC (statistical process control) in facilities, and the use of PPAP (production part approval process).
- **AS9100** [39] is an aerospace industry standard that is similar to that of ISO 9001, but is specific to the aerospace industry.
- **TL9000** [40] is the telecom quality system. It has the same basic structure and content as ISO 9001 with some specific processes unique to the telecom industry.

3.3 The Pilot Process

Assuming that your product is mature enough to start production (Chapter 2), you cannot jump from prototype to full production rate in a single leap. The production system and the product will rarely work perfectly on the first try – there are simply too many ways the process can go wrong. Tesla learned this when it tried to ramp production too quickly and continually missed production goals.

The structured sequence of incremental building and testing prototypes is called the pilot process. While companies call the pilot process by different names (Table 3.3), all pilots have the common goal of ensuring that when mass production starts, the outstanding issues with production, quality, and performance are minimized. The numbers of pilots needed will vary based on the industry, the complexity and cost of each pilot, the maturity of the design and manufacturing process, the number of design changes, and the promised delivery dates. We will use the terms pilot, pilot build, and build interchangeably.

The goal of each pilot build is to uncover design and production problems before mass production. Companies start by building a small number of units using a mix of production-intent processes (those you will use in mass production) and prototype processes (those which can produce similar parts but with a shorter lead time and limited fixed cost investments). The early builds are used to resolve concept issues such as software/firmware/integration issues or user interface problems. As issues are resolved, the factory builds more units at higher rates, ultimately using final tooling, fixtures, test procedures, workers, and processes. At each step, the products are subjected to rigorous testing and evaluation to ensure the products and processes are meeting their targets. For this book, we will break the pilots into three types: engineering verification testing (EVT), design verification testing (DVT), and production verification testing (PVT). EVT focuses on testing the product concept and engineering, DVT on the product performance as-built, and PVT on the production system's ability to produce at the desired rate.

Typically, for most consumer products, the build sizes grow by orders of magnitude with each phase. The following list describes the phases of building prototypes and pilot products, starting during product development and ending in mass production (an alternative framework – alpha and beta testing – is described in Inset 3.1). Figure 3.2 shows this same information graphically.

- **One prototype.** Going from an idea to the first works-like/looks-like prototype is the focus of the activities of the early phases of product development. Teams work to understand customer needs and begin to design a solution. Once the solution is determined, teams decide how the aesthetics, user interfaces, and functional parts will work together. During the concept through detail design, teams use prototype processes to build the initial units. They are often built using prototype electronics with 3D printed parts or machined parts. The aesthetics are achieved using expensive manual build processes.
- **Five prototypes**. Next, teams create a small number of prototypes, enough to do the necessary field testing and get feedback from customers and other stakeholders. These samples (works-like/looks-like) are typically made, before the factory is engaged, using low-volume or prototype processes. For example, laser cutting may be used instead of stamping. Prototypes made in small numbers should look and perform similarly to the final product but might not have the exact aesthetics or mechanical properties. Using these prototypes, teams can also get feedback on the producibility of the design from internal or external manufacturing resources (Chapter 10).
- **Ten engineering samples**. Product realization starts when the team transitions to using final mass production methods. The next phase, typically called engineering verification testing (EVT), uses a mix of

Table 3.3
Description, alternative names, and characteristics of each pilot phase

		Engineering verification testing	Design verification testing	Production verification testing
Description		Test the design with a combination of production-intent parts (both custom and off-the-shelf) and soft-tooled parts. The production-intent mechanicals are often not ready at this point. However, the functions, fit, and aesthetics can be simulated using 3D printed parts. Initial integration of software and hardware can be tested.	Test the functionality with production-intent parts. The team can tweak the tooling and identify design issues. These units are often used as certification units. Early packaging prototypes are also evaluated.	Test the performance of the production system. PVT pilots are used to adjust the production design, including worker training, standard operating procedures, and production rates.
Other terms for pilot phases	Toy industry	Engineering pilot (EP)	Design pilot (DP)	Production pilot (PP)
	Automotive	Pilot assembly	Pilot assembly	Ramp to production
	TRL	TRL Level 7: Prototype is near or at the planned operational system.	TRL Level 8: Technology is proven to work.	TRL Level 9: Actual application of technology is in its final form.
	MRL	MRL Level 6: Capability to produce a prototype system or subsystem in a production-relevant environment has been demonstrated.	MRL 7: Capability to produce systems, subsystems, or components in a production-representative environment has been demonstrated.	MRL Level 8: Pilot line capability has been demonstrated. The manufacturing system is ready for mass production.
Number of units		5–10	10–100	100–1,000
Use of the units/samples		Sales samples, photos for marketing, software testing, verification and validation of functional specifications, beta testing.	Certification, durability testing, reliability testing, marketing samples.	Final quality checks, aesthetic inspection, early sales, samples for promoters.
Design maturity		Form and function are complete. CAD has been fully defined.	Final details including tolerances and fits have been decided.	"Pencils down," only minor changes expected going forward.
Aesthetics		Prototype created (often hand-finished) with overall aesthetic look and feel designed.	Fit and finish being adjusted, feel of actuated parts tuned, improving weight balance, and finalizing CMF.	Final fit and finish specified. No changes beyond this point. *Golden samples* (examples of products with the right aesthetic value) have been approved and final specs have been communicated to the suppliers.

Table 3.3
(*Continued*)

	Engineering verification testing	Design verification testing	Production verification testing
Tooling status	Some long-lead tools are still being cut. Often, mechanical parts are made using prototype or soft tools. PCBA tooling is complete.	Tooling was done in steel-safe mode (Chapter 12) and not polished. The tool is being reworked to adjust for fit and finish.	Final tooling is complete and tools are polished.
Fixtures	Few fixtures are used. Prototypes are hand-built.	Some production fixtures are used in prototype fabrication.	Final production fixtures have been built and tested.
Packaging	Being designed.	Prototype packaging complete.	Full-production/shipping packaging complete.
Product quality tests	Functional verification has been done to ensure that the prototype meets technical specifications.	Durability and reliability testing is underway. Some certification testing has been started and has to be complete before MP.	Final aesthetics have been approved. QC testing is underway, especially verifying that the units pass any tests failed in DVT.
Testing equipment	Lab/bench test equipment is used to comprehensively test functionality of the prototype.	Some final production test fixtures have been built and used in testing.	Final production testing equipment has been completed and verified.
Assembly labor and production line	Prototype shops or in-house engineering build units by hand.	Prototype lines are used to build the units using highly trained assemblers.	The ultimate production team has been trained and is operating on the final production line.
Production planning	Early forecasts of numbers of needed parts have been done and ordering of long lead-time parts (material authorization) has begun.	Purchase orders (POs) for final production parts have been placed. Forecasts of numbers and scheduling of needed materials are close to finalized.	All purchases and deliveries of materials are planned for initial mass production. Production plan is in place. A rolling forecast for mass production is being created.

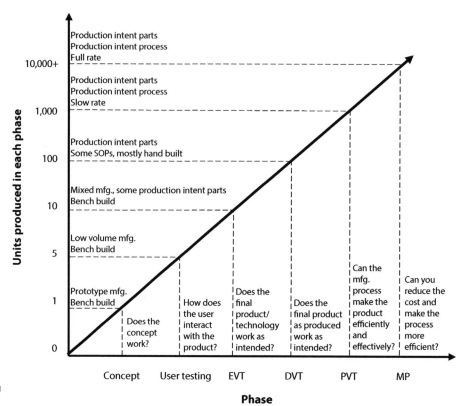

FIGURE 3.2 Going from 0 to 10,000

production-intent and low-volume prototyping. Sometimes the EVT is done at the final factory and sometimes it is done using low-volume manufacturers. Most of the part designs are the same ones that will be used in final mass production units but may be made differently, and the product meets most of the specifications. These units are used to evaluate the product design as it will likely function in the field. The units are also used to test firmware and software. The units are typically not very robust and might not withstand the normal wear and tear required of the final units.

- **One hundred design samples**. The DVT phase uses production-intent parts, but products are not built on final assembly equipment. This phase is used to ensure that the production-intent parts work as expected. DVT is used to support the certification process (Chapter 17). DVT units are also used to test durability and reliability.
- **One thousand production verification samples**. During the PVT phase, bugs in the production are worked out. The final production line equipment and workers are used, but at low enough production rates to catch and fix problems before large numbers of parts are wasted. For example, teams may discover that a hand-assembly process generates too many defects, a testing process takes too long, or the production line contains bottlenecks.
- **Ten thousand plus products**. In the ramp to mass production (MP) phase (Chapter 19), rates increase to final production levels. During this process, teams streamline processes, reduce waste, and improve production flow.

Inset 3.1: Alpha vs. Beta Testing

 Some organizations use the terms alpha (α) and beta (β) testing to frame how prototypes and early units are used to gather customer feedback (and many use the term incorrectly). Alpha and beta testing is typically used in software development; however, some hardware development teams use the terms as well. α and β tests are generally used to validate product usability with customers (Table 3.4). Alpha units are typically early works-like/looks-like prototypes that can pre-date the EVT runs and are used for usability testing and software development. Beta testing is usually done on DVT or PVT units that are used for flushing out any issues that arise when the products are made using production-intent processes and are assembled using the mass production techniques.

Table 3.4 Alpha vs. beta product testing

	Alpha	**Beta**
Pilot stage	Pre-EVT or EVT	DVT or PVT
Testers	Friends and family, early adopters, experts in the field	General users
Scope of testing	Open-ended: used to find the unknown unknowns	Controlled experiments to evaluate critical risks and unknowns
Output	Insights that drive significant design changes	Verification that the product is ready for mass production
Market visibility	Internal only	Used to create buzz for a product through early users

3.3.1 Steps in the Pilot Process

Teams cannot just show up at the factory and run a pilot. A significant amount of planning work is needed to ensure that everything is ready and the right data is collected during the pilot. Figure 3.3 shows the steps involved in preparing for and executing pilots.

Plan Sequence and Timing of Pilots
Early in the product realization process, teams carefully plan their pilot strategy. This includes setting the number of pilots, their purpose, how long each is expected to take, and the number of units (samples) built at each step (Inset 3.2). Any long-lead items needed for the pilots are ordered. Delivery deadlines for tooling, capital equipment, and testing equipment are communicated to the factory and suppliers. The purpose of the planning phase is to ensure that the right resources are allocated at the right time for each pilot.

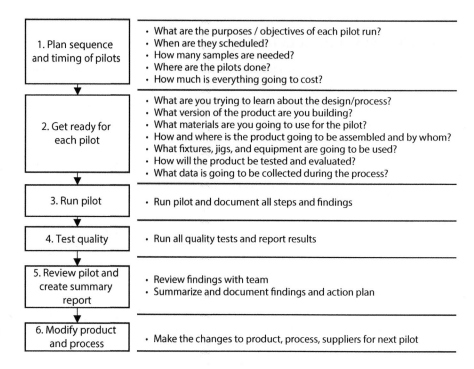

FIGURE 3.3 Steps in the pilot process

Get Ready for Each Pilot

Once the overall sequence and timing of pilots is determined, the team has to plan for each individual pilot run. The actual execution of the build is relatively easy; the hard part is getting ready. Piloting is like painting a room: most of the time spent is not putting the paint on the walls, but getting the room ready. You make sure you have all the necessary materials; you do not want to be running out of the house to buy brushes in the middle of painting. You patch and sand the walls to ensure a smooth painting surface and carefully place tape and drop cloths to prevent paint from getting onto the trim or floor. Rolling the color onto the wall only takes a few minutes, but the entire project would look like a mess without the hours of careful preparation you did beforehand.

Similarly, the better a team plans for piloting, the more efficiently the process will run, and the better insights it will give to the team. Many teams rush headlong into the pilot phase and then realize that they have not ordered enough samples, planned the tests correctly, or recorded critical information. Checklist 3.1 provides a list that teams can use to understand whether they are ready to do a run.

Run Pilot

The pilot run is typically conducted over a couple of days to a week (depending on the numbers made and the complexity of the product). Representatives of the engineering team, production team, and quality team

Inset 3.2: Planning the Size of Each Pilot

Pilots are used to debug the design and the manufacturing processes and produce a number of sample units. The units produced are used primarily for testing, but also by various business functions for other related activities. Part of planning a pilot is making sure you have enough capacity, material, time, and money to produce enough units to satisfy all of the company needs. In addition, you need to plan for some yield losses – units that do not work or have significant flaws. The sample plan lists the number of units that need to be built in each pilot. The whole organization needs to be surveyed so there is no last-minute "I need one too!!" Samples may be needed for the following activities:

- **Quality test types.** These units will be destructively tested and typically cannot be re-used.
- **Certification samples.** These units are sent to the certification testing labs.
- **Development.** The firmware and software teams will need samples to test the integration of the software with the hardware.
- **Investor samples.** Investors may want samples to keep them happy.
- **Sales samples.** Samples might be given to potential customers to test and evaluate.
- **Marketing samples.** Samples might be given to marketing to drive sales and generate marketing collateral.

NOTE: The above icon is used throughout the text to indicate where documents need to be created and managed.

observe the production, troubleshoot any design issues, and identify areas for improvement in real time. These observations should be comprehensively documented.

Test Quality
Some of the units will be tested according to the pilot quality test plan (Chapter 7). The quality tests will include evaluations of:

- **User feedback**. Users will be able to give much better feedback when they have a working product that looks and performs like the final product. Samples are used to identify whether the specification was correct or if changes need to be made to the product requirements.
- **Functional performance**. Every aspect of the product's functionality should be tested.
- **Aesthetic performance**. The aesthetic performance of the product should be evaluated to ensure it meets the aesthetic standards, including:
  - Fit and feel: are the gaps and fits correct?
  - Color: are color and finish consistent and do colors match?
  - Surface: does it have the desired finish without defects?
  - Weight and balance: how does the product feel?

Checklist 3.1: Pilot Planning Checklist

Do you have clear learning objectives for the pilot? Just building the product to build it does not take full advantage of the learning opportunities.
- ❑ What aspects of the design need to be evaluated through formal testing and user testing?
- ❑ What risks on the risk document need to be resolved?
- ❑ What quality testing is to be done in this phase?
- ❑ How many samples do you need for testing, customer samples, marketing, and other business functions?
- ❑ How close does the build differ from the final product?
- ❑ How are the samples going to be used?

What version of the product is going to be built? The design is likely to continue to be updated throughout the pilot planning phase. At some point ahead of the pilot, the team needs to decide what version is going to be built to ensure all of the materials are ordered correctly.
- ❑ **Bill of materials.** The BOM used to build the product needs to be finalized. Last minute changes in the bill of materials can disrupt ordering and planning.
- ❑ **Drawing versions.** The version of the drawing used in the actual pilot should be documented. If there is a problem, you need to know which version was built.
- ❑ **Which colors/versions** are being produced?

What materials will be used for the pilot? The team needs to make sure that materials are ordered in time for the pilot.
- ❑ **Manufacturing methods**. How is each part made? For example, while the part will be injection molded eventually, the part might be 3D printed for an early EVT.
- ❑ **Source**. Where are the parts coming from? Who are the suppliers? Are these the final suppliers? Are the suppliers sending you prototypes or final production-intent parts?
- ❑ **Cost**. How much will the parts cost?
- ❑ **Lead time and material readiness**. Will all of the parts arrive on time and be of the right quality?

How is the product going to be assembled? Builds might occur on a bench or on the final production line.
- ❑ **Bench, pilot line, or production line**. Where is the pilot going to be run and when? Will it slot in with existing production or run on a pilot line? When is the pilot scheduled?
- ❑ **Standard operating procedure (SOP), assembly processes, and manufacturing**. The manufacturing and assembly processes used to build the units should be selected and documented.
- ❑ **Staffing and labor.** Who is going to build the pilot products and what are their skill sets? Are you using expert assemblers who can finesse the assembly, or general workers with limited training?

What fixtures, jigs, and equipment are going to be used? Ultimately, the production line will need custom fixtures and jigs to support production.
- ❑ What manufacturing equipment needs to be ready for the pilot?
- ❑ If fixtures are not ready, how will the process be simulated?
- ❑ What material totes and other equipment is needed to protect the sub-assemblies and final product and the product moves through the assembly line?

How will the product be tested and evaluated?
- ❏ **In-line testing**. What tests, measurements, and evaluations will be taken during the build? To facilitate finding and fixing problems, early builds may be subject to more extensive quality testing than final production. Some testing requires custom equipment and fixtures – for example, ICT (in-circuit testing) on PCBAs. Is the equipment available, and if not, what will be used instead?
- ❏ **Quality test plans.** What quality control tests will be run on the finished prototypes, and are the testing equipment and fixtures ready? How are the samples allocated to the tests? For example, if the team is using 3D printed parts as stand-ins during EVT, it is wasteful to do the full suite of chemical contact tests on the product.

What data is going to be collected during the process? The team needs to plan what data will be systematically collected during the build process and how it will be collected.
- ❏ **What questions are going to be asked:** Process cycle time? Accuracy and completeness of SOPs? Time and motion studies to identify difficult-to-assemble products? Yield rates and scrap rates?
- ❏ **How will the data be collected?** What pictures, data, and notes should be taken for documentation purposes? Who is responsible for the observations and documentation?

- **Durability**. Some of the units will be subjected to stress tests such as drops, vibrations, high and low temperatures, water immersion, and chemicals (oils from hands or cleaning fluids).
- **Reliability.** Some of the functions will be repeatedly tested for their ability to perform over the desired life of the product.

Review Pilot and Create Summary Report

After the pilot is completed, a single report should be created for each pilot which summarizes all of the activities conducted and their outcomes, including in-line observations, recommendations for changes, quality failures, and other comments. The report ensures that all team members can review the results and know what they need to do before the next build. Also, the report is an excellent record and resource for subsequent products. Too often, lessons are not documented, risks are not followed up on, and changes are not implemented.

The review of the pilot must include careful observations made during the build: any part defects, assembly challenges, or inefficiencies should be noted and photographed, the problems should be explained, and a comprehensive root cause analysis should be done. The issues surfaced during a pilot can include:

- Aesthetic defects (e.g., warped parts, excess flash).
- Mechanical defects (e.g., buttons that do not fit).
- Assembly challenges (e.g., wire harnesses are difficult to route).
- In-line test failures (e.g., camera image tests fail because auto-focus does not work).
- Excessive cycle times on some stations (e.g., assembly takes too long and inventory builds up behind the station).

Inset 3.3: Product Realization as an Extension of the Lean Start-up Model

 The term "lean start-up" is used frequently when discussing how to develop innovative new products in an entrepreneurial setting. Initially described in Ries's book on the subject [27], lean start-up defines practices that help companies quickly develop products using frequent feedback from customers. While initially applied to software, its principles are being applied to hardware development as well.

Product realization is philosophically consistent with the lean start-up model. There are several fundamental lean start-up principles that, when followed, will make the product realization process much smoother.

- **Minimum viable product.** The MVP is the simplest possible version of the product that addresses the critical customer needs without containing all of the bells and whistles. Especially if you are launching a new technology in a new market, the fewer features and the less complex the BOM, the easier it will be to go through the product realization process.
- **Innovation accounting and actionable metrics** are metrics that teams can use to drive product improvements. Deciding what you are going to measure and how you will measure it before each pilot will help ensure that failures are identified and remedied quickly.
- **Build-measure-learn** emphasizes building several iterations of the product and rapidly improving the design. The build-measure-learn concept is also very consistent with the pilot phases in which the learning shifts from "does the customer want this?" to "does the technology work and can we build it?"

- Errors in standard operating procedures (or lack of SOPs).
- Material damage (e.g., the totes do not protect materials).
- Packaging issues (e.g., the product does not fit in the specified box).

Each issue or problem identified should include:

- A description of the problem or failure.
- A root cause analysis.
- A corrective action plan.
- Team member(s) responsible for addressing the problem.
- Timeline for correcting the issue.
- Tests to verify the fix in the next pilot.

The risk management document (Section 4.3) should be updated and the production schedule should be reevaluated to accommodate all the tasks involved in correcting the issues found during the pilot.

Modify Product and Process
Based on the plan for and report on the pilot, the various teams will need to implement improvements before the next pilot. The plan for the following pilot will also need to be updated to reflect any re-runs of quality tests or to verify that the changes made had the intended impacts.

3.3.2 Piloting Drives the Total Product Realization Timeline

Delays in piloting are virtually impossible to make up and inevitably delay mass production. It is very difficult to make up time in downstream pilot runs as the tasks in the pilot are often sequential (you have to test the product before you find a quality issue before you redesign before you change your tooling). When faced with delays, it is often tempting to cut out pilot runs, but this can lead to significant quality failures downstream.

Anything the team or factory can do to reduce the number of pilots and the time between them will create more buffer in the schedule and give room for delays. The overall schedule is driven by the number of pilots, whether or not you start your pilot runs on time, and how quickly you can turn around changes for the next pilot.

How Many Pilots You will Need

Most teams underestimate the number of pilot runs that will be needed to thoroughly debug the product and process. The number of pilot runs is highly correlated with:

- **Maturity of the design** (Chapter 2). If the team is still iterating on the product details, more pilots will be needed to test each design modification and debug the design. Also, the later the changes in product realization, the more expensive the changes will be because tooling modifications are typically costly and time-consuming.
- **Challenging target specifications that cannot be tested ahead of time**. For example, if low noise is critical to the quality of the product and there are a lot of moving parts, it is likely the product will need multiple iterations to ensure it is not too noisy.
- **Highly coupled subsystems.** If subsystems can be built separately and tested ahead of pilots, it is easier to get the whole product to work. For example, if motors and controllers can be tested outside the system, it will be easier to integrate them into the whole design. If all components must be built together or cannot be tested until the product is fully assembled, more pilots may be required.
- **Immature firmware and software**. The hardware, electronics firmware, and software integration problems only become apparent when the firmware and software are married with the real hardware. Integration issues may only be solvable through changes to the printed circuit boards. Re-spinning (redesigning the circuits and creating all of the design files) a board can significantly delay the next pilot.

Do You Start the Pilot on Time?

It is easy to miss the planned pilot timeline for relatively minor issues. Reasons for delays include:

- **Not identifying all suppliers.** Groups often assume that it should be easy to find a camera or motor with a specific performance and form factor. If procurement cannot find an appropriate component, it may be necessary to delay pilots to redesign the product to accommodate what is available.
- **Availability of materials.** You cannot start a pilot run until all of the parts are ordered with enough time to account for longer than expected lead times. Some suppliers will be better at meeting delivery schedules than others.
- **Availability of piloting capacity.** There may be limited capacity to do pilot runs within a factory because either of existing production or competition with other products for pilot capacity. If you miss your window, you may have to wait to get back in the queue, even if you are ready.
- **Availability of fixtures, tools, and testing equipment.** For production verification testing (PVT), the production line will need all of the final tooling and fixtures. These pieces of equipment are typically long-lead items, and any delay in their delivery will delay the pilot.
- **The responsiveness of the factory/CM.** Some CMs will be more responsive than others in getting ready for the next build.

How Long does it Take to Integrate the Changes?

After a given pilot is completed, the engineering teams, factory, and suppliers will implement changes to the design, process, parts, and procedures. The ability to turn around changes quickly is also critical to the overall pilot timeline. The time between pilots will be directly related to:

- **Speed of quality testing.** Designers cannot begin working on the next iteration until the quality testing on the current pilot is complete. Waiting ensures that the lessons learned from the prior pilot are reflected in the next pilot.
- **Responsiveness of your design partners.** Not everyone you partner with will be as invested as your team is in getting your product to market quickly. For example, an outdoor equipment company had outsourced the design of their electrical systems to a single-person shop to save on design costs. After EVT, the team needed to implement a number of design changes to fix a safety issue. However, the work was a low priority for the electrical design consultant because he had other clients. The pilot rounds were delayed because the team could not get a rapid turnaround on critical design changes.
- **Time to update tooling.** The longest lead time to get new parts is typically the time to re-cut the tool and reset the process parameters. If you have suppliers who have a quick turnaround, this will reduce the time between pilots.

3.3.3 Where to Do the Pilot Runs

A pilot can be run using a number of production models: flexible job shops, a dedicated pilot line, or on the actual production line that will ultimately be used for mass production. Each approach has its costs and benefits.

Job Shop

Early pilots (EVT) may be built in a job shop or maker space where the product is made by highly skilled workers. The cost of these builds is very high when compared to the final mass-produced product, but the volumes are typically low. The benefit of a job-shop approach is that you get a prototype that is both mechanically and aesthetically close to the final product. In addition, job shops do not typically require tooling, so the lead time for getting the materials together is relatively short.

Dedicated Pilot Line

The second option is to use a dedicated pilot line either in the factory or in a pilot facility. Dedicated pilot lines mimic the final production environment and facilitate learning without interrupting existing production. Pilot lines are characterized by:

- Dedicated space that can be reconfigured quickly for different products.
- Flexible fixtures, tools and equipment that can accommodate a range of products.
- Highly skilled workers and teams who can manage unexpected variations and adjust the parts to ensure a high-quality product.
- The products manufactured in these pilot lines are similar to the final product, but production speed and quality are not representative of those of mass production. Teams still need to ramp up production on final production lines to identify issues specific to mass production.

Some CMs offer pilot lines in the US (or Europe) and then transfer production to lower-cost facilities overseas for mass production. Running pilots near the development team and then transferring production to low-cost facilities has obvious benefits. First, more of the team members can be present at the build, and travel time and cost are reduced. Second, teams can better communicate with the build team during the pilot phase when communication is most crucial (e.g., fewer language or cultural barriers to inhibit communication).

Teams should also be aware that conducting the pilot and mass production in two different facilities can also have a few drawbacks. First, some of the tooling and parts necessary for final production might have to be built at an overseas factory, and shipped to the US for the pilot builds. Second, learning is hard to share between sites. For example, if workers find a clever work-around on assembly, they must train the overseas workers in the same method.

Production Line

Another option is to run the pilot on the production line that will be used for mass production. Typically, production lines are already in use. Pilots can be run when the line is not active or can be intermixed with existing production (mixed model production). For example, new automotive versions are typically produced alongside older versions on the same line. Pilot runs produced on mass production lines have the following characteristics:

- Manufacturing of existing product lines may need to be paused to run the pilot, products may be run off-shift, or the pilot runs may be intermixed with existing production.
- Teams use existing fixtures and equipment.
- The same workers can be used to run the pilot and mass production. Their experience helps them learn the new product.
- There are fewer surprises between the pilot and the production run.
- Typically used for products that are part of a family or are generations of the same product.

Sometimes the team does not have a choice if the company does not have a dedicated pilot line or facility. However, if you are going to select the contract manufacturer route, different CMs will offer different options. There are several costs/benefits of each approach, as well as ways of mixing them (Table 3.5). Some companies will do early EVT and DVT pilot runs in a pilot facility, and then PVT will be done in the final factory.

Table 3.5
Pros and cons of various pilot locations

Benefits of the pilot lines	Production line	Pilot facility overseas	Pilot facility in-country	Job shop
Easy for design team to engage in the pilot process		⇔	⇑	⇓
Ability to make rapid changes to tooling		⇔	⇓	⇑
No interruption to existing production	B	⇑	⇑	⇑⇑
Quality of the workforce	A	⇑	⇑	⇑⇑
Ability to find production issues related to labor, tooling, fixtures, etc.	S E L	⇓	⇓	⇓⇓
Ability to customize for customer samples	I N	⇑	⇑	⇑⇑
Fast transition from pilot line to mass production line	E	⇓	⇓⇓	⇓⇓
Ability to pilot very new technologies and new assembly methods		⇑	⇑	⇑⇑

⇔: Similar to baseline; ⇑: Some advantage; ⇑⇑: Significant advantage;
⇓: Some disadvantage; ⇓⇓: Significant disadvantage

3.3.4 Pilots Never Run as Expected

Ideally, a given EVT, DVT, or PVT pilot is well planned, all materials and tooling are at the right maturity for the pilot, and the build starts on time. However, reality is never ideal: tooling will not be ready exactly when hoped for, design changes will be pending, fixtures will not be built, or someone's flight will be canceled. As a result, pilots will always run slightly differently than planned. Also, extra pilots may be needed.

In the chaos of extra pilot runs, pilots not running as expected, or pending design changes, it is vital that the current state of the design – BOM, CAD, tooling plan, SOPs – is documented for each pilot. Several typical failures that occur during chaotic piloting include:

- It is too easy to forget "was the new battery configuration included starting in DVT3 or in DVT4?" and not be able to attribute functional or quality failures correctly.
- Running pilots for the sake of running a pilot wastes time without learning anything significant. Teams feel the pressure to run builds but do not have clearly defined goals. By rushing in and not planning, teams miss many learning opportunities.
- Not having quality test plans defined likewise wastes time. The samples sit around while teams decide what to test. As a result, problems are not found in time for the next build.
- Too often, organizations dismiss a failure in a quality test because "we have a pending design change to fix that." They then fail to re-run the test to validate whether the fault reoccurs with the new design.

By thoroughly planning pilots, understanding the goals, documenting carefully, and systematically managing the issues that are uncovered, teams can prevent the above problems from causing significant quality issues or delays.

3.3.5 Piloting New Manufacturing Processes

In some cases, the new product may include new and untested manufacturing processes. New manufacturing processes may have been run at a low production volume or in lab conditions, but need to be ramped simultaneously with product volumes. New processes can be especially challenging if increasing the scale of manufacturing requires the design and production of new capital equipment.

Failing to appreciate the complexities of simultaneously developing both a new product and new processes at full-production scale can lead to spectacular failures. For example, around 2013, GTAT was contracted by Apple to produce single crystal sapphire as a replacement for their glass phone covers. GTAT agreed to simultaneously ramp up a new large-scale production facility, increase yields, and grow much larger crystals than they ever had. Significant resources and capital were expended to build the facility and ramp production in a short period, but when the company was unable to deliver on its promised timeline, it overextended itself and eventually declared bankruptcy [5]. While there were many contributors to the bankruptcy, Chief Operation Officer (COO) Daniel Squiller attributed the failure to the inability to ramp production of large boules sufficiently quickly. They had produced smaller boules for other companies but the process changes necessary to produce the larger units took longer and cost more money than expected, leading to the liquidity crisis that ultimately caused bankruptcy [41].

Summary and Key Takeaways

- ❏ Manufacturing at scale is much harder than producing one or two units.
- ❏ The product realization process is part of the overall product development process.
- ❏ The product realization process has the goal of ensuring the product is ready to mass produce while meeting cost, quality, and schedule targets.
- ❏ The pilot process helps teams to move up the volume curve. The process starts with engineering verification testing (EVT), then design verification testing (DVT) and then, finally, production verification testing (PVT). Depending on the industry and the technology maturity, differing numbers of pilot runs will be needed to ramp up to full production.
- ❏ Different industries will impose different requirements on the product realization process.
- ❏ Planning for the pilot process is critical to successfully learning as much as you can.
- ❏ The process will never go as expected and delays are very difficult to make up.

Chapter 4
Project Management

'THE PATH AHEAD'

1. INTRODUCTION

2. ARE YOU READY TO START ?

3. PRODUCT REALIZATION PROCESS

MILE 2

4. PROJECT MGMT

Active project management is essential for all of the resource-intensive, iterative, and cross-functional processes involved in product realization. This chapter reviews four project management tools that your team will need: critical path management, delineation of roles and responsibilities, risk management, and enterprise data management.

Product Realization: Going from one to a million, First Edition. Anna C. Thornton.

© 2021 John Wiley & Sons, Inc. Published 2021 by John Wiley & Sons, Inc.

Getting a product from prototype to production requires the coordination of hundreds of people executing thousands of time-sensitive tasks on time and on budget with little room for error. Unlike being late with a prototype, there is often millions of dollars at stake when launching a new product. Entire companies can be made or broken not on the product idea but how it is executed. Product realization is also a process that has to be adaptive to unexpected changes; the whole purpose of the pilot process is to uncover and fix problems, but each of those changes can have a ripple effect throughout the process.

Failure to keep all these activities on track and manage the inevitable changes to them can result in:

- **Violating contractual obligations**. Sales contracts are often contingent on meeting specific delivery dates. Missing this window can not only risk cancelation of the existing orders but also risk ongoing sales.
- **Shortcutting the quality testing process**. It is tempting to sacrifice quality testing when a schedule crunch occurs, but this is almost invariably a mistake. Doing insufficient durability and reliability testing can leave the product at risk for failures in the field.
- **Missing seasonal trends**. Many products sell a majority of their units within fixed time windows: toys for the holiday season or grilling equipment for the summer season. Missing a target date can mean losing a sizable percentage of the entire year's revenue, and can give the competition a chance to get a foothold in the market.
- **Loss of consumer confidence**. Negative reviews can be hard to scrub from a product's image. If early missteps are made public, whether on tech e-publications or on social media, those poor reviews are permanent.
- **Insufficient cash flow**. Delaying the launch of a product can stress cash reserves. Teams often do not budget for unexpected extra pilot runs, additional samples, and the expenditure rate required to "keep the lights on" (e.g., office and travel expenses) past the expected launch date, so running late can mean running out of cash. This problem can compound because raising additional funds when you have just missed your deadline and do not have enough money for the first PO (purchase order) can significantly reduce your valuation. Investors can smell desperation.

Keeping everything on track typically requires the attention of a project manager. In an ideal world, the perfect team would not need project management. Everyone would be aware of crucial issues, teams would evenly divide the work, procedures and changes would be clearly communicated, and people would reach across functional boundaries to ask critical questions.

The reality, as you can easily guess, will always be less than ideal. Product realization is often too complicated for each individual to be aware of all aspects of the project. "I thought someone else was dealing with that," "That should have been easy," and "I didn't know that was due now" become familiar catchphrases when teams miss deadlines.

Project management is especially important in product realization when the tasks of all of the functional groups converge. The engineering teams are finishing up the designs and making late changes; the product team is getting the feedback from the user; operations and manufacturing are getting the production system ready; finance is tracking the money and managing cash flow; legal is managing the contracts with suppliers; and human resources (HR) is typically hiring new staff in preparation for the production ramp. Throughout product realization, there needs to be significant cross-functional communication. For example, the team responsible for quality testing needs to understand any changes to the specification document made by the engineering team and update their plans for the new requirements.

Active project management (PM) is critical to keeping complex activities like product realization on track. PMI [42], a leading project management institution, defines project management as "… the application of knowledge, skills, tools, and techniques to project activities to meet project requirements." A PricewaterhouseCoopers (PwC) study found that 97% of the companies surveyed believed that PM was critical to business performance [43]. Effective project management will help teams communicate the impact of changes, complete tasks well in advance of critical dates, and ensure that avoidable problems do not trip up the launch of a product.

This book will not cover all aspects of project management. There are numerous books already published that cover the subject in depth for any readers who are interested. This chapter will highlight a few project management tools that are critical to product realization:

- **Defining roles and responsibilities** (Section 4.1) ensures that teams are aware of their obligations and that the right skills are employed in a timely fashion. It allows teams to avoid excuses such as "I thought the other team was responsible for that," or "we didn't know we had to do that."
- **Managing the critical path** (Section 4.2) keeps the team focused on the sequence of activities that defines the overall schedule. Without knowing what is essential, people tend to do the work that brings them recognition, that is easily doable, or that they think is important. However, dull, hard, or tedious tasks are often most crucial to the schedule. Understanding the critical path helps teams focus on what the project needs rather than on what individuals prefer to work on.
- **Managing and tracking risk** (Section 4.3) ensures that potential roadblocks are addressed and cleared. Teams risk product failure when they neglect to notice potential issues or when they raise issues early in the process but neglect to follow up on them. Careful risk management systematically documents and tracks these issues and ensures that teams mitigate the risk and tie up all loose ends.
- **Managing enterprise data** (Section 4.4) ensures that all of the data needed to execute product realization is up-to-date and accessible. With hundreds of documents, it is easy for someone to post the wrong

version of a document, or for two people to unknowingly edit the same file. Enterprise data management (EDM) imposes discipline on how data and documents are stored, accessed, updated, and retrieved.

4.1 Roles and Responsibilities

When the idea for your product is first born, whether in a start-up or on a design team at a larger company, you'll probably be working on a small, collaborative team in which everyone does a little bit of everything: "I'm the lead engineer, but I also designed the logo and helped with customer validation."

By the time you're ready to launch a hardware product into the market, though, your team will involve far more people who can't all communicate easily with one another, so roles and responsibilities must be clearly defined. While a core product team may only have 5–10 people, the extended team can be much larger because it includes all of the factory, suppliers, and service team people who will touch your product. Before diving into product realization, it is essential to know what skills and capabilities will be required and to ensure that your core and extended team include all of these skills. It's dangerous to find out that you need a customer support skill set just before launch, leaving you scrambling to find a supplier at the last minute.

It may be tempting (and it's often most cost-efficient) to outsource many roles that are not in the immediate core capabilities of your team. Keep in mind, though, that outsourcing the role doesn't mean outsourcing the *responsibility* for the work. Someone within your organization needs to own and understand the outsourced work, assess the work quality, and integrate the supplier's contributions back into the core team.

The next two sections describe the typical roles and responsibilities in a product realization process and how to manage and track who is responsible for what. Keep in mind, though, that there are as many ways of organizing teams as there are companies. Depending on the history of the company, its size, the technical challenges, and the industry, each firm will have a small or large variation from what is described here. Individuals may have multiple roles or entire divisions may be dedicated to one set of tasks. A start-up may have a single person responsible for the entire supply chain, whereas a large automotive company may have entire divisions devoted to merely tracing the carbon footprint of their supply base. Your individual organization will likely differ in some ways from this template, but the jobs that need to be done will be similar.

4.1.1 The Functional Groups

All organizations – independent of technology, size, speed, or country – need roughly the same functional groups to launch a product. The organization chart defines how each of the groups reports up through to senior management. Each company may have unique names for their functional organizations or unique reporting structures. In almost any organization, the following are the functional roles directly involved with the product realization process.

Executive Team

The Chief Executive Officer (CEO), Chief Operating Officer (COO), Chief Product Officer (CPO), and Chief Technology Officer (CTO) typically set strategy and oversee the functions, but don't get involved in the day-to-day minutia of the product realization process unless there is a serious problem.

Project Management

Project managers coordinate across functional boundaries, track risks, and ensure that the teams stay on schedule. They sometimes report to a dedicated project management functional group or may report up through the engineering or product management organization. In some companies, what we refer to here as "project managers" might be called team leaders, project leaders, or project coordinators. Depending on the company, available team skills, and project needs, the project manager may provide technical input as well as ensure that cost, quality, and schedule goals are met. In others, a project manager only maintains the schedule, keeps and updates risk documents, and ensures that all tasks are completed on time.

Product Management

The product team members are responsible for managing relationships with customers. They identify customer needs, predict likely sales, and sell the products to the customer. During product realization, they act as the voice of the customer and make the final decisions on tradeoffs between cost, quality, and schedule. They test the pilot samples with customers and suggest modifications (to the product or the branding) between pilot phases. Product management includes the following functional groups:

- **Product owners** have primary responsibility for the final product definition, brand, and marketing approach. While they take input from many different parts of the organization, they make the final tradeoffs between cost, quality, and schedule.
- **Brand management** is responsible for the ways in which all aspects of the product (e.g., packaging, mobile applications, color, material and finish (CMF) specifications, and marketing) consistently communicate the brand to the consumer.
- **Marketing** works directly with the customers and analyzes global trends to understand customer needs, market dynamics, and how to target products to specific markets.
- **Sales**, not surprisingly, manages the actual sales of the product. In some organizations, the sales team may also be responsible for developing and communicating the sales forecast. In other organizations this is done by the product manager or the marketing role.

- **Customer support** team is responsible for interacting with the customer after the customer purchases the product. For example, they answer the phones when there is a complaint and help facilitate repairs. Sometimes these people are located in the quality group or operations group.

Engineering

Engineering the product is more than just defining the geometry of the parts. For a typical electromechanical system, the engineering team will include the following functional groups:

- **Research and development (R&D)** – sometimes called development engineering – develops product technology and ensures that it is mature enough for mass production. This team is often engaged throughout the pilot process to support troubleshooting and validate design changes.
- **Mechanical engineering** designs the mechanical parts (e.g., structures, gears, motors, and mechanisms).
- **CAD engineers** create the CAD models and drawings that define the geometry of the product and its parts. Sometimes the mechanical engineering and industrial engineering take on this role and do their own CAD.
- **Electronics engineering** designs the electronics including printed circuit board assemblies (PCBAs), displays, power management systems, and sensor systems.
- **Firmware engineering** writes the custom firmware for the electronics.
- **Software engineering** designs and writes software to run the product and provides the user interface (UI) of the device.
- **Application (app) design and development** creates apps that run on smartphones or laptops and communicate to the product. Apps need to work on a variety of operating systems and devices.
- **Industrial design (ID)** defines the look, feel, ergonomics, and hardware user interfaces for the product (buttons and switches). ID defines the CMF (color, material, and finish) specs for the product.
- **Packaging design** designs the packaging (e.g., gift box, inner packs, and inserts) to both hold the product securely during shipment and to communicate the brand of the product.
- **Finish and paint engineering** ensure that the coatings, finish, and paint meet both the durability and aesthetic requirements.
- **Continuing engineering** plans and executes design changes during mass production to reduce cost, improve quality, and fix problems found in the field.

Not all of these groups will be at the peak of their activities at all times during the product realization process. For example, R&D typically is very active in early product development and helps with the early engineering verification testing (EVT) pilot phases. R&D is available during the rest of the product realization process to provide analysis and troubleshooting. Continuing engineering usually becomes active as the product transitions into mass production when the customer starts giving feedback.

Manufacturing Operations

This team is responsible for the manufacturing equipment and facilities and for producing and delivering a quality product on time. Operations engages with the design team before the hand-off to the factory to ensure that the product is producible. This team owns the mass production process and includes the following functional groups:

- **Factory operations** manages the factory, the production lines, and the labor force. In cases where a CM is engaged, factory operations has primary responsibility for managing that relationship.
- **Manufacturing engineering** has responsibility for designing the tooling, fixtures, and material-handling equipment required for production. This team is also responsible for defining the SOPs.
- **Production forecasting and planning** teams make sure that the factory capacity can support the sales forecast. They ensure that materials are ordered to match the production schedule.
- **Industrial engineering/operations** teams ensure that factories have sufficient capacity, determine where automation is appropriate, optimize the production line to reduce the amount of work in progress (WIP), and ensure that the labor plan matches the production requirements. They may also design the totes and material handling equipment.
- **Distribution** ensures that the product gets from the factory to the customer. The distribution plan can include complicated shipping processes, multiple modes of transport, and navigating customs and duties in each country.
- **Reverse logistics** gets faulty products back from the customer, executes repair or disposal, and returns a working unit to the customer. This group often works closely with the customer service group to ensure that customers' complaints are handled correctly.
- **Supplier management** identifies and manages supplier relationships, from selecting the right suppliers to conducting quality audits of each critical supplier. This group works closely with the procurement team (see below under "Finance") on price negotiation and sales forecasts.
- **Material management** (can also overlap with purchasing and procurement) ensures the timely ordering of materials and manages the inventory. Material management communicates closely with procurement to place orders and coordinates directly with the suppliers about ordering and delivery.

Quality

Quality team members should be involved with the product development process from the start. If involved from the beginning, this team can help avoid choices that increase the risk of failing durability and reliability tests and increased scrap and rework. However, quality is often only invited to the table when the pilot builds start.

In some organizations, quality reports through operations, with the design-related quality team managed by the engineering groups. In other cases, quality is its own functional group reporting directly to the CEO, and is a parallel organization to manufacturing, engineering, and product management groups.

- **Quality engineering** defines the quality strategy for the product. The quality engineers work closely with the product engineering team to design products that will be robust to manufacturing variations. Quality engineering takes primary responsibility for defining the quality test plan and overseeing the Quality Management System (QMS). This team outlines the set of tests required to qualify the product for sale and shipment to the customer.
- **Regulatory compliance** is responsible for managing the certification requirements, testing, and paperwork. Regulatory consultants are often used in cases where there is no internal team or the product regulations are complex or unclear.
- **Test engineering** designs and develops custom test fixtures and procedures, and the members typically have an engineering background. For example, in the cycling industry, these engineers would be responsible for developing fixtures to test the reliability of a shifting mechanism. The test fixture would repeatedly shift the gears automatically and test performance over its expected life. Test engineering needs to work hand-in-hand with quality and design engineering.
- **Testing technicians** run the quality assurance tests and typically have associate degrees. Sometimes the testing technicians are associated with the factory and sometimes with the quality organization.
- **Quality assurance** ensures that the manufacturing processes done within the factory walls meet specifications. QA drives continual improvement to reduce scrap and rework. The personnel qualifications range from associate degrees to advanced degrees in engineering.
- **Shipment audit** is the check on quality done just before shipment, and is typically done by a third party who specialize in shipment audits.
- **Metrologists** are responsible for verifying the dimensional accuracy of fabricated parts. They have deep expertise in tolerance analysis and measurement methods such as coordinate measurement machines.
- **Continual improvement** is responsible for identifying problems in existing production, doing root-cause analyses, and working with various other teams to implement improvements. This team works closely with the continuing engineering team and sometimes reports up through operations.

Finance

Finance has a range of responsibilities including budgeting, managing cash flow, and ensuring that materials are ordered. The finance team is engaged throughout the entire process from initial product concepts through delivery and customer support.

- **Finance** manages the financial cash flows and expenses. Finance needs to coordinate with all other functional groups to understand the magnitude and the timing of cash outlays and revenue.

- **Procurement and purchasing** is responsible for ordering materials to ensure no delay in production. This group coordinates closely with operations and sales to understand the sales forecasts and when and how much inventory needs to be purchased. This group also negotiates cost reductions for volume purchases and sets up and maintains second sources for critical materials.
- **Order management** takes and executes the orders from the various product distribution channels. Order management works closely with sales and operations teams to ensure that the factory can fulfill the orders.

4.1.2 Assigning Roles and Responsibilities

The easiest way to ensure that all of the tasks are completed on time is to appoint someone (either an individual, a supplier, or team) to each deliverable and to hold them accountable. In start-ups, individuals may have ownership of multiple activities, whereas in larger companies, entire groups collaborate on single deliverables. The process of defining roles and responsibilities helps to identify gaps where activities are not allocated, to ensure individuals are not overcommitted, and to clarify who has ultimate approval on deliverables.

The RASCI (or RACI or RASIC) (**R**esponsible, **A**ccountable, **S**upporting, **C**onsulted and **I**nformed) matrix is one formal way of organizing and documenting these roles and responsibilities, as described in Table 4.1.

The RASCI document is a living document. Typically, the project manager updates the RASCI as the task list and team expand. Table 4.2 shows an example of a RASCI chart. The first column contains a list of the major deliverables and tasks in the process. The first row lists the people or roles involved in the process. Each cell identifies the responsibility each person has in each role – **R**esponsible, **A**ccountable, **S**upporting, **C**onsulting, or

Role	Description
Responsible	Who is doing the actual work? Who is responsible for the deliverable?
Accountable/Approver	Who is responsible for approving the deliverables? Who decides when the deliverable is not good enough? Who is ultimately accountable if something goes wrong?
Support	Who provides support during the implementation of the activity/process/service? Whom does the responsible party rely upon to do some of the work and subtasks?
Consulted	Who can provide valuable advice or consultation? Who is the subject matter expert?
Informed	Who should be informed about the progress of the deliverables or significant changes to the scope of the work?

Table 4.1
RASCI chart
definitions of roles

Table 4.2
Example of
a RASCI chart

	Team leader	Product eng.	Product manager	Quality eng.	Mfg. eng.	Industrial eng.
CMF files	I	I	A			R
CAD files	I	R	A	S	I	S
Quality plan	I	S	A	R	S	
Tooling plan	I	S	A	I	R	

Informed. The purpose of the RASCI chart is to ensure that every task/deliverable has one responsible resource (R) and one resource who has ultimate approval (A). The other three categories – supporting, consulting, and informed – define who will support the responsible group and who needs to be kept abreast of the results.

To build a RASCI chart, the project manager will typically execute the following steps:

1. **Define the major deliverables** and assign a row to each one. The rows should include the deliverables on the critical path as well as any preparatory or secondary activities. This list should consist only of major deliverables, not individual tasks (such as those outlined in a schedule or Gantt chart). For example, a quality control plan is a deliverable. There may be many tasks associated with creating a quality plan, including drafting the plan and reviewing it with suppliers, but these should not be included in the RASCI chart. Defining the RASCI at the individual task level significantly overlaps with project scheduling and can create duplication and errors between the two.

2. **List the resources and suppliers in the columns.** In larger companies, the columns should list the job description (e.g., lead mechanical engineer) or role rather than an individual's name. This ensures that the document will remain correct even if people move or change roles. For smaller organizations, in which individuals have multiple roles, it can be easier to assign columns to individuals.

3. **Draft the RASCI chart.** Typically, the table is drafted by the project manager or the product lead. Only one role should be tagged as "responsible" and only one role as "accountable" for each deliverable. This enables single-point accountability.

4. **Review with each person or group leader** to verify that people are correctly assigned roles, that they understand their roles, and have enough time to complete the work on schedule.

5. **Identify tasks where insufficient resources are allocated** and identify extra resources (or a plan to engage them). For example, it may become clear that no resource is allocated to designing

and managing the packaging. At that point, the team may begin to search for a vendor who could manage both the design and production of the packaging.

6. **Maintain the document** over time as the team and deliverables change. In keeping with the ethos of the document itself, only one person (usually the project manager) should be responsible for maintaining and modifying this document. The document should be available for review by the whole team.

4.2 Critical Path

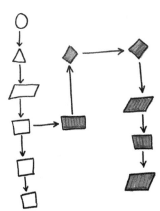

This book is only into Chapter 4 and has already described a complex and interrelated set of tasks. Precisely enumerating the product realization process and all of the tasks in a large and complex Gantt chart is a daunting, if not impossible, effort. Given all of the tasks to be done, you'd likely lose focus on those that ultimately drive the total timeline. To focus the team on what is most important to the delivery, teams can identify the critical path and actively manage it. This is not to say the other tasks can be ignored, but focusing on the critical path helps prioritize what is critical to the schedule. In the past, project management had the adage that projects become late one day at a time. However, because of the uncertain nature of product realization, you can be on track one day and a major problem can create sudden delays of weeks if not months.

The critical path is the sequence of tasks and resources that determine the shortest time it will take to deliver the project. The path is typically maintained as part of the project schedule through the Gantt chart or other scheduling documents. Often the critical path will change through the project as delays occur and additional iterations are needed. Feeder tasks are the activities that have to be finished before an item on the critical path can begin. In order to stay on schedule, teams must try to ensure that:

- **Every task on the critical path takes as little time as possible**. Taking longer to build tooling or to run a pilot directly impacts the overall schedule, and it is challenging to recover this time.
- **Feeder tasks are all completed on time.** For example, if materials are not ordered far enough in advance of the pilot phase, your team will have delays on the critical path.

In the past, project managers would outline the critical path manually, but almost all teams now use project management software to create their critical paths. There will be differences in the activities on critical paths depending on the industry and the complexity of the product. For most electromechanical product realization processes, the critical path typically starts with finishing the design, building tooling, then passes through the pilot phases and manufacturing phases, and ends with distribution. Figure 4.1 shows a typical critical path with a set of parallel feeder tasks.

A typical consumer electronics product's delivery schedule is determined by the time needed to sequentially execute the following tasks:

- **Complete the product design package** (Chapter 6). The product design must be close enough to final to get accurate cost estimates from the factory and to design the tooling. The design package (all of the CAD, drawings, bill of materials (BOM), specification documents, and CMF files) is complete only when manufacturing engineering reviews the files and ensures that they can make the product at the target cost. Getting the design package ready for hand-off to the factory is highly dependent on the work and attention to detail paid during the detail design phase. Completing the design package can be as easy as exporting the files and sending them to the factory, or can it take weeks to clean up the drawings to get them ready for hand-off.
- **Review all parts for manufacturability** (Chapter 10). Even though the design package is complete, the team will need to make minor modifications to enable manufacturability – for example, adding ribs, removing undercuts, and adding snaps – depending on the manufacturing processes used. Each type of process will have its own set of rules. If the engineering team neglected to take manufacturing complexities into account, this phase could involve significant redesign. It can be as fast as a day review or can take several months to iterate designs.
- **Design the tooling** (Chapter 12). This phase takes the CAD of the fabricated parts as an input to designing the tools needed to manufacture the parts. Tooling design is not as simple as making a negative of the part geometry and machining the cavity. For example, tooling design for injection molding includes accounting for shrinkage and warpage as well as cooling, material flow, parting lines, and ejector pins. Tooling design can take weeks to months.
- **Build the tools.** Cutting hard tooling is a very long process and can take 2–12 weeks depending on the complexity, size, type, and the number of tools. Cutting tooling is also expensive, so getting it right the first time is very important.
- **Produce the first parts from the tool** (also called first shots). The design verification testing (DVT) pilot run requires using production-fabricated parts. The time to get the first run done after the tooling is completed is a function of manufacturing and assembly equipment availability and time to produce the batch. This can take days to weeks. Until the tool is ready, first shots have been made, and the first part inspection (FPI) completed, the DVT process cannot proceed.
- **Execute the DVT and PVT pilot runs** (Chapter 3). The product will need to go through sufficient DVT and production verification testing (PVT)

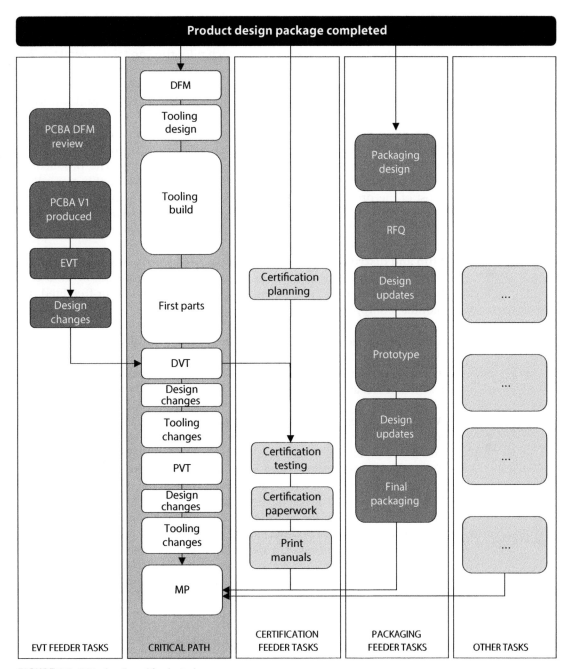

FIGURE 4.1 Critical path and feeder tasks

pilots and the associated testing to build confidence that the product is of the right quality and performance for mass production. Depending on the complexity and type of product, each pilot run can take several weeks to months. After each set of builds, the team must test the products and implement any needed modifications to the design and the tooling. Each build can take a week or two, with a month or so between builds if tooling modifications are needed.

- **Start mass production** (Chapter 19). Once the production is validated, then mass production can start. There is always a gap between starting mass production and finishing enough product to ship. The production system will likely operate slowly at first, even if PVT was able to resolve most of the production issues. Getting enough saleable product for shipment can take days to weeks depending on the total manufacturing lead time of the product and the production run rates (how many products a factory can build per unit of time).
- **Ship and distribute the product** (Chapter 16). Depending on the method of transport (sea, air freight, or trucking), the transportation time can be on the critical path. This can range between days if air-freighted, to months if shipped by sea.

You may have noticed that this list is not a comprehensive summary of all the tasks that need to be completed to get your product to market on schedule. Other tasks that feed into the critical path, called feeder tasks, can also delay the schedule because being late delays the critical path. For example:

- An EVT pilot uses prototype production parts and runs in parallel with tooling design and manufacturing. DVT cannot start until the EVT is complete, and the resultant design changes completed. If EVT is delayed, so is DVT.
- The DVT process usually must wait for tooling in order to produce the first parts. However, even if the tooling is completed on schedule, the DVT build may be delayed beyond the tooling build time if the materials are not purchased in time or there are late design changes.
- The packaging design and packaging pilot processes run in parallel with the product pilot process. Any delay in packaging can delay the ability to produce or ship a product, even if everything else is ready.
- Certifications are legally required to ship the product, and any delay in those can delay shipment.

4.2.1 How Long Will the Critical Path Be?

When planning how long product realization will take, the team will need to assess the timing of all activities on the critical path and set a target schedule. While some activities cannot be compressed (e.g., cutting tooling), others might be influenced by the target ship date. For example, a tight schedule might limit the maximum number of pilot runs a team can deploy. Teams can use questions like those in Checklist 4.1 to set the schedule.

Checklist 4.1: Factors to Consider when Developing a Product Realization Schedule

Product design package and specifications questions
- ❏ How long to get to a mature design package?
- ❏ Is the specifications document complete? If not, how long will any modifications take?
- ❏ If some or all of the design is being outsourced, how long to get a completed design?

Manufacturing relationship questions
- ❏ If the team is using a contract manufacturer (CM), is the relationship already formalized?
- ❏ If you don't yet know where the product is going to be manufactured, how long to finalize the production partners?
- ❏ How long will it take your production partner to review your design package?

Tooling design and production questions
- ❏ How manufacturable are the parts and how much redesign is needed?
- ❏ How long will it take to design the tooling?
- ❏ How long will it take to build the tooling?
- ❏ What fixtures are required, and when are they needed for each pilot?
- ❏ When are tools needed for each pilot and where can prototype tooling be substituted to get the tool off the critical path?

Material readiness questions
- ❏ What materials are needed, and from which suppliers? How long will it take to get the materials from each supplier?
- ❏ Are you having any custom parts designed by outside suppliers that will require testing and iteration?

Pilot process questions
- ❏ How mature are the product and manufacturing processes?
- ❏ Is there limited piloting capacity (i.e., do schedule or budget constraints limit the number of pilots that can be run)?
- ❏ How much redesign is expected after each pilot? Is it an update of an existing product with slightly modified aesthetics, or is it a completely new and untested product type?
- ❏ Are you running pilots to test new manufacturing processes? Can these be done in parallel or do they need to be done in sequence?
- ❏ Are you using an existing production line or are you waiting for the production line to be built?
- ❏ Is there a dedicated pilot facility?

Certification questions
- ❏ What certifications are needed, and how long will they take to acquire?
- ❏ How early can the certification process start? Can you start certification during EVT or DVT, or do you need to wait until PVT?

Order fulfillment questions
- ❏ What shipment method will you use to get your product from the manufacturer to the retailers/consumers, and how long will that take?
- ❏ How are you fulfilling orders? Are you shipping directly to the consumer from the factory, or are you shipping from a distribution center?

4.2.2 What Can Derail the Schedule?

It should be no surprise to you that the answer to the question "What can derail the schedule?" is "many, many things." Murphy's oft-quoted Law states that "Whatever can go wrong will go wrong."

As far as this author knows, there has never been a single product brought to market that did not have schedule hiccups that threatened the promised delivery date. Suffice it to say, anticipating these types of failures and ensuring enough buffers in the schedule will help reduce the impact of the unknown unknowns.

All of the things that could go wrong during product realization could fill an entire book, but below you'll find just a short list of the several common causes for delays (and for a lot of gray hairs and sleepless nights).

- **The product fails durability testing.** In one case, the designers of a product that plugged into the USB port of a car did not take into consideration that the products would sometimes be dropped. During standard drop tests, the product easily shattered. Unfortunately, the design didn't allow space to add reinforced snaps or screw bosses, and the housing had to be completely retooled. The retooling delayed the product for several months.
- **Less than capable workers in PVT**. In the case of a high-end audio product, the company tried to ramp to production quickly without training workers at the manufacturing facility. These workers were not careful in the assembly process, and they got glue on the outside of thousands of units of the product. The product was very high-end, and customers would not pay for a shoddy glue-stained item. The initial shipment of several thousand units had to be rebuilt, causing a delay of almost a month.
- **Failure to detect faulty systems**. The CM for an internet-enabled communication device claimed that it could do cellular testing on-site, but failed to detect that the antenna assembly in the factory was poorly soldered and prone to breakage. Many defective units were shipped from the overseas CM to the US, and the fault only discovered when the distribution facility tested the product in the US. Several thousand units had to be recalled, and it took the company another month to get the product to the consumer.
- **Critical components may not be available when you are ready to go into full production.** When a communications device company was close to launch, a memory chip that was central to the design of device suddenly became unavailable because a large international consumer electronics company bought the entire world's inventory of the component.

The PCBAs had to be redesigned to accommodate a different memory chip, and all the testing and certification had to be redone. This set back the product launch by several months.

- **Lead time for bulk orders may be longer than you think.** Teams assume that desired components will be available because they had purchased the prototype parts in small volumes from various distributors. One manufacturer discovered very late in the process that the lead time for a bulk purchase was around 180 days. As a result, the team had to scramble to find materials on the spot market for which they had to pay a premium.
- **Getting the packaging right can take months.** Teams often start the packaging design later in the process than they should. In one case, a cardboard packaging insert in the gift box did not adequately restrain the product, allowing the product to jostle and break. Only during vibration testing just before PVT was the problem identified. The redesign delayed the launch by two weeks.
- **Getting the right accessories selected, ordered, and tested can also cause problems**. One manufacturer of a home appliance didn't consider the wall wart (the device that connects a USB cable to the electrical outlet) when designing their electronic device, assuming that enough wall warts that met their aesthetic requirements would be readily available at a low price. The supplier they chose had a hard time supplying the part in volume with short notice. While there ultimately was no delay, it used up a significant amount of the team's limited bandwidth.

4.3 Risk Management

Merriam-Webster defines risk as the "possibility of loss or injury." In the context of product realization, risk is the chance that something might go wrong that impacts cost, quality, or schedule. Most teams start by planning assuming nothing will go wrong. However, teams also need to actively manage the risk of things going wrong by anticipating potential problems. Ultimately, much of the time and resources in product realization are spent reducing risks that the organization identifies, and many risks require cross-functional efforts to overcome them.

Hard, uncertain problems often get dismissed as "That should be easy." or "Amazon did it; we should be able to do it." Issues raised in meetings are not followed up. People who raise the potential problem assume that highlighting the problem means someone will solve it. Too often in product recall situations, someone in a post-mortem meeting will say, "I raised that issue at

Vignette 3: How to Stay on Schedule

Bryanne Leeming, Founder and CEO, Unruly Studios

Unruly Studios started developing what became Unruly Splats in 2015. In 2017, after 23 prototyping rounds and many tests with over 3000 customers to understand what features and capabilities were important to kids, teachers and parents, we knew the product was ready because people were ready to pay for it when they saw their kids and students playing on it and they saw how it worked. However, we didn't have a product-ready design. Our industrial engineer mocked up 14 different designs and we surveyed 75 parents and kids at a local STEM Fair. When we tallied up the votes, the consensus was unanimous and we had our final design. Now we had to get it into production. We wanted to promise a date we could achieve, so we started talking to factories.

Unruly was new to production, but we made sure our team and set of advisors had the experience and connections we needed. My co-founder, David Kunitz, had over 25 years of experience manufacturing toys at Hasbro and Mattel. Our other team members and advisors came from Mattel, iRobot, and the MIT Media Lab.

It was important to us that we have a sense of the cost and timeline to manufacture Splats before we launched the campaign because we were determined to have sufficient cash flow to deliver our product on time to customers. We met our manufacturer through a previous co-worker of David's from Hasbro. David's friend had a 30-year relationship with our factory in China. That introduction built a lot of trust from the start: we trusted them because of the personal referral and they trusted us because of their longstanding relationship. That trust helped us in many ways including getting very favorable payment terms (which helped significantly with cash flow) and a lot of attention and help. They were also very excited about manufacturing Splats as we were their first Bluetooth product; we both dove in to discover unknowns and work out the details together.

We had met the factory and got back-of-the-napkin quotes on what they thought it would take. From that we were able to set our price point and delivery timing. Factory representatives came to the US shortly after to visit and we had the opportunity to host their entire team at our office. We showed them all our functioning prototypes and shared the beta software app so they could play some games. We actually had their whole team playing a game of whack-a-mole on the product when they came to Boston, so we could make sure they knew what they were building. They loved it!

We launched the Kickstarter campaign in October 2017, and in November 2018 we delivered our first products to customers. The initial launch was successful because our product arrived to customers on time (a rarity for hardware crowdsourced products) and with minimal bugs. We attribute our success to two things. First, by doing iterations before we started production,

we reduced costs and improved reliability. Secondly, and perhaps the more important, was our team of advisors with strong supplier relationships.

To stick to our timeline, we had a lot of back-and-forth with our factory and were overnight-shipping every new prototype they created. Dave went over to China for the production run to make sure everything went smoothly and to approve the first Unruly Splats directly off the line.

At one point, we did notice some delays. Our CM had a lot of projects at once and we were not being prioritized. Because we had met in person both in China and in Boston, we were able to send a note to the owner and get back on track. It continues to be a great partnership.

The personal referral to our CM helped us immeasurably because we immediately started out with a trusting relationship. If our team of advisors had not shared their contacts with us, we might not have found a supplier so willing to work with a tiny customer. Our supplier is used to working with big name brands, but they took us on because they trusted our advisor and because they believed in our product. They even extended us favorable pricing, which is a boon to any start-up. We have since done an additional round of production 5 times the size of the original run, our product forecast is growing and we look forward to a long relationship with them.

Text source: Reproduced with permission Bryanne Leeming.
Photo source: Reproduced with permission from Unruly Studios.
Logo source: Reproduced with permission from Unruly Studios.

the start of the project." Teams can choose to bury their heads in the sand or they can actively manage the risks. By proactively identifying risks ahead of time, the teams can put together cross-functional plans to address those risks.

Here are two typical examples of risks identified during the product realization process and what the teams can do to address them:

- **Scenario 1**: A team producing a new IoT (Internet of Things) sports product is worried that their new supplier will miss a key delivery date for a critical electronic subsystem: the supplier keeps promising aggressive delivery dates, but isn't responding quickly to calls and is late with specifications and samples. The team is worried that the materials won't be ready in time for the next pilot. To reduce the risk that late delivery of parts will impact the schedule, each functional group works to reduce the risk:
  - *Supplier management* communicates with the supplier continuously about the timing and updates the team regularly.
  - *Engineering team* creates a backup plan to have a surrogate part that could be used to build the pilot samples to enable testing of the rest of the product.

– *Quality team* develops a streamlined process to evaluate incoming units and parts to ensure that incoming quality control doesn't hold up deliveries.

• **Scenario 2**: The industrial design team specifies an electroplated surface on a new kitchen appliance. The part is at a user's eye level, so any defects in the surface finish will be very apparent. The low-cost supplier hasn't shown the team examples of the surface quality. Also, the supplier does not have a rigorous inspection standard, though they claim that they will sort out any defective units. The team is worried that they will need to scrap a significant number of units, causing materials shortages. To ensure that the supplier delivers adequate quality the teams take these actions:

– *Quality team* develops more precise quality specifications with the supplier to ensure that no bad product is shipped.

– *Procurement* orders more materials to allow for additional scrap.

– *Finish engineering* evaluates alternative solutions to reduce the aesthetic impact of any defects.

Notice that in these two examples, the teams act before the problem has become critical. The teams simply vigilant about the warning signs of

Table 4.3

Risk tracking records and descriptions

Column	Description
Unique item number	Facilitates tracking and ensures that even if names are changed, the risk can be tracked across multiple documents.
Short name	A descriptive and unambiguous name helps teams understand which problem they are discussing.
Category	Clusters items into related issues. For example, a hardware project might include supply, process, aesthetics, mechanical, and electrical categories.
Description	Describes the issues with enough detail that other teams can understand it.
Date added	Identifies lingering risk items. Teams tend to add a number of risks at the start of a project that are left to linger on the list. If the items haven't been addressed, the teams need to decide if the item should be removed; or if the team has been putting off dealing with a difficult problem.
Due date	Ensures that teams track time-sensitive issues.
Status	Tracks the status: closed, tabled, or open. Some items may not be active but need to be revisited later in the project. Closed items provide a history of the decisions as well as maintain the possibility of reopening them. The history is also useful for next-generation products.
Owner	Ensures accountability. A single person/role needs to own each item to follow up and update the status of each risk.
Risk level	Enables teams to prioritize risk: low, medium, or high.
Status	Tracks the weekly updates on actions taken to resolve the risk. Every update is dated and attributed to an owner.
Next action item	Highlights what the next steps are and who is responsible for that action.

potential delays, and proactively designed around those risks. Unfortunately, at many organizations, when team members raise concerns, these concerns are not followed up on until they become delays. Good project management requires proactive planning and problem-solving. The ultimate goal of the risk management process is to not let these risks slip by and become critical.

Risk management processes require the project manager and team to track what risks are raised, the status of the mitigating action, and when risks are closed out. Ideally, the list of risks and their status is maintained in a way that the whole team can see it, and updates are made in a timely manner. For small- to medium-size companies, risk management lists are typically maintained in spreadsheets shared through the cloud. Some companies have tried to repurpose bug-tracking software for risk management, but the structure of these tools does not easily translate to hardware-related issues. Table 4.3 lists the information in a risk document, and Table 4.4 gives an example.

The project manager should conduct regular reviews and updates of the project risks. This can be done through individual discussions with team members or during regular project updates. During regular project meetings, teams should review each of the action items and focus on items that are high risk, have been open for a while, or have had no action on them. After new information for each risk item has been reported, the team can discuss and plan next steps. The tracker is updated and then used as the agenda for the next meeting. By formalizing the regular discipline of reviewing risks, groups are less likely to let key issues turn into costly delays.

Table 4.4 Example of risk tracking document

#	Short name	Category	Description	Date open	Due date	Status	Owner	Risk	Status	Next action and timing
1	Supplier delivery schedule	Supply chain	Uncertain if supplier X is going to be on time with their delivery. Risk of missing EVT	1/15	2/28	Open	Supply chain	High	Supplier is reassuring us they will be on time but are regularly late. No clear resolution	Engineering: reporting on possible replacement parts for EVT by (date) Supply chain: update by (date) with supplier and visit by (date) to assess actual timing

4.4 Managing Your Enterprise Data

The second law of thermodynamics applies to everything from fluid dynamics to teenagers' bedrooms to document management: everything moves toward chaos, and it takes energy to return things to order. It is easy to lose track of critical documents, the wrong version to be posted, or two people to be editing the same CAD file without coordinating.

Teams will create a large amount of information to support the product realization process and mass production, including the manufacturing processes, quality tests, certifications, packaging designs, forecasting, schedules, material ordering, and production schedules (Checklist 4.2). All of this information is needed to transition the product to the factory, ensure compliance, and build the product reliably. This set of data is broadly referred to as the *enterprise data*.

Some documents are "controlled," meaning that changes made to the document have to be tracked, have a formal release process, and have a method for approving any changes (Inset 4.1). Data and document management is complicated by the fact that documents produced by different functional groups are highly coupled. For example, a change in a part design requires an update to the tooling design, a change to the current BOM, and cost updates. A change to a sales forecast will ripple through material ordering, production planning, and inventory. All of these changes must be carefully documented and all affected documents updated. Older versions must be made accessible only for historical purposes.

As teams move from the early design phases to detailed design and product realization, the most significant change for most organizations is the increased focus on document and data-intensive deliverables. Every aspect of the product, processes, materials, plans, and ordering, has to be formally documented and the information verified through either the pilot or testing process. For the process to be successful, all of the information across departments and roles must be consistent and up-to-date. For example, to answer the question "What is going to be built?" teams need to create a specification document, a bill of materials, the drawing package, and the CMF documents. All of the information across the three files has to be consistent.

Document and data management is a critical aspect of product realization because it supports:

- **Single source of information.** If document versions aren't controlled, it is easy for different people to work from different or outdated data.
- **Comprehensive information.** Complete and accurate documents prevent misinterpretations or different interpretations from department to department.

Checklist 4.2: Document/Data List

The documents/data listed here are called out with the document icon throughout the text

Document/Data	Location	Document/Data	Location
Actual-to-actual (A2A)	Section 14.5.4	Pilot checklist	Checklist 3.1
Auditing plan/results	Section 19.4	Pilot quality test report	Section 7.2.3
Bill of materials (BOM)	Section 6.2	Pilot reports	Section 3.3.1
Cash flow model	Section 8.5	Preventative maintenance plan	Section 19.5
Certification documents	Section 17.1	Process flow	Section 11.1
Color, material, and finish	Section 6.3	Process plan	Section 11.4
Continual improvement plan	Section 19.2	Production forecast/plan	Section 15.1.3
Control plans	Section 13.4	Purchase order (PO)	Section 14.5.6
Cost model, COGS	Section 8.3	Purchasing and production activity control (PAC)	Section 15.1.1
Cost reduction plan	Section 19.3		
Customer complaint data	Section 18.4	Quality test plan, pilot	Section 7.3
Drawing package, electrical	Section 6.5	Quality test plan, process	Section 13.3
Drawing package, mechanical	Section 6.4	RASCI	Section 4.1.2
Enterprise data plan	Section 4.4	Request for information (RFI)	Section 14.5.1
Failure modes and effects analysis	Section 5.3.3	Request for quote (RFQ)	Section 14.5.3
		Return material authorization (RMA)	Section 14.5.9
Invoices	Section 14.5.8		
Labeling plan	Section 17.2	Risk management system	Section 4.3
Master production schedule (MPS)	Section 15.1.1	Sales and operations plan (S&OP)	Section 15.1.1
Master services agreement (MSA)	Section 14.5.5	Sales forecasts	Section 15.1.2
		Sample plan	Inset 3.2
Material authorization (MA)	Section 14.5.7	Schedule/critical path	Section 4.2
Material resource planning (MRP)	Section 15.1.1	SOPs, quality test, process	Sections 7.2.3 & 11.5
Non-disclosure agreement (NDA)	Section 14.5.2	Specification document	Chapter 5
		Strategic manufacturing plan	Section 15.1.1
Packaging design	Section 6.6.3	Tooling plan	Section 12.4

- **Rapid access**. The whole product realization process is very time-sensitive. Critical documents need to be accessible quickly. Cloud storage of documents with controls on who can access documents and who can modify them, is important.
- **Documents created before needed**. Project managers should anticipate which documents will be needed, and ensure they are completed on time. The process becomes too chaotic if teams are creating documentation retroactively and in a hurry.

Enterprise data management (EDM) is the collection of tools and processes used by an organization to create, track, maintain, and use their enterprise data. Organizations should create an EDM system before creating the data itself. It is easier to create an organized system from scratch than to go through and clean up a mess that's already created.

In its simplest form, the EDM might be a shared hard drive with quick email updates from one team member to another when a file is changed. While this might be simple to create, it's not simple to use: document traceability will be difficult, and the system will rapidly become chaotic.

Often, smaller or newer companies will develop their own slightly more robust EDM systems, perhaps developing their own software and asking employees to agree to certain procedures. While developing your own software is cheaper than purchasing software in the short term, it may cause time-consuming problems in the long term since it relies on the discipline of every team member to keep it organized.

When companies get large enough, they typically need to employ a purchased EDM system. Document management is too complicated and too many people are involved to depend on each person following the process to keep the system working. Moving from a hand-built system to a fully integrated EDM system is typically challenging, time-consuming, and painful. Depending on the company's current suite of software, their data needs, and budget (both dollars and resources), companies should choose the right-sized system that best balances cost and capability. EDM systems range from hand-built systems, to SaaS and cloud-based systems (Inset 4.2), to software installed on individual computers, to large enterprise-wide systems.

The term EDM covers a wide range of tools and software used to track and manage all aspects of a business's data. EDM systems fall roughly into three categories: dedicated workflow and document management, management of materials and production, and tools to manage customer interactions. This set does not include the wide variety of tools used to track and predict the financial flows within an organization (we will leave financial controls to those with finance degrees).

Inset 4.1: What is a Controlled Document?

Up-to-date information critical to designing and building your product needs to be available to the right team members. It also needs to be version-controlled and protected from unauthorized access or changes. You do not want all your CAD models on one person's local hard drive, not accessible to the team. Also, you do not wish for anyone in the company to be able to mistakenly delete or change critical files.

As early as 2001, engineers at General Motors knew of problems with ignition switches that ultimately resulted in 124 deaths and over 30 million cars recalled. The recall was made more difficult because, sometime during 2007, a change was made to the ignition parts to reduce the risk of failure, but the changes were made without notification or change control. As a result, later models may have had faulty switches installed when they went in for repairs, and it was hard to trace when the change was implemented. In addition, the change didn't meet GM's original specifications [44, 45]

Larger companies usually use software-based document management methods, but smaller ones can manage them manually. However, all document control processes share the following characteristics:

- **File naming conventions** and document numbers enable easy searching and ensure that team members know which file is the most recent version.
- **Storage** locations for documents need to be consistent so that when someone goes to look for a document, they can be assured of finding the most recent version.
- **Approval and change policies** define how leadership approves documents and how changes are communicated to the right functional groups. CAD and other documents will start in draft form and are frequently updated by multiple team members. Before production starts, the drawings need to be "locked" before being released to the factory or supplier. After the engineering team officially releases drawings, changes need to be formally approved by the design team and communicated to production.
- **Access policies** ensure that files are accessible to the right people and *only* the right people. Your IP (intellectual property) is precious and should not be purposefully or accidentally made public. It is unfortunately very easy with online services such as Google or GrabCad to make files public accidentally or to forward them to people who have not signed an NDA (non-disclosure agreement). Also, sales information or contract terms may be restricted to a small team to reduce the risk of competitors learning critical competitive information.
- **Retention and version control** ensure that all versions of documents are maintained, but that the wrong ones are not used by accident. For example, it may be necessary to keep multiple versions of drawings that represent engineering changes done during production. Teams still need access to older drawings if quality issues arise. However, you only want the latest version released to the factory.
- **Cross-document consistency** ensures that updates in documents are reflected across all of the related material.
- **Enforcement of enterprise data rules** ensures that everyone in the organization follows this crucial process. The second law of thermodynamics also rules data: everything will tend toward entropy and chaos. It is not enough simply to *have* a document management process; the process must be consistently and carefully enforced. This can be done through PLM (product lifecycle management) systems that regulate the adherence to processes, or through manual auditing and enforcement.

The first group of tools – workflow and document management – can generally be divided into two categories:

- **Product data/document management** (PDM) systems are used to manage critical documents, their versions, and the processes for updating them. They are essentially glorified filing systems.
- **Product lifecycle management** (PLM) systems manage both the documents and the processes around them. PLM systems typically have embedded PDM systems within them. For example, PLM systems define and control compliance with the procedures to introduce a new SKU (stock-keeping unit) into the sales pipeline. The software ensures that the right people are informed and that documents are approved correctly.

The second class of tools – management of materials and production – primarily focuses on documents related to materials, production planning, manufacturing, and supplier management. They are used to ensure the right materials are ordered, that the production system can meet customer demand, and that the current state of production can be tracked. Also, these systems provide consistent access of information across functional groups. For example, a change in a sales forecast will update the inventory requirements, ordering plan, and production release schedules. These software systems often have embedded PDM and PLM modules, and can, in turn, be categorized into groups:

- **Material resource planning** (MRP) focuses primarily on material control, ordering, and consumption rates to support production. MRP systems control when orders are released to the factory to ensure that deliveries

Inset 4.2: Software as a Service (SaaS)

Until the growth of SaaS (software as a service) and cloud-based software in the early 2000s, companies were forced to use painful home-grown systems or to spend thousands (if not millions) of dollars on enterprise software. The purchased systems often required significant customization before use and a large team within a company to manage it. SaaS software enabled companies to get the limited functions they needed at a low price without the IT (information technology) overhead.

The last decades saw the expansion of SaaS suppliers. The use of cloud services enabled the development and deployment of lower-cost tools that allow the enterprise to get a simplified version of the more extensive software at a lower cost without requiring extensive training to use it.

One popular example of this is Google Docs: this online service replicates the most commonly used 10% of the functionality of expensive word processing, spreadsheet, and presentations software at little or no cost, with the added feature of seamless online collaboration.

occur on time. They allow materials to be ordered with sufficient lead times while controlling the amount of inventory in the system.

- **Manufacturing execution systems** (MES) are used to track where the parts are on the factory floor in real time. MES are typically used for more sophisticated and complicated processes where it is critical to track the real-time status of the production floor (e.g., the thousands of parts used when producing cars).
- **Enterprise resource planning** (ERP) is used to manage all resources within a company (more than just materials). Many ERP systems have an MRP system as a module within them. For example, SAP, one of the largest ERP systems, has modules that will manage operations, financial accounting, human capital, corporate services, and real-estate management.

The third group of tools supports interactions with customers and includes:

- **Customer relationship management** (CRM) systems are used to track and analyze customer interactions. The simplest CRMs ensure that you are following up with customers throughout the sales cycle, while the more complicated systems manage the relationship through the entire product development and field support processes.
- **Warranty management systems** are used to support the customer after the product is purchased. These systems track customer interactions and outcomes as well as the costs associated with sending spare parts, replacing parts, or dispatching servicers to fix product on the customer's site. Warranty management systems are typically highly integrated with both the financial systems and the CRM systems (Chapter 18).

Summary and Key Takeaways

- ❏ The product realization process requires a large number of people and outside companies to execute highly interrelated tasks under extreme cost, schedule, and quality pressures. Proactive project management is needed to ensure a successful launch.
- ❏ The goal of project management is to balance the competing requirements of quality, cost, and schedule.
- ❏ Understanding and managing the critical path will reduce the risk of preventable delays.
- ❏ Teams need clearly defined roles and responsibilities, and they must identify critical resources (either in-house or outsourced) before the resources are needed.
- ❏ Teams need to identify and mitigate risks during the process. Active tracking of risks helps to guarantee that the team addresses and closes out all of the risks.
- ❏ Systems need to be put in place to ensure that team members can quickly find the most up-to-date data, that no one can change data without informing key stakeholders, and that all changes are approved.

Chapter 5

Specifications

PRODUCT PLANNING

5. SPECIFICATIONS

6. PRODUCT DEFINITION

CAD

7. PILOT-PHASE QUALITY TESTING

8. COST & CASH FLOW

A specification document outlines all of the requirements for the product. In addition to providing a guiding framework for product design, this document drives many of the quality and testing processes during the pilot phases. Ideally, a comprehensive specification document is completed long before the product realization process begins in order to reduce the likelihood that serious performance and quality issues will be discovered late in the process.

This chapter describes how the specification document evolves over time, what should be included in the document, and several approaches to ensuring that the document is complete.

Product Realization: Going from one to a million, First Edition. Anna C. Thornton.
© 2021 John Wiley & Sons, Inc. Published 2021 by John Wiley & Sons, Inc.

Every successful product must solve a customer need. If your team has not carefully understood and designed for your customers' needs, the product is unlikely to be a commercial success. Once the team has gone thorough customer discovery using frameworks such as design thinking, Christensen's "Jobs to be done" theory [46], or lean start-up processes (Chapter 2), you need to ensure that the product you design meets all of the needs you have discovered along the way. Additionally, you need to document the issues the customer many not have thought about, such as "What happens if the product is dropped?" or "Does the customer care that the buttons break after ten thousand pushes?"

To stay true to the customer needs that inspired the product while tracking all of the learnings along the way, teams must develop and maintain a specification document (spec doc). Without documenting the common view of the product goals, it is easy for team members to develop divergent views of what they are designing – like a childhood game of telephone.

Teams should start documenting customer needs as early as possible in the design process. Not only will this keep the customer needs top-of-the-mind, it will also help reconcile differing opinions on what the customers want. Too often, teams do not realize the importance of writing everything down before product realization starts. Or if a team does draft the specification document early in the development process, they often file it away and neglect to update it regularly throughout the product development process. When these situations occur, the spec doc does not contain the most recent information needed to drive the quality assurance, component selection, and supplier selection tasks at the beginning of the product realization process. In the worst cases, the teams never define the specs, and the document has to be created from scratch when the purchasing, factory, or quality organizations ask for it.

In different industries, this document that we will refer to as the "specification document" or "spec doc" might go by different names: the product brief, the requirements document, product requirements document (PRD), product specifications document (PSD) or the requirements specifications document. The specification document typically starts as a list of general requirements (voice of the customer) and evolves to contain quantified product specifications as the team refines the product definition. This document enables organizations to have a shared vision for the product and a quantified description of success. It is used to communicate that vision to suppliers and partners, and it provides a framework for verifying and validating the design during the pilot process. The document(s) typically contain both qualitative (requirements) and quantitative (specification) information about the product. While these two words are often used interchangeably, each has a very specific meaning.

- **Requirements** are non-solution-specific statements about the goals of the product that capture the needs of the customer and the broad goals for the product. They are typically open-ended statements such as "The user wants the product to be light" or "The unit needs to last a full day without charging." Requirements are general and don't constrain the design space but are used to guide the design process.

- **Specifications** are typically associated with a design solution and unambiguously define the detailed and quantified requirements. For example: "This product should weigh no more than 100 g and ideally should be less than 80 g."

To add to the confusion, there are often multiple specification documents for the typical electromechanical product, including:

- **Firmware specification** defines the permanent software programmed into a read-only memory which bridges between the hardware and the software [47].
- **Software specification** defines the software functions and user interface.
- **App specification** describes how mobile apps will work.
- **Part/component specifications** define the function, interface, performance, and operating conditions of each part. Purchased part specifications are often available on a supplier's website.

Before defining what a spec document should be, we need to define what a spec document is not:

- **A list of critical parts**. A specification document is NOT a Google Doc with a list of the most expensive components. Some teams create a document with a battery, a Wi-Fi interface, a Bluetooth chip, and a list of functions, and call it a spec document. The simple list is at best a preliminary bill of materials or a shopping list for the Shenzhen electronics market.
- **A color/material/finish (CMF) specification.** The CMF document defines what the product will look like and rarely contains any engineering requirements or specifications. The CMF is critical to communicate the branding, final finish, and aesthetics, but is not comprehensive enough to guide the rest of the design or the product realization process.
- **A brochure or marketing material.** Marketing documents, used to communicate the product to potential customers, are not comprehensive or quantitative enough to support engineering decisions.
- **A document created by a single person.** One individual might spend weeks writing and editing a specification document. It might then be put in a team file folder, never to be seen again.
- **One-and-done.** Some teams spend a significant amount of time creating a spec document in the early stages of the product, but never update

the document as the process moves along. As a result, by the time the design goes to production, the specifications document is out of date or irrelevant.

Having a comprehensive specification document is critical for the product realization process for a number of reasons:

- **The spec guides the product realization process.** A large part of the engineering verification testing (EVT), design verification testing (DVT), and production verification testing (PVT) process is the verification of the target product performance. To verify the performance of the product, the engineering and product teams need to quantify and document the expected behavior of the product.
- **The product realization process informs the spec document.** A large number of design changes will occur during the piloting process. Open-ended user testing will find new needs, and durability and reliability testing will find weaknesses with the product. The design changes need to be consistent with any specifications already defined. If a change to the specification is required, then the engineering, product, and quality teams need to understand the other impacts of that change. For example, if an additional button is added to the design for a new toaster, does that change the shape of the packaging? How does that affect the overall target COGS? How will the new button be tested for durability and reliability?

5.1 Integrating with the Product Development Process

As pointed out in the introduction, the specification document is developed and modified as the product development process progresses. Before the product is designed, the voice of the customer needs to be understood. The team may start with vague customer needs such as "I want to communicate effortlessly" and "I want to travel with my product." Throughout the customer discovery phase, the engineering and product teams translate the voice of the customer into requirements for the product, and then the teams translate the requirements into quantified specifications. For example, "I want to be able to travel with my product" becomes "The product needs to have a long battery life," which is translated to "The product must work for at least six hours and ideally more than eight hours without charging."

The product team doesn't write the specification document in one go. Ideally, teams start recording the voice of the customer as they refine the product concept and user profiles. As the product is detailed, the team continually updates it with more information. For example, early in the concept generation phase, teams may have a product brief and knowledge of the

FIGURE 5.1
The spec doc
supports all of the
stages of product
development

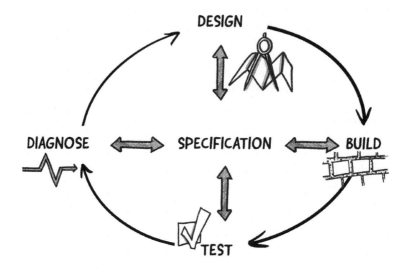

stakeholders, but the industrial design team may not finalize the details of the aesthetic criteria until later in the detail design phase.

When the specification document is completed, it becomes a central document in the product realization process (Figure 5.1). The engineering teams have the responsibility of ensuring the design is consistent with the specification document. Manufacturing engineering will select processes and control them to ensure conformance. The quality teams will verify and validate the specifications throughout the pilot processes and production. Finally, customer support will help customers whose product does work as expected.

5.2 Parts of the Specification Document

The format and content of the software, firmware, application, and specification documents are relatively standardized across multiple industries. A quick search returns almost 100 books on the topic. On the other hand, there are relatively few standards on how to write a specification document for a hardware-based product. While some industries have unique requirements, most specification documents share a standard set of elements. All will have quantified product specifications, a product brief, and durability requirements. Unfortunately, many specification documents are less than comprehensive – they focus on the qualitative specs or only describe how the product should perform in use. Below is the list of the types of information that should be included in a specification document to ensure that it comprehensively describes the product. The next

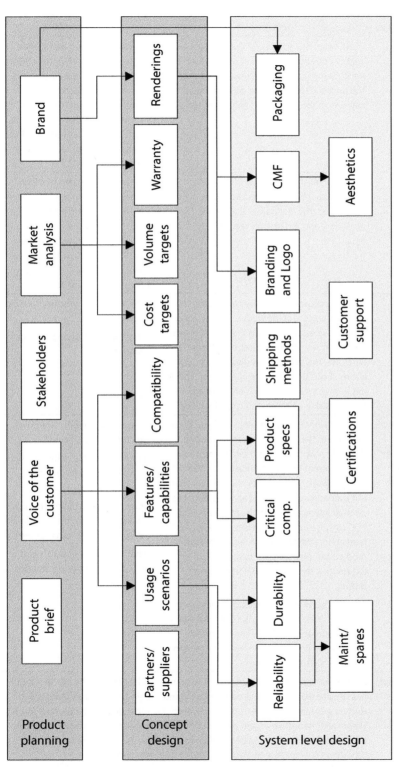

FIGURE 5.2 Parts of the specification document and when they are initially drafted

Table 5.1

Specifications
identified during
product planning

Category	Description	Example
Product brief	*A text-based description of the product goals, targets, and customer.* Videos, photos, and sketches may accompany this product brief. This section is typically owned by the senior member of the product management team.	"We are delivering the smallest possession tracker on the market. The product will be easy to install and use while having enough battery life to ensure that the product can be powered over its expected life. The product will be attractive enough for it to be an accessory in itself."
Market analysis	*The size of the market and which types of customers are being targeted.* Both the beachhead (the first market), as well as secondary and tertiary markets, are described.	"We are first targeting midsize technology companies that have 100+ employees in a single location. The revenue of the businesses needs to be at least $150 K per person. We believe there are at least 10,000 of these companies within the US."
Stake-holders	*All of the people/organizations involved in the purchase and use of the product and their needs.* The consumer of the product does not have to be the person who purchases or initiates the purchase. This section may also contain information about customer personas. Table 5.2 has a list of typical stakeholders, all of whose needs need to be taken into consideration.	"The user of the product is a child between the ages of 9–10." "The purchaser is the parent or grandparent."
The voice of the customer (VOC)	*Captures what the customer has told you.* These requirements are often subject to interpretation as to how to address them but are the "raw" information from the customer. Most of the VOC will be gathered by the product team, but everyone on the team can contribute based on their interactions with the customer. The VOC can also include promises made by marketing. During product realization, the early product samples are used to gather further information through validation and user testing. The VOC provides the touchstone for making decisions. Whenever there is a question about which approach to take during design, the team should test their ideas against what the customer says they want.	"The product needs to help me execute my work faster." "I need to be able to throw it in my backpack without worrying." "I have trouble getting my grandmother to use electronics. I think it is because the buttons are too small and she can't see well." "I hate the cords in my current device, they always get tangled."
The brand	*Defines the personality of the product.* The brand manager and product leaders design the brand to elicit a specific response from a customer. The aspects of aesthetics and performance as related to the brand should be listed. The product may be part of an existing brand or creating a new brand.	"The brand of the product should communicate ruggedness and functionality. The product should appear to be reliable and able to be used in a variety of environments."

Stakeholder	Role
Initiator	Recognizes the value of a product
Gatekeeper	Gives information to the decision-maker
Decider	Makes the purchasing decision
Influencer	Influences the users to make the purchase
Purchaser	Pays for the product
User	Uses the product

Table 5.2
Types of stakeholders

three sections will describe the information documented during the three phases of product development: product planning, concept design, and system-level design. Figure 5.2 shows the information defined in the specifications document, when they are typically documented, and how they relate to each other.

5.2.1 Specifications Drafted during Product Planning

Early in the product planning phase, the team is still formulating the overall product strategy and early concepts: the product managers and sales are talking with customers, the research and development team is developing the technology, and the product team is testing initial MVP (minimum viable product) prototypes with customers. The initial sales projection and the size of the market are estimated. Table 5.1 outlines the information typically defined at this stage.

5.2.2 Specifications Drafted during Concept Design

During concept design, the engineering teams generate multiple ideas which the whole team evaluates. At this point it is important to get many perspectives on the concepts. For example, branding may identify that the product doesn't have the space for a logo and customer service may identify issues with maintaining the product.

The specs are still typically qualitative, but they contain more detail about the product's functions. As the product team learns more about the customer needs (e.g., how the product will be used, what the cost targets are), they refine the concept (or throw it away and start over). Table 5.3 gives the list of specifications defined at this stage. Imagine how the Juicero debacle might have been avoided if the engineering team had evaluated multiple design concepts that would have better met the cost and functionality targets. Who knows, they may have decided just to sell bagged fruit as a hand strengthener as well a source of fresh pressed juice!

Category	Description	Example
Features and capabilities	*Lists the actions that the product should be able to perform.* Typically, these are defined by the product owner, but the whole team may contribute. It is critical to keep the feature and capability list from changing at the last minute. The MVP (minimum viable product) features should be highlighted.	"The product will turn on automatically when someone enters the room."
Compatibility requirements	*Defines how your product interacts with other products or systems (or doesn't interact).* The compatibility is based on factors such as market, market strategy, supply chains, and competition.	"The bike water bottle must fit 90% of all standard bike cages without interference with the top tube."
Usage scenarios	*Defines all of the conditions under which the product is reasonably expected to perform and survive.* The scenarios are typically text descriptions. They are the inputs to the durability and reliability specifications described in the next section and Chapter 7.	"The product will need to be able to sit in a car for several days without degradation to battery, function, or aesthetics."
Warranty	*Defines the warranty period and coverage of the warranty.* Also, this section will describe the expected total life of the product along with the target return rate. (Chapter 7 and Chapter 18)	"The product will have a 1-year warranty with an expected usage of 3-5 years." "The overall return rate will be less than 5% with less than 3% of the returns due to quality, reliability, and durability issues."
Partners and strategic suppliers	Lists the people and organizations who, while outside your company, are integral to the development of the product.	"Acme batteries will supply our battery, and we will co-develop a custom-shaped battery to satisfy performance and geometry requirements."
Rendering or sketches	*Shows the look and feel of the product.* Where the brand describes the personality of the product in words, the sketches show it in pictures.	Figure 5.3 which shows a CAD rendering of a passive amplifier for a smart phone.
Cost targets/ price target	*Provide product cost and price targets to ensure that the business plan is viable.* The agreed-upon cost targets for BOM, non-recurring engineering costs (NRE), COGS, and distribution costs should be documented (Chapter 8).	"Cost of goods sold at Launch will be less than $15." "Cost of goods sold after six months of production will be less than $10."
Volumes/ Forecasts	Define how many SKUs will be sold in each period.	"10,000 units per month for the first 6 months and a 10% growth each month thereafter"

FIGURE 5.3
Example of rendering
of a passive speaker
amplifier
Source: Reproduced
with permission from
Lemeng Shao

5.2.3 Specifications Drafted during System-Level Design

The specification document should be mostly developed by the time the team finishes system-level design. During this phase, the overall concept is completed. Enough is known about the technical solution that quantitative product performance targets can be defined. These targets will be updated and changed as the team learns about the customer, technology limitations, and manufacturability. Table 5.4 provides a list of specifications that are defined at this stage.

5.3 Gathering Information

Specification Document

The scope of the specifications document is quite large and encompasses more than just the performance specifications. Looking at the list of what needs to be included can be overwhelming. So how can you be sure that your team has included everything? It is easy to fall in love with your idea and solution without asking the questions "Did someone else try this?" "How did customer react to other similar products?" "Am I thinking like the customer rather than an engineer?" and "How can it fail?" While each product is unique and teams will draw on different resources to populate the spec document, the following can help teams elicit customer needs, identify specifications, and ensure that the document is comprehensive:

- **Template and checklists.** Use a specification template to make sure you have covered the basics in a specification document. The template can be derived from the contents of this book, may be created based on another product, or can be found online.

Table 5.4
Specifications
set during
system-level
design

Category	Description	Example
Product specifications	*Quantifies Measurable targets.* Specs can include metrics such as size, weight, and performance requirements. These are typically defined by the engineering team but are validated by the product owner. The specification should consist of a target value, and a minimum and maximum acceptable value. It should also indicate whether the team should target nominal (the middle of the range), low end (LTB – lower the better) or high end (HTB – higher the better).	"The product will be no heavier than 30 g with a target weight of 25 g. Ideally, the product should be as light as possible (LTB)"
Critical components	*Lists preselected critical components and platforms.* These are typically specified by the engineering teams. Listing critical components is essential for coordination across functional teams; for example, firmware design needs to know what platform will be used, and purchasing needs to understand whether critical components have long lead times. Note: this is not the entire BOM but just the *critical components* all teams need to know about.	"Will be based on the Snapdragon Mobile 835 platform."
Certifications, regulatory, compliance	*All of the tests and certificates required to legally sell the product.* These include the regions where the product will be sold and a preliminary list of certificates (Chapter 17).	"The units will be sold in USA, Canada, and Europe." "The product will require Bluetooth certification."
Durability	*Quantifies what stressors the product will be exposed to in each of the usage scenarios.* Stressors can include temperatures, vibrations, dust, water, loads, use, and abuse (Section 7.3.3). The durability requirements are derived from the usage scenarios and are typically defined by the quality team with consultation from engineering and the product owner.	"The product can be dropped from a height of 30 cm onto a wooden floor with minimal aesthetic damage and no functional damage."
Reliability	Defines an average number of times each feature needs to work without failure during the warranty period. The reliability requirements are typically defined by the quality team with consultation from engineering and the product owner (Section 7.3.4).	"Button A will be actuated with a force of 5 N for 10,000 cycles during the warranty period."
Packaging and accessories	*Defines the final product and packaging as purchased by the customer.* It will include a list of all of the packaging – for example, gift boxes, inner packs, and master cartons – as well as all of the accessories, manuals, inserts, and other pieces in the gift box (Section 6.6).	"The packaging will include a gift box with a sleeve. The contents will include a screwdriver and installation template."

(Continued)

Table 5.4
(Continued)

Category	Description	Example
Branding and logo placement	*Location of any logos or branding information.* Branding and logo placement are usually described through renderings of the industrial design. Teams often forget to think through where the logo and branding features will be added and then run out of space. The branding is coordinated between the brand management, product owner, industrial design, and finishing engineering. For example, SRAM's product design enables prominent branding on their SRAM XX1 Eagle Crankset (Figure 5.4). One competitor's product design, in contrast, has limited space and a much less prominent logo.	"The logo should be placed on the front surface of the product with the deco wrapping around the top and bottom."
CMF	*Defines what color, material, and finish are going to be used for each visible surface.* The CMF is typically maintained in a separate document (Section 6.3). It is defined by the industrial engineering group along with the product owner.	"The handle will have a soft-touch paint finish of Pantone tone 2478 CP." "The product will come in three color schemes: white, black, and rose gold."
Aesthetics	*Defines the acceptable quality of each of the visible and aesthetically important surfaces.* The aesthetic requirements can include color matching and limits on defects such as flash or debris (Section 7.3.2).	"The finishes on all exposed surfaces will be high gloss with no more than one dust inclusion per square inch."
Shipping requirements	*Details how the product will be distributed through the various sales channels.* Also, it will identify what stresses the product is expected to experience during transportation. (Chapter 16).	"The product will be shipped by ocean freight in master cartons stacked on pallets. Likely to be LTL (less than truckload) shipments." "Shipping is subject to all of ISTA-2A except the stacking test." [48]
Maintenance and spares	*Defines how the product will be maintained and how worn parts will be replaced.* These requirements are defined by the whole team with significant input from customer support (Chapter 18).	"The product will have a custom rechargeable battery that can be removed." "The gift pack will include a single replacement battery." "Additional replacement batteries will be sold through the company's website."
Customer support	*Defines how customers will get help and information from the company about their product.* The specification will include information about what pathways exist for communicating (e.g., toll-free numbers or web portals), what service providers will be used, and how returns will be handled (Chapter 18).	"Support will be managed through distributors who will return a defective product to us for reimbursement."

FIGURE 5.4
Branding and logos
of the SRAM XX1
Eagle Crankset
Source: Reproduced
with permission
from SRAM

- **Other internal specification documents**. Specification documents from other products in your company can be used as a starting point for creating a new specification. Products from the same company often share similar performance, certification, and aesthetics requirements saving you time.
- **Storyboarding each aspect of the life of the product.** Documenting and agreeing on how the stakeholders use the product will help the team identify implicit specifications – those that may be in people's heads but need to be written down. Scenarios might include purchase, unboxing, setup, use, storage, transportation, recharging, recycling, and repair.
- **Reviewing promises to the customer.** The marketing collateral (the collection of media used to support the sales of the product) will have explicit and implicit commitments. For example, marketing videos or other materials may show a product in use outside or by children. These promises to the customer have broad implications for quality testing and certifications which may not have been explicit in the spec document.
- **Reviewing other similar/proxy products.** Other existing products will help set the customer's expectations for your product. Understanding the explicit and implicit promises to the customer can help you define what the customer is likely to expect.
- **Failure modes and effects analysis (FMEA)** is an effective way to anticipate the circumstances in the usage or manufacture that could contribute to failures. The spec document should define the requirements needed to avoid these failures (especially safety-related ones).

5.3.1 Storyboarding

The customer's experience with a product is much more than just using it for its primary purpose: for example, for a smartphone, making a call or taking a picture. The biggest mistake that most teams make in building a specification is to focus solely on the primary use of the product, but ignore all of the other stages in the life-cycle of the product. Think about how much abuse your phone takes. It has to sit in your back pocket or purse, it gets left in the car, and it gets dropped in puddles.

For example, if you were designing an insulated coffee press, what would that product need to do? Sure, it needs to make coffee. But if that's the only thing you designed for, you might end up making a coffee press that's as frustrating to use as the one the author recently purchased: it makes coffee well, but the design team did not think through what would happen when the press was put in the dishwasher or left to soak in the sink (very obvious steps in the life-cycle of any well-used coffee maker). The over-molding on the outside of the press is not watertight, so water collects between the over-molding and the body of the carafe. As a result, when the carafe sits on its side in the cupboard, it leaks water everywhere. Maybe the product came with warnings not to put it in the dishwasher, but the manual is long gone and users typically never read it in the first place (especially if you are an engineer and think you know better).

Storyboarding is a process that teams can use to document how the stakeholders interact with a product both during the actual function of the product as well as throughout the life of the product (e.g., purchasing, transportation, setup, connection, maintenance, and disposal). By laying out every step in the life of the product, the non-obvious needs of the product become apparent. Storyboarding can be as simple as a set of boxes and arrows defining the steps, or as complex as fully rendered cartoons depicting the user experience.

Look around your house and you will easily be able to tell which products were designed with the entire user experience in mind and which were not. The author recently purchased two products within the same month: a set of wireless speakers and a robotic vacuum cleaner. The producer of the speaker system did a great job at streamlining the user experience. From the packaging and instructions that made setup a breeze, to its intuitive app that automatically updated, it was a frictionless system.

The robotic vacuum, on the other hand, was clearly not designed with the user's whole experience in mind (or at least this user). In the vacuum's defense, it had a difficult task, contending with all the hair shed by two very furry retrievers. The vacuum was designed to run autonomously, and to signal its human owners whenever it required maintenance with a loud audio prompt. Because of the significant workload of vacuuming all that dog hair, the vacuum required a fair amount of maintenance. It is set to run at night, so the only ones who hear the audio prompt are the dogs, who bark at it. The vacuum cleaner is often dead in the middle of the floor the next morning. The audio prompts are often vague and buttons are poorly labeled, so it's often hard to tell what needs to be fixed or how to fix it. Locating the manual on the website is very challenging, as the exact model is not clearly marked on the product. Finally, finding the replacement parts jammed in one of several junk drawers requires significant digging. The product is effective at keeping floors clean, but the maintenance process is frustrating. In this case, doing a storyboarding of this experience would have highlighted several needs: the need for more obvious and consistent labeling and for better web/app support to troubleshoot what's wrong with the device. It would also be much easier to store spare parts on the device itself and to warn the consumer of imminent maintenance needs (as is done with printer ink cartridges).

Here is an example in which thoughtful storyboarding created a seamless customer experience: a company was producing a wall-mounted device and knew that customers cared about the product being perfectly level. By storyboarding the product installation, the design team decided to include a small level and mounting location in the plate. Also, the holes and slots were marked to make it very clear which holes were used to screw the device to the wall. There was a debate on whether to include a screwdriver in the kit, and whether the pain of having to find a screwdriver outweighed the cost of including one. By talking with customers, the team decided that the benefit of adding the screwdriver didn't outweigh the cost, extra packaging, and additional shipping weight.

Storyboarding should be done for each aspect of the product's life. Table 5.5 lists typical usages seen by consumer products, though depending on the industry or product, the list of scenarios will, of course, be different.

5.3.2 Customer Promises

The next step in the specification documentation process is explicitly documenting what promises the marketing, sales, product teams, and the website make to the customer. By means of marketing collateral, conversations, and tradeshows, the marketing and sales teams may have explicitly or implicitly promised your customer certain capabilities and features. Miscommunication about these promises is especially prevalent in organizations in which marketing and engineering teams do not communicate well. These situations can lead to costly redesigns. If, just before the product is going into production, the marketing team suddenly mentions "but we promised the customer it would be half the size!" the product may need to be delayed or marketing has to explain why they can't meet the promise.

Web sites, videos, or other marketing collateral provide implicit promises that must be incorporated into the design of the product as early as possible. For example, if the marketing film shows the product used outdoors, the team must explicitly define a UV resistance and maximum operating temperature. Otherwise, the product may yellow or overheat and cause a warranty claim. In another case, a marketing video might show a child using the product. If the spec does not specify the age range for users, the

Scenario	Description
Shipment, handling, and distribution	How is the product shipped? Will it be repackaged? Does it have to be tested or updated before being sent to the customer? For example, if a product needs to have its firmware updated after being shipped to the distribution center, the product packaging should be designed to accommodate this.
Purchasing	What is important to the customer during the purchasing process? Where is it purchased: in-store or online? How is it brought to the customer's location? For example, with clothing or wearable products, the consumer may want to see and feel the actual product (or a store sample) rather than just a picture on the outside of the box.
Unboxing, setup and install	Is the unboxing experience relevant? What does the customer need to do to install the product? It should be easy to take out and as painless as possible to set up. The printed materials should match in color and feel and fit with the product branding.
Each of the major functions	How is the user going to use each aspect of the product? Are all of the functions intuitive and easy to use? What happens if the user makes a mistake, and what are the potential incorrect uses?
Failure of each of the primary functions	What happens if some aspect of the product fails? For example, if the unit has an error in the electronics, how does it tell the user how to respond? Do the users need to dig through their junk drawer for the manual or is it obvious what they need to do? Can the product self-repair or self-diagnose?
Maintenance	Does the product need to be maintained by the user? Can it be cleaned, and with what substances? For example, many household cleaning products can do damage to certain materials and finishes.
Outdoor/indoor use	Where is the product likely to be used? What are the environmental conditions of each? For example, if the user operates the product outside, what impact will dirt or grass have on the unit?
Transportation	How is the product transported or carried? How often will the user be moving or transporting the product? Designers should assess the risk of factors such as accidental drops and the effect of temperature fluctuations.
Storage when not being used	Where is the product stored? How is it stored? For example, items stored in basements or garages are more likely to experience high humidity and temperature fluctuations than items stored in kitchens or bedrooms.
Recycling	Can the product be recycled? Can they send it back to you? What certification requirements do you have for recycling your product? Can you repurpose the product?

Table 5.5
Use scenarios

packaging isn't labeled as unsafe for children, or the product doesn't adhere to the appropriate ASTM toy standards [30], the company may be exposed to potential recalls and liability for injuries.

The engineering teams and product teams can better align the explicit specifications with these implicit promises by:

- Reviewing the specification document with everyone who is discussing the product outside of the company. Each relevant person should sign off on the specification document early in the process to confirm that all implicit and explicit promises have been captured.

- Setting rigorous communication controls on who can discuss what about a new product with whom outside your company.
- Reviewing all marketing materials, videos, and websites before they are shared externally to verify that all operating conditions, features, and user profiles match the specification document.
- Re-reviewing all marketing materials after they are posted to ensure nothing was missed.

5.3.3 Learning from Other Products/Competitive Analysis

When creating the specifications document, teams do not need to learn everything about their customers from scratch; they can and should learn from similar products already on the market. Your customers will have expectations based on their experiences with other products. As Bridget van Kralingen, Senior VP of IBM Global Services, said, "The last best experience that anyone has anywhere becomes the minimum expectation for the experience they want everywhere" [49].

Competitors' products will contain design features that customers love, as well as make mistakes that cause dissatisfaction (and at worst, recalls). Reviewing field performance and customer feedback for related products can give you insights into what your customers may explicitly or implicitly value in your product.

If you do not have a direct competitor, look at products that have similar users, user profiles, or location usage. For example, for a smart device in a wall plug the team might look at plug hardware and intercoms as well as other smart home devices. A headphone with biosensors might benefit from a look at headphones as well as biodata products such as Fitbit. For this section, we will use the scenario that you are developing a battery-powered portable smoothie blender. Customer expectations – ranging from UV (ultraviolet) resistance of materials, durability expectations, and noise requirements – can be gauged by researching proxy products such as wired (non-portable) smoothie makers and battery-powered portable appliances.

Once you have a set of products to research, many public sources can be used to find the implicit expectations set by other products, including:

- **Expert product reviews** in trade journals or sites such as C|Net and Consumer Reports can give you a sense of which metrics reviewers value most. These specifications are probably no surprise to the team, but it is helpful to track them over time as new competitive products come on the market.

- **Product reviews on social media or internet commerce sites** can give you insights into what makes customers happy (5-star reviews) or very unhappy (1-star reviews).
- **Federal recall databases** such as the Consumer Product Safety Commission (CPSC) list [50] all recalled products in the United States and the causes for recall. Reviewing this database can give insights into the mistakes that other companies made that have led to significant failures.

For the hypothetical blender example, the information sources and the customer needs they identify are shown in Tables 5.6 and 5.7. These needs should be added to the specification document. The design team should be especially aware of any safety-related issues.

5.3.4 Failure Modes and Effects Analysis (FMEA)

To verify that you have a comprehensive specifications document, your team should also identify any potential failures and the impact they could have on the customer. FMEA is a standard systematic process that teams use to catalog possible failures in a product, the root causes of each failure, and the ultimate effect on the customer. The FMEA owner – typically either the quality engineer or project manager – ranks the issues according to the likelihood of the failure, the severity of the failure, and the ability to detect the failure. The method is widely used and is often a requirement when designing medical devices and automotive and aerospace products. There are many online articles and books that discuss FMEA in much more depth than can be covered in this book. Table 5.8 shows an example of a very simple FMEA.

Source		Customer needs
Formal reviews of smoothie blenders		• High end finishes • Wattage • Capacity • Accessories • Small size • BPA free
Customer reviews on e-commerce sites	Features customers liked	• Quality of the blend • Time to blend • Ease of cleaning • Pulsating mode • The quality of the material
	Features customers didn't like	• The motor failing (smell and burning) • Cracked casing • Poorly fitting lids and leakage • Poor fit to the base • Hard to clean • Noise

Table 5.6
Professional and consumer reviews

Table 5.7

Reviewing CPSC recalls can help identify critical safety specifications [31]

CPSC recall	Primary root cause	Suggested specification
"The blender poses a laceration risk if consumers pour or invert the pitcher after removing the lid while the loose stacked blade assembly is still inside the pitcher."[51]	Loose assembly	The blade must be locked to the bottom
"A piece of the blender's mixing blade unit can break off during use, posing an injury hazard."[52]	Metal fatigue	The blade must not fatigue over life of product
"The rechargeable battery packs can overheat, posing fire and burn hazards."[53]	Battery hazards	Batteries must not overheat under maximum temperature, load, and charging conditions
"While placing the cup on or off the base of the blender, the blender can be inadvertently turned on, activating the blade. This can pose a serious laceration hazard to consumers."[54]	Accidentally switched on	The on switch must have a lock-out to prevent the system from powering on when in an unassembled state

Source: CPSC Recall List.

5.4 Managing a Specifications Document

Depending on factors such as company size, product complexity, or regulatory requirements, a specifications document can have a variety of different forms and formats. Some companies use software to manage the information, others use a text document, while still others use an Excel-based format. Some organizations split the information into a generalized requirement document that captures the voice of the customer, and a specification document that captures the details of the specific design concept. Others combine the information into one document.

Typically, small companies or companies that produce simple products use either a text-based document or a spreadsheet. Larger companies that are producing highly complex products (e.g., an aircraft) typically use dedicated software tools. For example, the military, aerospace, and automotive industries use enterprise requirement management tools to manage complex hierarchical requirements and ensure traceability and accountability throughout the supply chain and production system. The software-based specification tools take the form of both locally installed software products and (more recently) SaaS cloud-based products. Many of the affordable SaaS model tools are designed to support software development, but, at the time this book was written, there were limited options for hardware development.

If your team is building a spec document from scratch, there are two typical forms: text-based and table-based, but neither solution is universally recommended. The choice of text vs. table is often based on the preference and prior experience of the team members.

Table 5.8 Example FMEA

Process/ feature	Potential failure mode	Potential failure effects	Severity (1 mild – 10 severe)	Potential causes	Occurrence (1 seldom – 10 often)	Current controls	Detection (1 easy – 10 hard)	Risk priority number (RPN)	Action taken
Battery charge	Over-heating	Burns or fire	10	Defect in the battery chemistry	2	Sensor to check for excessive temperatures which cuts power to the charging circuit	1	20	Add quality checks on the battery to ensure correct assembly
Battery charge	Battery will not charge fully	Not enough run time	2	Incorrect charge cut-off value	1	Full charge of the battery checked on a sample of product	5	10	SPC of the battery charge measurement station to track changes over time
Blending	Too much ice clogs the blade	Incomplete blending	5	Incorrect usage by the customer	3	Manual and training	5	75	Checking if the motor power and blade can handle a range of blending scenarios

Regardless of the format, your specifications document must include:

- **A cover page** with information about the document. The information on the cover page prevents confusion between this and other documents and between different versions. Information should include:
  - Product name
  - Revision date and version number
  - Author
  - Approvers
  - Unique document number
- **Revision history** to identify what changes to the document were made and when (it can be hard to find the changes otherwise). The first page typically has a table called a revision block. The revision block itemizes any significant changes to the document since its initial approval. Each revision is documented with dates, a summary of the revisions, and any approval signoffs made and by whom.
- **The product name, revision date, and revision** on each page of the document.
- **A way to group the specs** into categories to facilitate review.
- **Unique numbers for each specification.** Numbers should not be reused if specs are added or deleted.

To manage the specification document, the EDM system (Section 4.4) should ensure that the specification document and updates are:

- Available to all approved team members.
- Access-controlled to prevent leakage of IP outside of the organization.
- Reviewed and approved by the relevant functional groups.
- Communicated to the whole team and do not just reside in one person's local directory.

In creating and maintaining the specification document, the following guidelines should be followed:

- **Be quantitative if possible**. It is always better to have a measurable specification rather than a vague one.
- **Build and own as a team**. The spec document is a contract – agreed to by the whole team – that defines what success is. If only one person is responsible for generating the spec document, the team may not be aligned on the goals of the product (and most likely won't wade through a 20-page document).
- **Keep it organized and numbered**. Assigning a unique number to each specification will help with traceability and change management. The unique index can be linked to other documents (e.g., quality test documents).
- **Make the document change-controlled**. After it is first completed and released, no changes should be made to the text without approval by the right people.

- **Link the document to other product realization documents**. The pilot quality plan, production quality plan, and shipment audit plans should be derived from and consistent with the specifications document.

Summary and Key Takeaways

❏ Specification documents are critically important because they force teams to define successful outcomes, clearly communicate the design goals to the internal and external teams, and serve as the basis for the quality test planning.

❏ Specification documents include a wide range of information beyond just the performance specifications.

❏ Specification documents are living documents that should be continually updated and maintained while also being carefully controlled.

❏ There are several tools that teams can use to populate a comprehensive specification document, including storyboarding, FMEA analysis, and evaluating competitors' products.

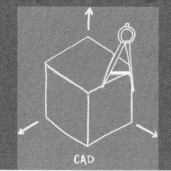

Chapter 6
Product Definition

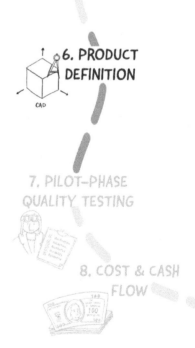

A complete and comprehensive definition of the product is necessary for designing a production system, selecting suppliers, ensuring product quality, and avoiding delays. The product definition is not just a CAD model of the product but includes the complete drawing package, definition of the finished goods including packaging and accessories, the bill of materials (BOM), and color, material and finish (CMF) documentation.

Product Realization: Going from one to a million, First Edition. Anna C. Thornton.
© 2021 John Wiley & Sons, Inc. Published 2021 by John Wiley & Sons, Inc.

In the last chapter, we discussed the importance of detailing the product specifications, and refining and updating those specifications as the product moves through the product design process. However, the factory can't build a product based solely on a specification document. The engineering and design team needs to create a complete definition of the product including the drawings of the parts, the BOM , assembly information, packaging, and CMF definitions. The specification document defines what the design team *wants* to design and the product design package completely defines *what* the design team creates. Later chapters will describe how the team determines *how* the design will be built.

The product design package is more than just the CAD model of the product geometry. Young engineers are taught that if they create a CAD model, they have designed a product. However, beautifully rendered CAD give a false sense of security that the parts can actually be built and assembled into the final product. In addition, the CM needs to know more about the finishes and tolerances which are typically not called out the CAD model.

The factory, CMs, and suppliers require a clear and unambiguous definition not just of the parts and materials needed to make the sellable product but also the CMF and packaging. Defining the product is even more complicated when several versions of the same product are available to customers. Differences between SKUs (stock-keeping units) (Inset 6.1) can include colors, level (low, medium, and high-end models), unit counts (sold in ones, twos or fours) or country of sale (different power adapters and manuals).

The clearer and more comprehensive the product design package, the smoother the product realization process will be. A missing component on a BOM can delay production, and inaccurate CMF specs can lead to unexpected aesthetic outcomes. Suppliers will build parts based on their interpretations of the drawings, and if you leave off a critical dimension, the components may not fit. If generic materials are specified, material substitutions can change the performance.

The product definition is used to:

- Estimate and track COGS and non-recurring engineering (NRE) costs (Chapter 8), both early in the process and throughout product realization. The geometry of the parts and the BOM can be used by estimating software or sent to suppliers to get more accurate parts quotes (Chapter 14).
- Assess the manufacturability of the parts (Chapter 10).
- Design and build tooling and fixtures to fabricate the parts (Chapter 12).
- Define the processes to manufacture and assemble the product (Chapter 11).
- Define how the quality will be assessed (Chapter 13).
- Order and manage materials (Chapter 15) during production.

Inset 6.1: SKU

 A SKU (stock-keeping unit, pronounced */skju:/*) describes a unique product that can be sold. Sometimes SKU refers to the alphanumeric code assigned to each unique sellable product version. Careful and thorough product definition can ensure that different SKUs in a product line are handled correctly. A single product realization process can launch multiple SKUs with the same core technology:

- Each color product is allocated a unique SKU number.
- Products for different regions may be given unique SKUs because of differences in packaging, accessories (e.g., electrical plugs), or manuals.
- Products may be sold in multiples. A single product will have one SKU number, a package of 10 will have a different SKU number.
- Products can have minor variants to sell at different price points (e.g., higher-end car models have more expensive audio systems and seats).
- Products with different pre-loaded software may have different SKUs (e.g., Tesla includes range-extending software on some models).

Managing multiple SKUs in a single production line is resource-intensive and can lead to errors including wrong labeling, incorrect packaging, and mismatched parts or colors.

While a majority of the mechanical and electrical drawing packages will be shared across all SKUs, each SKU may have unique drawings. Each SKU typically has a unique CMF, BOM, and labeling. Also, the quality test plan may have different tests for different SKUs. For example, different finishes will have different aesthetic requirements.

The product definition is captured in several interrelated documents:

- **Bill of materials** (BOM) is a comprehensive list of all the parts, systems, and materials required to build the product (Section 6.2).
- **Color, material, and finish** (CMF) is a document that defines the aesthetic requirements for each of the product SKUs (Section 6.4).
- **Mechanical drawing package**. A 3D CAD model of parts is not sufficient for manufacturers to build the parts; the factory needs a formal drawing package. A formal drawing package includes a fully dimensioned set of drawings with tolerances, datums, material specifications, and surface treatments. The package consists of drawings of the individual parts and of the total assembly (Section 6.5).
- **Electronics design definition** includes all the files that describe the printed circuit board assemblies (PCBAs), wiring harnesses, connectors, and wiring paths (Section 6.6).
- **Packaging design documents** define the boxes, accessories, and inserts required to transport the product and to communicate to the customer what they are buying (Section 6.6).

6.1 Types of Parts

Before describing the information that goes into the product definition, we have to enumerate and explain the types of parts and elements that go into building a product: fabricated, custom, commercial off-the-shelf (COTS), electronics, consumables, and raw materials. Each part type has unique definition methods, supply chain, and typical lead times. For example, if your product uses COTS (commercial off-the-shelf) parts, you can specify those by listing their suppliers, their manufacturers, and the manufacturers' part numbers. Most COTS parts are available on short notice. On the other hand, if your product requires specially fabricated parts, those will need to be defined using formal engineering drawings and typically have long lead times for tooling and fabrication. Table 6.1 summarizes the categories that we'll use for the rest of the book.

6.1.1 Fabricated Parts

 Fabricated parts are typically custom-made for a single product or product family. Usually, the engineering team has the responsibility for detailing the design of, and manufacturing processes for, fabricated components. The two most frequently used processes for creating fabricated parts in hardware products are injection molding and machining; however, the range of possible manufacturing technologies to produce fabricated parts is extensive and can include forging, casting, stamping, sheet metal bending, etching, laser cutting, and thermoforming [55].

Part category	Description
Fabricated parts	The parts that are custom-fabricated, including injection molded, cast, and forged. These typically require producing or purchasing custom tooling.
Custom parts	Parts based on a conventional technology but customized to the exact specifications of a product (e.g., motors, gear boxes, batteries, and LCD panels).
COTS parts	Parts such as screws, springs, and standard batteries. They can be generic (from any manufacturer) or a manufacturing part number from a specific manufacturer can be required (i.e., assigned).
Electronics	Electronics parts including PCBAs, PCBs, flex-circuits, wiring harnesses, connectors, and sensors.
Consumables	Materials typically bought in bulk and only a fraction used on each product. They often do not show up in the CAD or assembly drawings and can include glue and tape.
Raw materials	Bulk materials used to produce fabricated parts. These materials, such as stainless steel bar stock for turning, and plastic sheet stock used for thermoforming, are typically purchased in larger quantities; the quantities purchased don't match the actual materials used in final parts because you have to account for the scrap generated during processing.

Table 6.1

Categories of part types typically found in a bill of materials

Fabricated parts share the following characteristics:

- **Engineered** to meet functional, assembly, mechanical, and aesthetic requirements.
- **Unique** to the product or product family being built and not shared with other companies.
- Require **tooling** or a CNC program.
- May require **secondary processes** such as painting, electroplating, or heat-treating.
- Often require a non-trivial **setup process** for each production run and are made in batches to amortize the costs of setup times.
- Often outsourced to a **third party** for tooling and production. Fabricated parts often require dedicated equipment that is too expensive to maintain in-house.
- Require **DFM** (design for manufacturing) to ensure that the parts can be built effectively and efficiently with high quality and low cost.

The fabricated part definition usually takes the form of CAD and drawings and should include:

- Fully specified drawings of the finished part including tolerances, surface finishes, and material specifications.
- Detailed descriptions of post-processing requirements.
- Drawings for each step in the manufacturing process. For example, for forged components, there will be both as-forged and as-machined drawings. Each will have dimensions and features specific to the step in the process.

Each fabricated part must go through a multiphase design and production process:

- **Preliminary design** defines the basic geometry of the part as envisioned by the designer. The initial draft of the part geometry occurs during the detailed design phase of the product development process.
- **Review for manufacturability of the basic design** (DFM) ensures when the overall part is designed it can be manufactured. For example, are you specifying large asymmetric hollow parts made out of stainless steel? Typically large hollow metal parts are made by deep drawing but you are limited to roughly symmetrical parts (Chapter 10). It is easy to design parts that are hard to make and it is hard to design parts that are easy to make. If engineers are ignorant of how processes work, they can end up requiring complex multi-action tooling, re-fixturing, and significant post-processing; all of which drive cost.
- **DFM for tooling** drives a second set of minor design changes to enable tooling to be designed. For example, if a team is using injection molding to create a housing, the tooling designers might include ribs to prevent warpage or a draft angle to enable easy removal of a part from the tool (Chapter 12).

- **Tooling design** defines the geometry of the molds used to make the final parts. Tooling engineers start with the part geometry, assembly requirements, and aesthetic requirements, and add manufacturing features such as parting lines and gate locations (where tools separate and where molten plastic is injected into the cavity). Tooling engineers also design the tools to optimize heat transfer for heating and cooling the tool, fluid flow to ensure complete filling of the mold, and the mechanical behavior of the tool. For example, the tooling designer will design the gates and sprues, cooling channels, and ejector pins needed to run the injection molding equipment with the minimum cycle time.
- **Tooling build** may be the longest lead-time item in product realization. Tooling is typically cut out of hardened steel, and machining the cavities can be very time-consuming. Usually, the tooling is built in *steel-safe* mode (i.e., the mold cavity is smaller than needed, so changes require only material removal rather than the more difficult task of adding metal) without the final surface finish and details.
- **First shots** are the first parts fabricated from the production equipment and are used to check for functionality, dimensional accuracy, and manufacturability.
- **Pilot runs**. The factory will run several pilots on each fabricated part in conjunction with the engineering and product teams. Each pilot will result in slight tooling modifications to ensure precise dimensions, functional performance, and correct aesthetics.
- **Final tool modifications**. At the end, the tool-maker will apply surface treatments, heat treatment and polishing to create the right surface finishes for the part and to make the tooling ready for mass production.
- **First part inspection** (FPI) is conducted by the part manufacturer to validate that the part conforms to all of the dimensional, functional, and finish specifications.
- **Batch runs**. Parts are produced in large quantities to amortize expensive setup costs across multiple parts.

Despite everyone's best efforts to get the design and tooling right on the first try, the pilot process will invariably find issues with the fabricated parts. Problems include:

- **Dimensional accuracy**. All fabricated parts have some inherent unit-to-unit variability. Pieces that look like they fit in CAD with the right clearances may not actually fit in real life.
- **Parts can change shape** as loads are applied or when internal stresses are released after manufacturing. Both of these contribute to dimensional variation. Also, materials shrink and warp during fabrication of the part.
- **Material and mechanical properties**. The manufacturing processes can create local material inhomogeneity and defects. For example, a certain

polycarbonate part was injection-molded, and the factory did not use a heated knife to cut off the sprue (the channel used to bring material into the cavity). Manually breaking it off introduced micro-fractures that, with regular use, grew and ultimately caused the parts to crack. Unfortunately, changing to a different material would have required significant design and tooling changes. Instead, the team had to strictly control the sprue removal process to prevent the issue from recurring.

- **Aesthetics**. How a fabricated part looks is often different from the designer's vision. Sink marks, color variations, surface defects from dust in electroplating, or swirl marks from machining might be more aesthetically problematic than expected. Unfortunately, they are usually only discovered when the first parts come off the line.

The quality and cost of the fabricated parts and tooling is strongly correlated with how much the design team understands and incorporates the rules of manufacturability. While many problems can be solved during piloting, it is best to avoid as many as possible by designing the part correctly in the first place.

6.1.2 COTS – Commercial-off-the-Shelf Parts

COTS are those parts that are bought directly out of a supplier or distributor catalog, such as screws, O-rings, and threaded inserts. They are purchased as-is and often come with specification sheets that define the guaranteed performance parameters. COTS parts are specified in the BOM by either a manufacturer and a part number or a generic description of the part.

It is important to be as specific as possible when selecting COTS parts. Even a slight change in the types of screws used, for example, can have dramatic impacts on short-term and long-term performance (Inset 6.2). In one instance, a company had been purchasing screws for an outdoor product, but had not specified that these screws had to be stainless steel (SS). When the supplier did a cost reduction, they changed to non-SS screws, causing a rusting issue when the product was exposed to rain and salt air. At all times you should assume that the factory will replace the COTS part with a lower-cost part if possible. If the specification is too general, the purchaser has the flexibility to purchase material that may not work as you intended.

Design engineers can decide to specify either an assigned or a generic COTS part. For example, you can specify a generic AAA battery, or you may require a specific part number (an assigned part) from a particular vendor. Assigned parts are often specified when better performance is needed and are almost always more expensive. Without changing the supplier, it is hard to get significant cost reductions (cost-down) on assigned parts. Where performance is not as critical, the team can save money by specifying a generic COTS part.

Assigned COTS parts also carry a higher risk of supply shortage or, in the worst case, obsolescence – also called end-of-life (EOL). In one case, a motor factory went on strike and was unable to produce their motor; in another, a large consumer electronics company purchased the entire world's supply of a specific memory chip. In another case, an engineering team had to redesign the product housing because a specific battery was no longer available, and the housing was designed to exactly fit that geometry. No other suppliers had off-the-shelf solutions that had the same envelope. The risk of not being able to second-source COTS assigned parts is much higher than teams typically anticipate. Almost all of the companies the author has worked with have had to manage around either EOL parts or shortages of assigned COTS parts at some point during piloting or production.

6.1.3 Purchased Custom Parts

Custom parts may need to be specified where the off-the-shelf versions do not sufficiently match the needs of your design, and your company does not have the skills or bandwidth to design or manufacture the technology. Custom parts are needed when the existing COTS offering does not physically fit your design (e.g., battery form factor), does not meet the functional requirements (e.g., exact power and speed), or requires a unique interface (e.g., a unique keyway).

Custom parts are typically purchased from suppliers who specialize in one technology or product family (e.g., batteries or bearings) and can use their expertise to customize an assembly or part unique to your product. Custom parts range from fully customized parts built specifically for your product to custom modifications and changes to a supplier's existing product line.

On one end of the range of customization, product teams may partner with other companies to produce unique, fully customized subsystems. The outside partners typically have deep expertise in their technology and create unique solutions for the product team. The partner may invest in developing new design or manufacturing technologies specifically for the partnership. These parts are difficult to cost ahead of time because both companies are investing in the product, and both are sharing in the profits and the risks. It is almost impossible to re-bid these parts or bring in second sources because of intellectual property (IP) and contract issues. Because the design of your final product is highly dependent on the design of the customized subsystem, changing suppliers typically means a significant re-design of the whole subsystem. Your final product may even advertise the partnership between the two companies as a selling point, further complicating any rupture in the partnership.

At the other end of the customization spectrum, teams may specify parts that are slight variations on a supplier's existing product offerings (e.g., changing the connector on a battery or the form factor on an LCD screen). Suppliers may offer customizations from a pre-set array of options. For example, if you are producing a robotic product, and you need a custom bearing with a unique combination of inner and outer diameters, a supplier could produce these for your company without having to create custom tooling. Typical examples of custom parts used in electromechanical products include:

- LCD screens with custom sizes, colors, and bezels.
- Cameras in different geometric configurations.
- Batteries with custom capacity, geometry, and connectors.
- Seals and gaskets that are custom-cut for your product's unique geometry.
- Motors with a variety of configurations, gearing, and connectors.
- Gearboxes that are typically uniquely configured for your application.

Depending on the complexity of the changes needed, the cost for custom parts can be incrementally higher than the COTS part, or exorbitantly more expensive. If a custom part is needed, you should talk with the supplier as early as possible to determine the significant cost drivers and how incremental changes in size, shape, power, or performance might dramatically increase cost (Chapter 8).

Suppliers typically publish comprehensive specifications for their products, and these specs enable you to identify whether they have the necessary capability. The specifications include the part envelope (they usually do not provide detailed CAD for all of the subparts), connectors (e.g., geometry, power), performance, and operating environment (e.g., maximum operating temperature). Also, the supplier may provide durability, quality, and reliability specifications that can be used to compute your product's reliability (Chapter 7).

Custom parts may have a significant lead time and large minimum order quantities (MOQ). Companies are often surprised at the amount of inventory they need to purchase. Also, the supplier may require you to pay for unique tooling, certification, or verification testing, so there may be significant expenses beyond the costs of the custom parts themselves. All of these expenditures can impact cash flow. Finally, it is challenging to switch suppliers or find a second source if there are issues.

6.1.4 Electronics

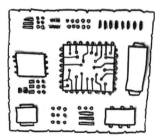

Most products are not purely mechanical but include some mix of mechanical and electrical parts. To fully define a product, electronics engineering needs to specify custom control boards, wiring, and sensors. The design of the PCBAs can be complex and time-consuming. Companies are often lulled into a false sense of security because getting an Arduino to function is relatively easy. However, designing the boards, possibly including complex antenna configurations, power management, and charging circuitry requires sophisticated technical expertise.

The design and fabrication of the production-intent PCBAs are often completed during the engineering verification testing (EVT) phase before final versions of the mechanical components are ready. At that point, the team can test the integrated operation of the PCBAs, firmware, and software using prototype mechanicals. For example, the team can test whether the button actuations trigger the correct software responses, but the button can be 3D-printed.

During the design of the PCBAs and electrical systems, teams often fail to take into consideration the following issues:

- **Test pads**. ICT (In-circuit test) requires locations to be set aside on the board for the test probes to interface with the PCBA (Chapter 13). Test pads and connectors may also be needed if you have to program the microcomputers before shipment. It is hard to add test-pads into the boards after the fact. Teams should think through how the boards will be tested and programmed and ensure there is sufficient space set aside for the test pads. If you need access to those test pads after assembly, you may want to provide access through the case or housing.
- **Structural support.** Teams must design supports to prevent the boards from undergoing stress that would cause damage in the short term (from drops) or long term (from vibration or fatigue). Designers often forget to leave room for the screws or heat staking of the boards to the mechanical structures.
- **Connectors and wiring**. The design engineers often fail to account for the space taken up by connectors and wires. Many teams discover too late that these connectors or wiring do not fit, and workers cannot assemble the product. For example, in a portable electronic device, the minimum bend radius of a flex-circuit was much larger than expected and couldn't fit in the housing as specified by the industrial designers. The entire housing had to be expanded to accommodate the minimum bend radius, delaying production for several weeks.
- **Component selection for reliability**. Some electronic components are highly sensitive to thermal or humidity extremes. The parts should be selected to perform in the range of expected operating conditions. For example, some components may be sensitive to moisture and require conformal coating. However, it is important to note that the lead time on more robust components can be longer and the parts are typically more expensive.
- **Packaging and temperature ratings**. You want to be as specific as possible about the temperature and packaging specifications. Parts are not interchangeable and can have dramatic impacts on assembly and reliability if the wrong parts are ordered. For example a transistor MMBT2907A is available in an SOT-23-3, SC-70-3, and SOT-223 package. While the function is the same, the footprint is different. If you order the wrong package, the part won't fit on the board.

- **Specifications for any post-processing**. The boards may be either potted (i.e., filling electronics with either a solid or gel) to reduce sensitivity to moisture and vibration, or conformal coated (i.e., spraying or painting a thin polymer film) to avoid the damaging effects of saltwater and humidity. If your product requires water resistance, be sure to comprehensively define the method and materials for the conformal coating. For example, if spray or dip coating are not explicitly mandated, the factory may use a manual brush methods which suffer higher defect rates and result in a less-than-perfectly-waterproof circuit board.
- **Programming**. Some microcomputers may need to be flashed with software.

6.1.5 Consumables

Consumables are materials that are bulk-purchased from a supplier or distributor. Examples of consumables are solder, wires, glue, scratch-resistant covers, and labels. Consumables are needed to assemble your product, but often are not considered important enough to show up on the assembly drawings or the engineering BOM. As with COTS parts, it is critical to specify the properties and materials of consumables so no unwanted changes are made. Small changes in adhesives can have dramatic quality impacts. Some generic versions of an adhesive can have very different viscosities and adherence to some materials.

Consumables typically don't show up in the engineering BOM but do show up in the manufacturing BOM. In addition, consumables are typically bought in bulk for a manufacturing site and kept in and dispensed from general inventory. Finally, the materials are typically used in fractions of a purchased unit. Consumables are defined in the same way as COTS: either as a specific manufacturer plus part number or as a generic description.

6.1.6 Raw Material

In a manufacturing BOM (described below) it may be necessary to specify the raw material used to make the part. A couple of examples include bar stock of aluminum used to make turned parts, or unmixed urethane to produce low-volume urethane cast parts. Raw materials can be either generic materials or specific to a manufacturer. When you are defining amounts of raw materials in the BOM, remember you need to account for offcuts or waste (e.g., scrap shavings from a turned part) that won't appear in the final product. Be sure to order not simply the amount needed in the product, but the amount including any offcuts or waste.

Vignette 4: Don't Assume You can Source Exactly What you Want

Jonathan Frankel, Founder & CEO, Nucleus

Nucleus was founded in late 2013 when the founder, Jonathan Frankel, wanted to install an intercom in his house and was quoted only expensive, analog systems instead of the modern, wireless system he was expecting. The system was billed as an "Anywhere Intercom" – an intercom that could instantly, over Wi-Fi, call any other connected unit in the world, whether it was upstairs or in a different country.

The Nucleus team hired a leading industrial design firm to create a sleek, modern look for the device. In addition, the Nucleus team required that (a) the device be thin enough to hang on a wall, (b) while also possessing a camera with a wide enough angle such that it could see most of the room in which it was hanging, and (c) have a built-in physical privacy shutter for the camera. The ID firm created a concept that met all three requirements by including a camera that rotated on a hinge; rotating it down or up would expose more of the room, and rotating it 180° would effectively become a physical privacy shutter.

The camera became the main feature around which the rest of the form factor was designed. However, when Nucleus took the design to their CM – one of the world's largest and most sophisticated manufacturers, who had also invested in the Nucleus Series A round – the CM told them that there was no camera that existed that was thin enough to fit inside the rotating panel that the ID firm had designed. The Nucleus team then spent months examining two alternatives: one, finding or designing a thinner camera to fit the original design, or two, creating a thicker rotating panel, which in turn would necessitate adding ~20% of thickness to the entire body of the device. After determining that neither of these alternatives were feasible or palatable, the Nucleus team needed to re-design the entire ID to incorporate a realistic camera.

The delay set us back several months. We learned the hard lesson not to assume we can get any technology in any form factor. If we had done some preliminary research into camera suppliers ourselves, or requested that our ID firm identify an actual part, we could have saved several iterations on the industrial design and several months on the project.

Text source: Reproduced with permission from Jonathan Frankel.
Photo source: Reproduced with permission from Nucleus Life.
Logo source: Reproduced with permission from Nucleus Life.

6.2 Bill of Materials

The factory needs a complete and accurate list of all of the parts and materials required to assemble, test, and ship the final product. This list is called the BOM. It should consist of everything from the injection-molded parts to the tape used to close the box and the fraction of the bottle of Loctite required to glue parts together. Remember you can't build a product without first ordering everything! You don't want to have your million-dollar project on hold because you forgot to order the right tape.

There are two types of BOMs:

- **The engineering BOM (E-BOM)** contains all of the parts that appear in the final product and its packaging, but not materials used during manufacturing that are not part of the product itself.
- **The manufacturing BOM (M-BOM)** contains all of the materials required for the manufacturing of the product, in addition to those listed in the E-BOM. For example, often, assembly lines use temporary plastic protection during assembly to prevent scratches or contamination from dust or oils. This plastic film is installed during the early stage of production and is removed before packaging; therefore, it would be part of the M-BOM but not appear on the E-BOM.

It is critical to have a comprehensive and up-to-date M-BOM for the following reasons.

- **Cost of material estimation**. The BOM enables the team to estimate the material costs. Many people are surprised at how much the costs of little parts, connectors, and wires add up.
- **Defining your supply chain**. Each part should have a supplier or distributor identified. Whenever possible, your team should try to consolidate, using the same supplier or distributor for multiple parts to reduce management overhead and costs. A quick count of the unique suppliers will highlight early on how complex the material ordering will be. Also, the blanks in the BOM will drive the to-do list of the purchasing team.
- **Material planning and ordering**. The BOM specifies a lead time for each part, indicating how far in advance the purchasing team needs to order materials. The BOM forces teams to identify the long lead-time parts, helping to avoid costly delays.
- **Tracking change orders**. The BOM should be kept up to date and consistent with any production change orders. This ensures that the factory builds the most up-to-date version of the product.
- **Quality control**. Each part in the BOM should have an associated link to either the drawing or manufacturing specification. This enables incoming quality control to ensure that the parts are correct as they come in the door (Chapter 13).
- **Cost-down/second-sourcing activities**. The BOM gives you an accurate picture of all of the material costs in your product along with how the part is purchased. This data is critical when trying to take cost out of the

Inset 6.2: Generic, Assigned, or Consigned Parts

 Parts in the BOM are typically assigned one of three designations: generic, assigned, or consigned. The differentiation among the three drives how parts are ordered and can have a significant impact on costs, lead times, quality, and cash flow.

Generic enables the factory to purchase the parts from any vendor. If a specific manufacturer part number is indicated, the generic designation tells the factory that equivalent parts are allowed without approval. For example, a BOM might read "#8 x 1 in SS Phillips Drive wood screw." This would enable the factory to buy parts that match that specification from any supplier.

Assigned parts require a specific part to be purchased from a specific manufacturer and sometimes through a specific distributor. Parts are assigned to ensure quality, and typically these assigned parts are more expensive than generic because you have less opportunity for cost-down. Parts can also be given the "assigned" designation when the company has negotiated a price with a specific supplier. For example, you may negotiate a much lower price on one component but commit to buying four other parts from them at a slightly higher price. The BOM data for an assignment will contain the data shown in Table 6.2. Making the Serial SRAM chip an assigned part would preclude the part from being replaced by an equivalent from Microchip or other generic memory chip manufacturers.

Consigned parts are typically limited to a handful of lines in the BOM (if any). Consigned parts are purchased not by the contract manufacturer but by the product development organization. For consigned parts, you take responsibility for ordering, paying for the part, and ensuring that the part is delivered to the factory on time. Consigned parts can make it challenging to manage forecasting and material handling because there are two ordering systems supplying the same production line. The risks of ordering errors and parts shortages are very high. Parts may be consigned for several reasons:

- The supplier may not want the low price to be shared with the CM.
- The markup on consigned parts is very low or zero. Consigning saves on the material markup on costly parts (Chapter 8).

Proprietary parts may be provided pre-programmed with firmware to protect intellectual property from being appropriated by the CM or their other clients.

Table 6.2 Data in the BOM for an assigned part

Part description	Manufacturer	Manufacturer's part number	Supplier
Serial SRAM memory, 256 kb, 3.0 V	ON Semiconductor	N25S830HAT22I	Avnet

product. For example, the team can change a part from an assigned to generic, or find a new supplier who can provide a part at lower cost.
- **Legal and regulatory requirements**. In highly regulated industries, government bodies such as the Food and Drug Administration (FDA) and FAA (Federal Aviation Administration) require traceability of the materials installed on each product. Especially in the case of aircraft, tracing the exact version of the BOM for each product and the ability to trace the

source and design control of each part is critical to getting the flight certificate for the airplane. The inability to determine the precise BOM of early Boeing 787 production units led to some of the initial delays.

- **Tracking factory costs**. As design changes are made, parts and costs may change. The purchase order sent to the factory should indicate which BOM version is being referenced so you are charged the right amount on your PO. For example, if you have done a major effort to reduce costs by replacing parts, you want to be charged for the new BOM costs, not the old.

6.2.1 What Information is Captured in the BOM?

The BOM contents will vary depending on the type of product and the company, but the fundamental data being captured will be the same. Boeing

---

Inset 6.3: Part Numbering

Every part in your company has to have a unique number. If numbers are inadvertently reused, it is possible to order the wrong part. In addition, part numbering helps with change management either through revision numbers or through assigning new numbers.

How you number the parts in your product depends on the standards set by your company, the complexity of the product, and the software you are using. Typically, if you are designing a numbering system, you want to design the schema to be expandable and readable, and which will live beyond your first product. The options for numbering range from unintelligent to intelligent:

Unintelligent part numbers are assigned in sequence. It is not possible to learn anything about the part from the number. However, the method is simple, and it is easy to avoid assigning the same number to different parts. An example is EP101286, EP101287, EP101288, and so forth.

Intelligent part numbers enable the type of part to be determined just from the part number. When defining an intelligent numbering scheme, it is essential to think into the future because changing numbering systems is very difficult. Examples are INJ-00043-01 (INJ for injection molding), OTS-00571-06 (OTS for purchased parts), ELE-06154-11 (ELE for electrical), etc. Each identifier comprises a category, a unique number, and a version number. In general, the rules that will make life easier are (1) do not start a PN with a 0 (Excel does not do well with identifying numbers vs. text) and (2) keep the numbers the same length and pattern.

Your part numbers may also change as design revisions occur. Companies generally use the "3F" rule (fit, form, and function) to determine when change requires a new part number or can be updated with a revision number. If there is any major change to fit, form, and function, then the part needs a new number. In addition, when an individual part gets a new number, the parent assemblies also need to be renumbered to reflect the change.

Finally, there is a lot of debate about aligning part numbers and drawing numbers, but typically the two don't use the same scheme but reference each other (i.e., a BOM may list a drawing number along with the part number and the part number may be included on the drawing).

will describe a generic screw in a BOM in the same way a start-up company will. At a minimum, the BOM should include the following:

- **Part number**. These numbers should be unique for each version of the part. For example, if you change material in an injection-molded part or if you use a second source for a part, a new and unique part number should be assigned (Inset 6.3).
- **Name**. This short text uniquely describes the part, e.g., *front frame 1.*
- **Description**. A longer description that can help differentiate between similar parts, e.g., *left-hand front frame with PCBA mounting features.*
- **Quantity**. The quantity of each part in the parent assembly or SKU (can be a fraction as well). For example, if you buy Loctite by the bottle, it might be 0.01 g of a 5 g bottle (i.e., a single bottle will be used to make 500 products).
- **Cost per part**. This may simply list the cost to purchase the part or it may be broken down into the labor, material, and capital equipment costs. The breakdown of the costs depends on the part type and the company preferences. The cost is often a function of the MOQ (described below).
- **Extended cost**. The cost per part times the quantity (the total of all the extended costs is the total cost of materials).

Often the M-BOM will also contain information about:

- **Lead time**. How far in advance the part needs to be ordered to ensure timely delivery.
- **Minimum order quantity (MOQ)**. How many need to be ordered at a time to get the quoted price.
- **Manufacturer/supplier**. Who is providing the part or sub-assembly.
- **Manufacturer's part number**. The manufacturer may have a unique part number which enables you to place the order, as they won't know your part numbering scheme.
- **Ordering method**. Assigned, consigned, or generic (Inset 6.2).
- **Comments and notes**. The assumptions behind the data and any additional information should be maintained. Typically, this information is used internally and not given to the factory.

Additionally, BOMs can include the following.

- An image of the part for easy identification.
- References to COTS and semi-custom part specifications.
- References to CAD and drawings.
- Cost model data and assumptions.
- Web links to product information.
- Materials and finishes.
- A bounding box that defines the overall size of the part.
- The overall weight of each part.
- Trace width and mounting method (e.g., through-hole vs. surface mount technology (SMT) for electronics).

6.2.2 Indented BOMs

In products with a large number of parts, the lines in the BOM can be grouped into subassemblies or subsystems. These subsystems and their constituent parts are managed using an indented BOM, which makes the BOM much easier to read and maintain. BOMs maintained in software have the hierarchy embedded in the data structure. When using spreadsheets, indented BOMs are not literally indented; rather, a number in the far left column indicates the level of the BOM. For example, a number "2" in the first column will have a set of rows with "3" below it. That indicates that all of the parts labeled with a "3" in the aggregate create the part labeled with the "2." For example, for the M-BOM in Table 6.3, the product has two major parts – the bracket assembly and the packaging – which are indicated by a level "1." The level 2 parts (rod, bracket, and adhesive) are the parts of the bracket subassembly. Because this is an M-BOM, while the bracket is made solely of the urethane casting material, it uses several disposables (mixing cups and stirring sticks) to make the part. These parts are labeled with a "3."

There are several challenges with an indented BOM in Excel, including doing the cost analysis, sorting, and organizing of the BOM. It is relatively easy to incorrectly sort the file and mess up the indenting, and calculating costs is not as simple as adding up the columns. For complex indented BOMs, teams may need to purchase BOM management software because the cost of maintaining the Excel sheets quickly outweigh the cost of buying a dedicated software system.

6.2.3 Software to Manage a Bill of Materials

A company can choose between several BOM management methods, depending on their needs.

- **Cloud-based spreadsheets**. For start-ups and other small organizations, cloud-based spreadsheets (such as Google Sheets) are often used as the first place to develop a BOM. Google has third-party add-ons that provide BOM templates. While cloud-based documents are easy to use for early collaboration, they can make it difficult to manage sharing, version control, and change control. The engineering team, along with operations, will need to decide on the BOM format. Ensuring that the BOM is complete and correct can be difficult as there are no rules imposed by the software.
- **Excel spreadsheets**. Most CMs in China use Excel spreadsheets to manage the BOM. It is easier to control access in Excel than in cloud-based spreadsheets, but it is difficult to ensure that everyone has the right version and that changes are communicated correctly. The Excel documents have the same formatting and consistency issues as the cloud-based spreadsheets above.
- **SaaS tools**. There are several SaaS (software-as-a-service) BOM tools that manage the BOM, store hierarchical BOMs, and can manage additional

Table 6.3 Sample indented M-BOM

Level	Part number	Part name	Description	Type	Quant	Cost per unit	Extended cost	Assigned/ generic	Supplier	Mfg part number	LT (days)	MOQ
1	A01-ASY-0001	Bracket assembly	Assembled rod and brackets	Assembly	1	$33.52	$33.52					
2	A01-MAT-0001	Rod	Hard Anodized 6061 Aluminum Rod - 3ft	Raw materials	0.17	$26.00	$4.42	Generic	McMaster Carr	6750K13 - 3ft	3	6
2	A01-CON-0001	Adhesive	Loctite Instant Adhesive 495	Consumables	0.01 bottle	$28.00	$0.28	Generic	Home Depot	xxx-111-zzz	3	5 oz.
2	A01-FAB-0002	Bracket pieces	Individual pieces	Fabricated	4	$6.70	$26.82					
3	A01-MAT-0002	Casting urethane	Smooth Cast 300	Raw materials	0.15 lb.	$5.96	$0.83	Assigned	Reynolds	Smooth Cast 300 - Gallon	2	5.4 lbs.
3	A01-MFG-0001	Mixing cups	8 oz plastic cups	COTS	2	$0.05	$0.10	Generic			2	500
3	A01-MFG-0002	Stirring stick	Wooden stick	COTS	2	$0.01	$0.02	Generic			2	500
1	A01-PGK-001	Packaging	Box and inserts	Semi-custom	1	$1.50	$1.50	Assigned	TBD		10	500

analysis tasks. Many of the BOM SaaS tools can automatically populate costs, MOQs, and lead times from suppliers. The systems also enable organizations to analyze cash flows, identify where costs are high, and collect other fixed-cost data.

- **Enterprise software**. Large organizations manage their BOM within their existing enterprise software – enterprise resource planning/manufacturing resource planning tools (ERP/MRP) (Section 4.4) – to ensure the right materials are ordered at the right time, and to manage and automatically communicate any changes. However, embedded BOM tools are cost- and resource-intensive; they are typically implemented only when the investment costs are justified by the labor and time savings and error reduction. Also, major CAD systems usually have embedded BOM tools that can link CAD models and other production information.

Choice of a BOM system depends on the maturity and size of the company as well as the maturity and complexity of the product. Larger companies often opt for enterprise BOM systems, while smaller companies or teams (or those making less complex products) may opt to use free cloud-based solutions. Table 6.4 outlines the pros and cons of each type of tool. The following factors should be considered when selecting which approach to take:

- **Cost to purchase and install**. Comprehensive enterprise software can be costly. Teams need to balance the cost and cash flow required vs. the functionality it provides.
- **Learning curve**. The product development team may not have the bandwidth to be trained on complex BOM systems, whereas almost everyone will be familiar with Google Sheets or Excel.
- **Shared access**. Early in the design phase, teams will want to have shared access to BOMs, be able to simultaneously update the document, and make changes without burdensome permissions and change processes. They may want the flexibility of a cloud-based system with few controls. However, with a large team, an enterprise system enables changes to be made in a controlled way.
- **Regulatory**. Highly regulated products (such as aircraft and medical devices) require BOM traceability and version control. Teams producing these products might be more inclined to use enterprise software with built-in version control.
- **Change management**. At some point, the BOM will need to be locked (Inset 4.2) so that no changes can be made without the right approvals of the factory and design engineering. After putting the document under change control, updates to the BOM in enterprise systems can be very time-consuming; but there is a greater chance the change will be managed without hiccups.
- **Complexity management**. Complex and deeply indented BOMs are hard to manage in a flat table.
- **Analysis capability**. Teams may want to do what-if scenarios – lead times vs. cost, or MOQ vs. cash flow – to identify key drivers of cost and cash flow. Enterprise BOM software often has this capability built in, flat spreadsheets do not.

Decision factors	Cloud-based spreadsheet	File-based spreadsheet (Excel)	SaaS	PLM/ ERP
Learning curve	●	●	●	○
Ease of use and flexibility	●	●	●	○
Collaboration	●	○	●	●
Change management	●	○	○	●
Complexity management	○	○	●	●
Analysis capability	○	○	●	●
Cost to make a change to the BOM	●	●	●	○
Integration into the supply chain	○	○	●	●
Cost to purchase and install	●	●	●	○

● – Good ○ – Poor

Table 6.4
Pros and cons of BOM management tools

- **Integration with the supply chain**. Transitioning the BOM to the factory may require the export of the BOM to the factory's systems. Ensuring that the BOM is consistent with other data management systems (ERP, MRP) can be time-consuming, so teams may want consider the benefits of using whatever BOM system the factory already uses. Having the BOM and ERP system in the same software suite can reduce the overhead of transferring files, ensuring version control and reducing errors.

6.2.4 When Does the BOM Become a Controlled Document?

Early in the design process, the BOM will be frequently revised and updated. It will be necessary for many team members to be able to update the document without a formal and cumbersome change management framework. However, once the product is ready to go into the pilot phase, it is necessary to lock the BOM and impose a formal change management. When the BOM becomes a controlled document, a set of constraints are placed to require that changes are approved and communicated. In a large company, these procedures are typically already in place and may be enforced through enterprise software. In a smaller start-up, it will be necessary to impose this discipline manually (Section 4.4).

The BOM typically becomes a controlled document just before the first parts are ordered for the pilot build. Locking the document prevents teams from making last-minute changes that are not reflected in the pilot builds. Change control in BOMs is multilayered:

- **Change approval**. Any changes to the BOM need to go through a formal change approval to ensure that the relevant functional groups can approve and implement the changes and the changes will not have any unintended consequences.
- **Change communication**. When the operations team begins to plan for piloting, any changes to the BOM must be communicated to the whole

group to guarantee that the supply chain is informed, materials are ordered on time, and old orders are canceled.

- **Version control**. Old versions of the BOM need to be kept and be traceable in case root cause analysis (RCA) is needed in the future. During piloting, changes in the BOM may have impacts on the validity of prior quality testing. When piloting uncovers quality issues, it is helpful to know whether the parts conform to the current BOM version (and will impact future runs) or if an older part was used and may have been the culprit.

6.2.5 Who Owns Your BOM?

Many team members mistakenly assume that when they engage with a CM, the team will own their product's BOM and can know precisely what is in it and how much everything costs. Depending on your contractual relationship with your vendor, contract manufacturer, design partner, or broker, you may *not* have a legal right to see the detailed BOM. In some cases, you may have access to a simplified BOM without the vendors, part numbers, or costs (to prevent information from being leaked to another CM). If you hire an OEM (original equipment manufacturer) CM (Section 14.3.2) to design and produce your product to your specifications, they may own the product CAD and/or the BOM. Depending on your contract, they may not be obliged to share the CAD or the cost breakdown with you. They may want to lock the design information to reduce the risk of you bringing the product to another vendor.

It is critical, when writing contracts, to ensure you have full access to the full and detailed BOM along with the cost of each individual part (called the fully-costed BOM) of your product and the major subsystems for several reasons:

- **Change approval**. It is critical to guarantee that the team approves *all changes* to the BOM. Changing from a high-end bearing supplier to a low-cost one might reduce the cost but may impact quality. The factory or CM might not be aware that the bearing selection was critical to the durability and noise performance.
- **Cost oversight**. Understanding how your product is being costed can help to keep costs down. Transparency ensures accuracy. The author recently was asked to review a BOM for a medical equipment company for potential cost savings. She found that the CM (who fought hard to prevent access the fully costed BOM) had "inadvertently" moved the decimal place several points on the Loctite and had set the cost of the screws and bolts in the parts based on the lowest MOQ (even though they were buying them in bulk), among many other minor errors. These small errors resulted in about a $50 change in the COGS which, over 10,000 units, came out to about a $500,000 annual overcharge. Once notified of these discrepancies, the customer was able to reduce future COGS and get reimbursed for the overcharges.

- **Transfer learning to another factory**. You may want to second-source or move your production to another factory at some point. Not having access to the CAD or the BOM can make that change much more expensive.
- **Root cause analysis (RCA)**. If there is a failure in your product, knowing what parts were used and where they were sourced from is critical to conducting an effective RCA. For example, some crystals are very temperature-sensitive. If you do not have access to the BOM, it can be hard to help the factory identify that an inappropriate part was originally specified. The factory may not be aware of how the product is used, and you may not understand what is in the product, which makes it very difficult to find and fix a problem.

6.3 Color, Material, and Finish (CMF)

The CMF document comprehensively defines the color, material, and finish of the product. As with all specifications, the more detailed the information, the more likely it is the product will meet the desired standard. If the team shares the CMF with the factory early in the process, the factory can identify potential quality issues before the design is locked. For example:

- While soft-touch paint (often used in products that are handled) has a beautiful look and feel, it is easily stained with oils, and if it is not correctly applied, it delaminates and peels.
- Some metalized paints can affect antenna performance and are not suitable for devices with Wi-Fi or cellular communication.
- A mirror finish or bright finish can result in a high scrap rate because it is easy to see defects in the surface: for example, a strict finish requirement on an electroplated part of a device for the home resulted in an 80% scrap rate for one of the parts, causing the factory to have to order additional parts and to delay production.
- Paint colors from different processes or different suppliers (Inset 6.4) may not match as intended (as mentioned earlier, a fundamental rule is "you cannot match white plastic with white paint").

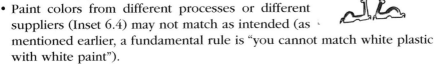

Typically, the CMF has a fully rendered drawing of each SKU with a table defining the color, material, and finish of each part. Figure 6.1 shows an example from Nucleus, a home video intercom system. The data in a CMF spec includes:

- **Color** is typically specified by the Pantone number [56]. In the case of Figure 6.1, the color is a simple Pantone Black. Pantone is a private company that has become the de facto industry standard for colors.

Inset 6.4: Quick Tips on Color Matching

 Colors are always going to be hard to match. Even if your product uses the Pantone system, the consumer's eye will pick up very subtle differences. Here are a few tricks if you must match colors in a product:

- Have one factory make all of the product or all of the color-matched parts. The factory can color-match each batch.
- Use different textures. The viewer's eye will adjust for different color tone and assume it is the texture, not the color.
- Separate the two parts with a contrasting color band.
- Create a shadow line using a chamfered edge or other geometric feature. The user will think the difference is due to a shadow, not the paint.

If you must match colors, you will need a method to measure whether they are the same. A frequently used color-matching standard is set by the International Commission on Illumination along with ISO. Colors are typically defined by a unique set of values XYZ, RGB, CYMK, or L*a*b*. The L*a*b* system, standardized under ISO/CIE 11664-4:2019(E) [57], plots the three color values for each pair of colors measured by the color testing equipment into a hypothetical 3D space. The distance between the color points, called Delta-E, measures how different the colors are. The eye can only pick up differences of greater than 3.

- **Material** defines what the surfaces are made from. More specific information about which formulation and vendor would be included in the drawings and BOM.
- **Finish** defines the quality and texture of the surface. Depending on the manufacturing process, there may be standards used to identify the finish. For example, in Figure 6.1 the camera cover has a finish of MT11010 which is a MoldTech [58] standard with a light texture with a depth of 0.001 in. The polycarbonate finish is a "polished" finish which is required for optical clarity.
- **Additional specs such as deco/branding** define any additional surface treatments. For example, the cover has a red sticker added and the part containing the camera lens is selectively back-painted black to only let light through the lens.

Rarely do engineers get training in secondary processes, finishes, or deco methods. As a result, the CMF is often a major stumbling block during product realization. It is tempting to over-specify finishes and put high tolerances on every surface, but remember that your customer is only going to see the outside surfaces. When the author first started making furniture, she spent significant time finishing the bottoms of the tables to the same quality as the tops. After wasting considerable time, the author realized that she only needed to carefully finish the tops, and give just a quick finish to the bottoms.

Camera lense

Color: Clear
Material: Polycarbonate.
Finish: polished
addtional: Backpainted black.

Camera cover

Color: Pantone black
Material: ABS.
Finish: MT 11010
addtional: Red dot graphic applied

All other camera housings

Color: Pantone black
Material: ABS.
Finish: MT 11010
addtional: none

FIGURE 6.1 Example of a color, material, and finish document. *Source:* Reproduced with permission from Nucleus

On manufactured products, parts will have A, B, C, and D surface finish designations:

- **A**. The surface is seen.
- **B**. Surfaces the user can see with deliberate effort.
- **C**. The surface is not seen and is not functionally critical.
- **D**. The surface is functionally critical.

The part specifications will define the aesthetic standard of each of these surfaces. Many standards exist for specifying finishes. The people specifying the finishes (typically the industrial design team) need to be familiar with the finishing specs and how they drive costs. For example, injection-molded parts typically use mold-finish standards set by the Plastics Industry Association (SPI) [59]. Three commonly used injection molding surface finishes include:

- **Tool marks visible**. This finish is for the underside or non-visible sides of products. If you change the batteries in a remote control, you'll see the lines and grooves that weren't machined or polished out on the underside of the cover.
- **Gloss level**. For parts without texture, this specifies how glossy the parts will be when they come out of the mold. Gloss level is defined by the process used to polish the tool. SPI [60] is the typical standard used and includes surface specifications such as:
  - SPI-C1: 600 grit stone, low polish
  - SPI-B1: 600 grit paper, medium polish
  - SPI-A2: Grade #6 Diamond Buff, high polish
- **Texture**. The surface can be sand-blasted or machined to give a rough surface or polished to give a glossy surface. Mold suppliers will provide

samples of different textured finishes that you can select from. For example, in Figure 6.1, the cover has a texture of MT1101.

Most other processes, such as electroplating and anodizing, also have finish standards. Being as specific as possible and using the right industry terms and standards will increase the chance that the product turns out as expected.

6.4 Mechanical Drawing Package

The mechanical drawing package defines the geometry of all the parts, tolerances, surface finishes, and materials, as well as how they are made (for example, an as-forged drawing and an as-machined drawing of a forged and post-machined part). The engineering drawing is based on the CAD model, but they are *not* the same thing. The CAD model defines the surfaces and edges of the physical geometry. However, the geometric model of the part on its own is not sufficient for communicating the design and manufacturing intent to the factory or suppliers; the more detailed mechanical drawing is needed for this.

Drawings follow a set of rules that enable an unambiguous definition of the geometry, as well as many other characteristics of a part: e.g., surface finishes, tolerances, materials, and feature criticality (Inset 6.5). While the current CAD models allow for easy translation of 3D geometry to 2D geometry, significant work is required to ensure that the drawing is complete enough to enable the manufacturer to produce the parts accurately. Typically, the designer has to add information manually and lay out the dimensions on the drawing. Drawings are controlled documents and typically have a formal release process that includes approvals and revision control.

Inset 6.5: A Short History of Drawings

The standards of how three-dimensional geometries are represented in 2D have a long history, starting with Gaspard Monge, the father of descriptive geometry, in 1765. The UK was the first to adopt technical drawing standards because the Industrial Revolution required the ability to manufacture multiple identical parts by different tradespeople. 1840 saw the advent of the blueprint process, which allowed for rapid and accurate reproduction of drawings so each person on the floor could have an identical copy. 1901 saw the definitions of drawing standards migrate across industries.

Since then, standards organizations have continued to update and release changes to drawing standards to reflect new technologies. Today, there are several standards, including the ASME (American Society of Mechanical Engineering) drawing and GD&T Standards [61, 62]. There are entire books dedicated to how to create engineering drawings.

A detailed mechanical drawing (an example of which is shown in Figure 6.2) contains:

- **Drawing of the geometry**. A scaled line drawing of the orthographic and isometric views of the part take up the majority of the space on the drawing.
  - *Multiple views of the geometry*. Typically, three isometric views of the part, all to the same scale, are included (top, side, back) along with an orthographic projection.
  - *Dimensions*. Drawings should fully specify every critical dimension. Dimensions should be measurable on real parts using standard equipment. For example, measuring the distance between a hole and curved surface can be very difficult and expensive compared with measuring the distance between two parallel surfaces.
  - *Datums*. Datums are common reference planes that define a baseline from which dimensions are measured. Typically, datums are aligned with either key surfaces, critical features, or assembly location features.
  - *Tolerance* gives a numerical range of how big or small a feature can be while still being acceptable. Features and dimensions on the same part may have different tolerances because of assembly or performance requirements. The box at the bottom of the drawing often specifies the blanket tolerances that will apply if no specific tolerances are called out for a given dimension. There are two primary methods for tolerances: coordinate and GD&T (geometric dimensioning and tolerancing). GD&T is very important where assembly tolerances are critical to the functionality or safety of a product. GD&T increases the chance that the functional goals of the product are aligned with how the parts are manufactured and measured.
  - *Assembly drawings*. Show how the parts go together in an assembly.
- **Drawing information**. The information block contains the part number, drawing number description, who generated the drawing, release date, and version. In addition, there is information about the size and scale of the drawing (i.e., 1:1 or 1:4). The drawing scale allows unspecified dimensions to be measured directly off the prints, although now most people go back to the CAD files.
- **Assembly bill of materials**. If the drawing is of an assembly, it will contain a bill of materials with part numbers and references to drawings of sub-assemblies and parts.
- **Notes**. Additional information about post-processing and other data is added in the white space of the drawing.
- **Surface finishes**. Surface finish and roughness can be specified for both aesthetic and functional requirements.
- **Material**. Defines the raw material that should be used (Inset 6.6).

Typically, the engineering team has to manually translate the CAD into engineering drawings (also called prints or blueprints). Even with advanced CAD systems, it is often not as simple as pressing a button to create the 2D representation of the geometry. Most CAD software companies are

FIGURE 6.2
Example of
a drawing.
Source: Reproduced
with permission from
Adam Kotler

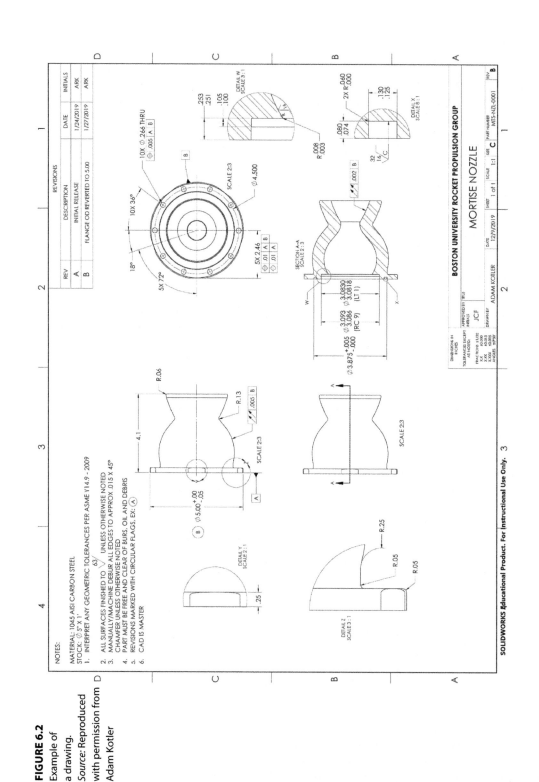

Inset 6.6: Material Specification

Being as specific as possible about materials (while giving the manufacturer the option of reducing costs by replacing with generic materials if possible) is critical both to ensuring ongoing quality and the ability to drive cost down over time.

For example, a team may decide to use Delrin for a bearing surface in their product to decrease friction while ensuring good tensile strength, creep resistance, and toughness. When defining the material, the engineer can take one of two approaches:

- **Be more generic and specify Acetal family**. An Acetal copolymer may be an acceptable replacement for DuPont's Delrin. Depending on the manufacturer, it may be less expensive than the name brand Delrin. If the design is not dependent on very specific material properties, the generic may be acceptable.
- **Be specific about the exact Delrin the product needs**. You will need to define specific manufacturers, formulations, and treatments if your design is sensitive to changes in any of the material properties. There are more than 30 different formulations for Delrin which allow you to select very specific material properties (e.g., manufacturing method, viscosity, and toughness, filler, and UV resistance).

developing technology to enable engineers to fully define a digital model and skip the 2D drawing, but the model-based definitions (MBDs) are not universally used, and most manufacturing organizations still want to see formal drawings. Table 6.5 describes the three methods of representing geometry and their pros and cons.

Purchasing, tooling, manufacturing engineering, factory operators, and quality engineering teams all use drawings during the product realization process for a variety of purposes:

- **Purchasing**. If the part is going to be made by an outside vendor, the more specific and complete the drawings, the easier it is to get an accurate quote.
- **Designing tooling**. For example, datums are used to set fixturing surfaces and parting lines of the parts. The surface finish requirements define the tool polishing, and the A-B-C surface designations indicate where gate and ejector pins can be located.
- **Defining process selection**. Tolerances, feature sizes, and feature locations drive the selection of manufacturing processes (e.g., 3-axis or 5-axis CNC), process planning (tight tolerances require more finishing passes), and specifying fixtures (if the part needs to be held during processing).
- **Determining whether a part is acceptable**. Drawings are used to verify whether the part produced is acceptable. Quality control will measure the part according to the dimensions defined on the drawing. If the features all fall within the tolerances, the part is considered good. Also, dimensioning and tolerances will determine how the part is measured. Dimensions that are hard to measure with traditional methods may require expensive coordinate measurement machines (CMMs). These issues are discussed further in Section 13.1.

Table 6.5 Pros and cons of CAD vs. drawing vs. MBD

		CAD	2D drawing	Model-based definition
Definition		Geometry is represented as digital 3D geometry. The geometry may not be fully specified and may not have tolerances, materials, or surface finishes specified.	A 2D drawing is a representation of the part geometry including dimensions, tolerances, materials, and surface finishes.	Dimensions and tolerances are specified within the 3D model along with notes on material, finishes, and other production notes.
Pros		A 3D model can be quickly specified. The CAD model can be used for digital manufacturing techniques such as 3D printing.	A 2D drawing is universally understood. The part or assembly is completely specified.	A MBD contains all of the details found in a 2D drawing without losing 3D information. A MBD creates a single "truth" and prevents the CAD and drawing from being in conflict.
Cons		A CAD model doesn't include tolerancing, materials or other notes required to fully specify the parts.	A 2D drawing can miss information and/or make certain features ambiguous. It is hard to accurately define complex surfaces. Data can be lost in translation from a CAD 3D model to 2D.	MBD is not universally used in the industry yet. Ensuring that the parts are fully dimensioned and specified can be difficult.

It is critical that the drawing package fully defines every detail the design team cares about, no matter how small. The author has seen numerous recalls on products that were correctly designed, but failed to specify all critical design details, causing quality failures. On one drawing package for a hydraulic-based product, a chamfer was not dimensioned and a recall resulted. The primary supplier, on its own, had added a chamfer with the correct dimensions. When the team switched to a secondary source, though, that second source used the same drawings, but didn't chamfer the part. Because the chamfer was not directly called out on the drawing, it was missed in the FPI and the sharply-edged part made it into production. The sharp corner caused wear and tear failures that led to a recall.

6.5 Electronics Design Package

In addition to the mechanical drawing package, your team will typically create an electronics drawing package that specifies, in detail, the circuit boards, flex-circuits, sensors, and wiring. The mechanical drawing package typically will have a placeholder for the electronics that defines its physical envelope so you leave enough room inside the housings. However, the actual electronics are defined using standards unique to electronic products. Printed circuit boards (PCBs), stencil masks, components and their placement, and connectors all have their own unique design definitions. The data to fully specify the PCBA includes:

- **PCB definition.**
  - *Gerber or ODB++ files* define the top, bottom, and internal copper layers, top and bottom solder masks, and top and bottom silk screens. Finally it includes the fabrication print which shows the board dimensions, hole sizes, thicknesses, plating, routed features, etc. Flex circuits are typically described as part of the fabrication print. The files either define a single board, or a panel which contains several individual boards (which are separated later using a router or by mechanically breaking the panels apart).
  - *Router file.* Defines the location of the holes and internal routing. This is typically generated by the PCB manufacturer from the fabrication print.
- **PCBA definition**.
  - *Top and bottom assembly prints* which show the location of each component.
  - *Test pad location list.* Shows the location of the test points; typically provided as an Excel file.
  - *Component BOM* itemizes the parts used to populate the board including a variety of integrated circuits, resistors, diodes, capacitors, and crystals. The component list can be maintained in a separate file or as part of the product BOM.
  - *Component placement list* specifies the position of each component on the board. It can also be called the parts list, centroid file, XY file, position file, or pick-and-place file.

In addition to these files, the electronics definition may include files that describe:

- **Flex circuits**. These are custom circuits assembled on a flexible thin film. The flex circuits take up less space than traditional rigid PCBs, and can be routed around the mechanical structures.
- **Wire harnesses**. These are pre-assembled bundles of wires and connectors. Whenever these are included, they should be specified by separate wire harness drawings.
- **Wire routing**. This drawing defines how wires are routed around the mechanical elements: the position, angle of rotation, etc. Wiring takes up an amazing amount of space and incorrect routing can lead to wires wearing and breaking.

6.6 Packaging

Packaging is the first thing customers experience when they purchase your new product (Inset 6.7), but sometimes the last thing that teams think about, often to the detriment of the product image and the delivery schedule. Packaging protects the product, it holds all of the additional materials

Inset 6.7: Unboxing Experience

"Unboxing" as a designed experience is a relatively new phenomenon. We all remember the feeling of opening presents on our birthdays when we were kids. The excitement of opening the wrapping and getting the toy out was often more pleasurable than playing with the actual toy. Studies have shown that the reward-seeking brain centers are stimulated by packaging and by anticipation of unboxing. That stimulation can trigger impulse purchasing. The importance of the unboxing experience is further enhanced by the increasing number of unboxing videos on YouTube and other social media sites.

While the general trend has been toward more elaborate unboxing, its role in purchasing decisions is constantly evolving. DotCom's eCommerce report [63, 64] identified a change: "Elaborate unboxing experiences and branded packaging, while wildly popular just three–four years ago, seem to be declining in importance for today's eCommerce consumer." You always want to keep ahead of trends. You don't want to spend a fortune on packaging if the customer doesn't value it.

(accessories and manuals), and it communicates your branding. Sometimes the packaging is as simple as a cardboard box with a sticker, while at other times it is almost a product in itself. Even your car is shipped with plastic covers to prevent scratches and damage during transit.

Design teams often underestimate the time and complexity involved in designing and producing a product's packaging. Long before your product goes into mass production, the packaging design must be fully specified and tested, even if you are just using an off-the-shelf cardboard box. Teams often start the packaging design too late in the process, resulting in rushing the packaging procurement and delaying product delivery to customers.

Also, organizations often do not understand the actual cost of packaging. Branding teams become enamored with Apple-quality packaging and complex unboxing experiences. Packaging designs are costed too late in the process, and the company is stuck with high costs with no time for iteration and cost reduction. A new audio company producing a $300 new set of headphones lost all of its profit margin because an overenthusiastic designer specified a packaging that ended up costing over $15. They didn't realize until too late the cost of their decisions, and by that time they couldn't redesign the packaging to meet a lower cost target. To ensure that the packaging is designed correctly for the product's needs, it is critical to understand the primary drivers of cost and complexity in packaging (Inset 6.8) as early as possible.

Each part of the packaging is defined using unique standards set by the packaging and paper industries. It is also important to ensure you have defined all of the parts of the packaging. The following sections describe the basic elements in a packaging design and how they are designed and defined.

Inset 6.8: Packaging Lessons Learned

When designing packaging, teams fall into a number of common potholes. This list is far from comprehensive, but should help avoid some of the issues other companies have faced.

- **The simpler the better.** The fewer the parts, the less there is to mess up. Keep packaging simple.
- **Packaging always costs more than you expect.** Start with a packaging budget and the key specifications it must meet, and then build from there. If you start with the highest-end packaging, you are likely to be late and over-budget.
- **Avoid complex patterns that wrap around a box.** If you have ever tried to get a wrapping paper pattern to line up when wrapping a birthday present, you'll understand this problem. It is hard for patterns such as waves or lines to match perfectly around a seam.
- **Things move in packaging.** Materials will slip and slide out of their cavities. Try to ensure that parts are either deep inside trays or are taped down.
- **Figure out your accessories early.** Adding a bag of screws late in the design process can require re-design and add significant cost.
- **Think through the unpacking process.** Ensure that the product can be lifted out of the packaging without damaging the product or injuring the customer's back.
- **Size the packaging to fit on a pallet.** You will get much more efficient packing and reduced shipping costs if you can ensure that the master cartons fit on a standard pallet without overhangs or wasted space.

The terminology and jargon in packaging are extensive, and could fill a book in their own right. In this section, we will outline just a few of the most commonly used terms to enable you to better communicate with your packaging supplier.

6.6.1 Box Types and Terminology

Packaging, at its most basic, usually involves a box. Boxes can be made of a variety of materials, have a range of configurations, and can be as cheap as several cents to as expensive as tens of dollars. Boxes can come to the factory either pre-assembled or in flattened form. The advantages of pre-assembly are that the boxes are typically of higher quality with a lower likelihood of being damaged, and they do not require assembly labor. However, the cost of shipping them to the packing location can be very high because of their volume, and storing inventory can use up precious floor space in your warehouses.

Since there are myriad types and configurations of boxes, the International Fiberboard Case Code [65] was developed by the packaging industry – this uniquely labels five classes of packaging and establishes codes for flat-board and inserts. This standard defines over 300 different types and configurations of boxes, and your team should consult this guide when choosing packaging.

Packaging suppliers can customize these boxes for you in several ways from single-color-stamped, to four-color printing of the stock, to case-made boxes with turned edges. In case-made boxes, high-quality graphics are printed on paper and wrapped over chipboard that has been glued together. These boxes don't have seams and don't look assembled. They are typically used for higher end products.

6.6.2 Packaging Elements

The packaging involves more than just the box that the customer sees, of course (Figure 6.3). To get a product to the customer safely, five types of packaging are typically required:

- **Point-of-purchase (PoP)** displays are the structures used to highlight a product in a retail location.
- **Gift box** is term used to describe the box the customer sees on the shelf. It contains one product and is the primary element of the unboxing experience.
- **Inner packs (also called inner cartons) are boxes** that contain a single or limited number of the same SKU in their gift boxes. They can be used for shipping direct to a consumer or may be combined in a master carton.
- **Master cartons** are the boxes that contain multiple inner packs and are used for transportation from the factory to the distribution center, and from the distribution center to the retail outlets or shipping centers.
- **A pallet** is the grouping of many master cartons into a consolidated structure that can be lifted by forklifts and stacked in containers.

Point-of-Purchase Displays (PoP)

If your product will be sold in a retail setting, it may be necessary to design PoP displays. PoP displays can be as simple as the custom-labeled trays that hold several products at the checkout counter (boxes of candy in the racks just at the eye height of your 5-year-old). They can be as complex as large end-caps (the end of an aisle in a retail location used to display product

FIGURE 6.3
Relationship between the gift box, inner pack or carton, master carton, and pallet

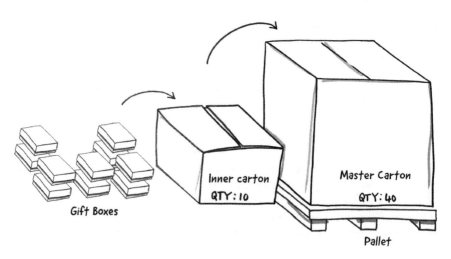

that is being promoted) with working products the consumer can try out. All PoP displays need to be consistent (in style and shape) with the gift box design and the overall branding and marketing of the product.

Gift Box

The gift box is the packaging that the customer sees and takes home. It can be as simple as a Mylar bag with a hanging tag (bags of screws at a hardware store) or a complex expensive package containing accessories and providing an elaborate unboxing experience (such as Apple's packaging for the iPhone). For products that are purchased by a customer and are small enough to be held in the hand, the packaging will typically be in the form of a box that is easy to pack and stack, whereas custom packaging shapes tend to be more expensive. The elements of the gift box might include, from the outside in:

- **Dust cover**. The outside gift box may be shrink-wrapped in plastic to inhibit damage and reduce theft.
- **Sleeve**. Some products have an outside sleeve that slides over the outer box. While this increases costs, it reduces the printing on the outer box. The sleeve can include SKU-specific information (matched colors), country-specific information, and other information that may need to be updated regularly. Since sleeves are generally easier and cheaper to redesign than outer boxes, sleeves can save companies money if the branding and labeling has to be updated. Sleeves also reduce the types of boxes in inventory needed to support multiple SKUs.
- **Outer box**. This provides structural support. It can be as simple as a standard cardboard box with single-color printing, to custom-cut cardboard boxes with cellophane windows (often for toys), to elaborate case-made boxes with ribbon hinges, fitted lids, embossing, and other aesthetic details.
- **Inner box/inner packaging**. There may be a box inside the outer box to hold parts of the product or create a unique unboxing experience.
- **Inserts**. In most cases, the product will need supportive inserts (trays, platforms, dividers, etc.) to prevent damage and movement. Inserts can be made out of a wide variety of materials (e.g., cardboard, foam, plastic, and molded pulp).

In addition to the product, the gift box will contain several items that need to be specified and designed. Remember to account for the cost of producing these materials and for the space that they will take up inside the packaging.

- **Aesthetic and customer experience materials**. Products that emphasize the unboxing experience will have envelopes, doors, and additional elements that enhance the experience.
- **Accessories**. Many consumable products will come with spare parts, cables, installation materials, or other accessories. These need to be secured so that they do not move and damage the product.
- **Manuals**. Virtually all products come with manuals, warranty documents, and quick-start guides. These documents need to be written and

typeset along with precise specifications of paper type, colors, sizes, and binding methods. The earlier these are drafted, the better.

- **Tags and labels**. The packaging will require labeling to support inventory management and warranty returns, and to satisfy certification requirements (Section 17.2). RFID (radio frequency ID) or other anti-theft tags may be integrated into the packaging to automate inventory taking and prevent shrink (retail term for product theft out the front and back door).
- **Scratch proofing**. If the product has a glass or other delicate surfaces, a removable film may be applied to reduce the chance of scratches or contamination from dust or oil in transit or unpacking.
- **Damage protection**. Foam inserts may be added to packaging to keep critical parts from vibrating or moving around during shipment.
- **Cases**. If the product needs protection during transport by the consumer, a case will be provided. For example, expensive noise-canceling over-ear headphones typically come with a hard case. Sometimes the cases are integral to the packaging. For example, when buying a power tool, the product usually comes in a hard case packaging (usually there is a sleeve or a sticker with the product information) which serves as the gift box, safe transportation around the job site, and storage of all of the accessories.

Inner Pack

Gift boxes are often packaged in multiples in an inner pack for shipment. Inner packs both protect the gift boxes and can be used to distribute products in smaller quantities to customers or distributors. Also, these inner packs allow for multiple SKUs in a master carton to be easily separated and uniquely marked. The inner packs are typically made of corrugated cardboard with labeling. The gift boxes may be wrapped in additional bags to prevent scratching and damage before being inserted into the inner pack.

Master Carton

The master carton is the box used to ship large quantities of product from the factory. The master carton should be of sufficiently high-grade material (typically corrugated cardboard) to ensure that the product will not be damaged during shipping (Section 7.3.7 discusses testing packaging for these conditions). These boxes may be custom-specified or stock boxes (standard sizes). If the product is particularly delicate, the master carton may contain internal support structures or foam blocks. The master carton is typically printed with the shipping information and any required hazard warnings (such as batteries) as well as the SKU and other product information.

Pallet

Most products are shipped on standard-sized pallets. The standard pallet is 48" × 40". As the team designs the packaging, specifying the pallet size is critical for several reasons:

- When shipping by truck or by container, the shipping company will charge by the load. If packaging isn't designed to make full use of the pallet space or if the height of the boxes don't fully fit a container, your effective cost per product to ship will increase.

- If the product is light enough to be stacked but the pallet is not fully loaded, it is not possible to double-stack the pallets, wasting valuable warehouse space.
- The master cartons will be shrink-wrapped to the pallet. If you do not entirely fill the pallet, it will be easier for the master cartons to slide during transport, potentially causing damage.

6.6.3 Packaging Design Documentation

The packaging design documents will typically be created by a packaging designer and will be communicated directly to the packaging manufacturer. Most of the time, the core engineering team won't need to access the packaging files and won't have the software to view the files. Methods to specify the packaging design include:

- A wide range of software packages that are used by packaging designers to fully specify the geometry, folding pattern, and graphics.
- The accessories and purchased parts will typically be specified using the same methods as for COTS parts.
- Manuals and inserts will be defined by the graphics files and a specification for the paper and print quality.
- Shrink wraps are defined by the type of process used, material used and material gauge.

Most of the time, the approval and review of the package design will be done first using renderings and then hand-built packaging samples.

Summary and Key Takeaways

- ❑ The product definition includes more than just the CAD files: it includes the drawing package, BOM, CMF, and packaging specifications.
- ❑ A detailed and accurate BOM is critical for all the steps in the product realization process.
- ❑ There are several formats and software systems that can be used to manage the BOM. Each organization must consider its own needs when choosing a BOM system.
- ❑ Factories need a drawing package to translate the design into a product. The more accurate and complete the drawing package, the less likely mistakes are to happen.
- ❑ The CMF defines the color, material, and finish specifications for a product.
- ❑ Each SKU of a product line will have its own CMF and BOM.
- ❑ The electronics need to be comprehensively defined using industry standard methods.
- ❑ Packaging cost can be high. Understanding the goals of the packaging and how it relates to marketing, margins, and total cost is critical. Making these tradeoffs as early as possible will get you the right packaging for your brand and price point.

Chapter 7
Pilot-phase Quality Testing

PRODUCT PLANNING

5. SPECIFICATIONS

6. PRODUCT DEFINITION

CAD

7. PILOT–PHASE QUALITY TESTING

8. COST & CASH FLOW

Product quality is about getting the product, as produced, to meet customer expectations and work correctly under all expected conditions over the life of the product. Though teams will do their best to design quality into the product, they can only test for most aspects of quality after they have a production-intent product. This chapter focuses on identifying the quality requirements, documenting them in a quality test plan, and then systematically testing products throughout the prototyping and pilot phases. Chapter 13 describes how quality is tested and controlled during product realization and full production.

Product Realization: Going from one to a million, First Edition. Anna C. Thornton.
© 2021 John Wiley & Sons, Inc. Published 2021 by John Wiley & Sons, Inc.

The term *quality* applies to a wide range of characteristics of a product and the processes used to make it. Because the word quality has such a variation of definitions in different spheres, we want to be careful about how we use the term in the realm of piloting and product realization (Inset 7.1 gives a short history of the term quality). When you hear "product quality," you might be picturing the difference between a beautifully designed, lightweight, and sleek laptop, and an off-brand clunky one. You might be picturing the difference between a robust SVU that can withstand a collision, and a tin-can car that crumples when tapped in a fender-bender. You can also be thinking about a product that lasts way past its warranty or one that fails early in life.

One of the major activities of the pilot phase is ensuring that the product meets customer quality expectations. Prior to piloting, the team should have spent significant time defining what the customer expectations are and documenting them in the specifications document (Chapter 5). However, most problems won't become apparent until the product is built and is put through its paces. Throughout the product realization process, the teams will need to test whether the product is achieving the quality goals set by the product development team.

However, it is important not to think of quality testing as a one-time effort that only happens during the piloting phase; quality testing should start as soon as physical prototypes are built. Engineering and works-like prototypes are used early in the process to test whether the technology can achieve its goals, then looks-like prototypes are used to test customer reactions, and finally works-like/looks-like prototypes test functionality with customers.

Most of the early pre-pilot testing is informal and open-ended, focusing on the validation process which answers the question "does the customer want what we are building?" Except in highly regulated products, initial testing tends to be

Inset 7.1: Historical Definition of Quality

 The Oxford English Dictionary [66] indicates that the word is probably derived from a combination of French, Greek, and Latin, and only relatively recently has the term meant excellence and superiority.

People started applying the term quality to product development and manufacturing with the growth of modern industrial practice. In the late 1800s, new terms emerged, such as quality mark, quality-tested, and quality producers. The growth of large-scale production post-WWII and the quality practices of Deming, Juran, and Taguchi [67–69] furthered the discipline. Industry added to the lexicon with a wide range of quality-related terms such as quality assurance, quality management system, quality circle, total quality management, House of Quality, quality control, quality controller, and quality management methods.

more like mad-scientist experimentation than rigorous formal experimentation. Early testing is used to guide product development, rather than being a hard test that can stop or cancel a project. Failures are seen as learning points rather than as emergencies (as they might be later in the process).

Once the product transitions into the product realization phase, the testing becomes much more formal and structured, generally falling into three phases: pilot testing, production testing, and field testing. These phases will share some testing protocols, but each has unique goals.

- **Pilot phases**. At each pilot stage – EVT, DVT, and PVT (engineering, design, and production verification testing) – the quality, engineering, and testing teams and the product owners assess the pilot samples (Chapter 3). Testing ensures that the design, supplied parts, manufactured parts, and production systems can consistently deliver a satisfactory product. This chapter will focus primarily on the pilot phase testing.
- **Production**. Once the product is in full production, it will be tested for quality to filter out sub-par products and drive continual improvements. Chapter 13 will review production testing methods and approaches.
- **Field testing**. If the product has on-board monitoring, performance can be continually checked, or scheduled diagnostic tests can be run, and data sent back to the company. Returned products should be analyzed for failure modes, customer satisfaction, and performance degradation (Chapter 18).

The first section of this chapter will specify how we will use the word *quality*, and then the rest of the chapter will describe the quality practices used throughout product realization.

7.1 Definition of Quality

When people talk about quality, they are typically referring to one of two categories of product quality.

- **Quality as designed**. The design quality is the vision of and goals for the product. Once a product is in production, these decisions are relatively fixed. When thinking about design quality the team will answer questions such as:
  - How does the product meet the needs of the customer?
  - How aesthetically appealing is the product?
  - What features are included and what materials are selected?
  - How is the product expected to perform?
- **Quality as produced**. The production quality is how well the manufacturing system is able to deliver the performance specs defined by the

design team. Small variations in the manufacturing processes will cause the product to perform less than optimally. For example, uneven soldering quality can lead to some devices failing. Ideally, the product is designed to be robust to the variation in the manufacturing process (robust design), but the reality is you can only find some problems after the product is produced using mass production methods. Questions to ask about manufacturing include:

– How well does the manufacturing process ensure that the product meets the outlined requirements?
– How much variation is there from product to product?
– What defects are introduced by variations in the manufacturing process?

The characteristics of each quality classification are illustrated by the differences between two refrigerators (also called white goods) shown in Figure 7.1 – one very high-cost and one lower-cost – to illustrate the difference between quality as designed and quality as produced.

7.1.1 Quality as Designed

The branding, price, materials, features, size, cost, noise, form, and finish are some characteristics consumers use to differentiate between high-quality and lower-quality products. Teams developing new products will generally make decisions about these factors early in the design process and, of course, capture them in the specifications document (Chapter 5). As the design evolves, these decisions about design quality are embodied in the

(a)

(b)

FIGURE 7.1 Two refrigerators. The one on the left (a) is significantly more expensive than that on the right (b)
Source: (a) Reproduced with permission from Todd Taulman, Dreamstime.com
Source: (b) Reproduced with permission from Georgsv, Dreamstime.com

Table 7.1
Design quality
aspects
and examples
from a refrigerator

Quality type	Definition	Refrigerator example
Industrial design	How does the product look and feel to the user?	• Stainless steel vs. plastic outer surface • Freezer on top vs. freezer drawer at the bottom • Ergonomics
Features	What features and functions does the product include?	• Multiple drawers • Ice makers • Humidity control • Water purifiers • Smart refrigerators with a monitoring app • Heavy-duty drawer rails
Performance	How well does the product perform for each of the features?	• Energy usage • Volume/usable storage • Noise level • Temperature control • Cabinet depth
Durability and reliability	How well is the product designed for durability and reliability?	• Electronics well-insulated from humidity and temperature fluctuations • High-quality bearings to minimize wear and noise • UV protection added to plastics to prevent yellowing

Table 7.2
Production
quality aspects
and examples
from a refrigerator

Quality type	Definition	Refrigerator Example
Industrial design	Does every product consistently meet the aesthetic standard?	• Are there defects in the surface finish due to inconsistent processes? • Are there uneven gaps between the doors due to variation in the door assembly process?
Features	Do all of the features work as intended?	• Do smart systems connect correctly the first time or is the antenna not functioning? • Are the drawer rails installed so that they are correctly aligned and the drawers open smoothly?
Performance	Does every product meet specifications?	• Is one refrigerator noisier than another due to compressor differences? • Do some refrigerators run hotter than others because some mechanisms are lubricated less than others?
Durability and reliability	Has the team minimized manufacturing variations that could lead to customer complaints?	• Are the screws torqued correctly, so that they don't loosen and rattle over time? • Are wired connections secured to prevent intermittent failures caused by vibration from the compressor?

component specifications, bill of materials (BOM), drawing package, and color, material, and finish (CMF) document.

While there are many aspects of design quality, they can be roughly grouped into four categories: industrial design (including user interface and aesthetics), features, performance, and durability/reliability. (Inset 7.2 describes the difference between durability and reliability.) Table 7.1 delineates what quality means in each of the categories, using the example of refrigerators.

Some of these quality metrics don't need to be formally verified during the pilot phase. For example, it is easy to check whether the height and depth of the refrigerator are within spec, without having to build a prototype. Other specifications, though, can only be verified after the unit is built using production-intent parts. For example, all of the contributors to noise (such as drawers that scrape against each other or a clunking ice-cube-maker) may not be obvious until the initial pilot builds are tested.

Inset 7.2: Reliability vs. Durability

These terms are often used interchangeably, but they have distinctive meanings.

Durable products can survive significant occasional stressors and still perform as expected with only minor aesthetic damage. A durable product functions after being dropped, hit, baked in a car on a hot day, or doused in coffee. For example, we expect a durable car to be able to survive Boston winters: we expect its suspension to withstand the ice-carved potholes (as long as they aren't deeper than the car's wheels), its bumper to survive minor parking "kisses," and its windshield wipers to contend with an inch of ice. Durability testing evaluates the ability for a product to withstand occasional stresses.

A **reliable product** performs the same way on the first day as on the last day of use. We often use the term reliable when talking about cars. If a vehicle is described as reliable, we would expect all of its features to work even after 10 years – functioning power windows, ignition, air conditioning, etc. Reliable products can survive the normal wear and tear of day-to-day use without their performance degrading (much). Reliability testing subjects the system to repeated action/motions to identify whether the performance degrades unacceptably.

It may not be possible to simulate all durability and reliability conditions. However, we all know that products tend to break down under thermal and vibratory stressors. **Accelerated testing** checks the device's ability to withstand higher than normal stressors and is used to find inherent weaknesses in the product.

Table 7.3 Types of testing

Durability testing	Reliability testing	Accelerated testing
One-time substantial but reasonable stressors that the product should survive	Repeated actions expected to lead to normal wear and tear	Extreme stressors that highlight weaknesses in a product

7.1.2 Quality as Produced

Once the design is defined, the manufacturing process and supply chain then needs to deliver the expected quality. The supply chain, the manufacturing process, assembly, and transportation will all introduce variations that could impact the customers' perceptions of quality. For example, the industrial design team might specify brushed stainless-steel doors. To achieve the target finish quality, the buffing process has to be selected and controlled, the assembly process has to ensure the doors don't get warped or damaged, and the packaging team has to specify films that prevent damage during transport.

Other examples of manufacturing variation that may impact quality include:

- Small dimensional variations in machining a bearing can lead to vibration that creates a noisier product.
- Solder defects can result in controllers not working because an electrical connection isn't reliable.
- Inconsistency in tightening screws can cause rattling.
- Variations in injection molding can lead to visible surface defects.

The quality-as-produced issues can be grouped into the same categories as for quality-as-designed (Table 7.2). Variations introduced in manufacturing can have impacts on aesthetics, features, performance of the system, and durability/reliability.

Ensuring production quality is a multistage and iterative process. Having a comprehensive specification document, which includes ranges of acceptable performance, can help design teams make material and process selections with production quality in mind. For example, if the quality of a bearing is critical to the noise performance, it may be necessary choose a well-known but expensive domestic supplier instead of going to Alibaba and buying the cheapest bearings. If a product requires high gloss surface finishes on a plastic part, manufacturing engineers may select expensive injection molding machines and time-consuming tool polishing.

Once the manufacturing processes and materials are selected, the processes need to be tuned and standardized to ensure the right quality over time. For example, cycle times and temperatures might be optimized for an injection molding process to assure consistent dimensions (Chapter 11).

Finally, it is not possible to design out all of the potential sources of variation. Many of the issues caused by production will not become apparent until the product is actually built using production-intent materials and processes. A quality test plan is needed to verify and validate that the product meets the design targets before it is released to the market.

7.2 Quality Testing

Most books on product design focus on how to select and define features, the look and feel of the product, and performance characteristics to meet customer requirements. This book assumes that you already have a brilliant design and prototype for your product, so we will focus on the second piece – how you ensure that your production systems deliver products that meet all the specs.

A large part of the product realization process is to verify and validate (Table 7.4) that the design and production systems perform as intended. The titles of the pilot phases – EVT, DVT, and PVT – reflect the importance of testing during the pilot phases.

The starting point for quality testing is the specification document. The specification document outlines the product goals, and the quality testing ensures that the product meets those goals. We typically talk about quality testing as having two parts: verification and validation. Validation answers the question "did you design the right product?" Verification, on the other hand, is used to evaluate whether your product meets the specifications and product design you have defined.

Both kinds of testing are needed to ensure that the product meets or exceeds customer expectations. For example, if the customer says, "the product takes too long to charge," then the design and production team need to know whether they specified the target charge time correctly (validation) or if the specific cell they got from a supplier is faulty (verification).

Often, teams wait until late in the process to begin quality test planning. As you might imagine, the teams that postpone quality test planning are usually the same ones who don't build a comprehensive spec document. Unfortunately, not planning early for quality testing creates a lot of schedule and cost risks. For example, a team might design a refrigerator

Validation	Verification
• Did you design the right product?	• Did you build the product right?
• Are the specifications right?	• Did you deliver the specifications?
• Does the customer like what you made?	• Do all of the units behave the same?
• Is the industrial design right?	• Are the parts of high quality (no flash or sink marks)?
• Is the user interface intuitive?	
• Is the "feel" right?	• Is it durable? Is it reliable?
• Can it be set up quickly?	• Is it free of failures or errors?
• Is the user using it in the expected ways?	

Table 7.4
Verification vs. validation

with a low-cost fork wire connection rather than a locking header to save on space and cost, only to discover too late that vibration causes the connector to work loose. By developing both the spec document and the related quality test plan early in the design process, teams can avoid being caught by surprise by a product that fails a durability and reliability test late in the piloting process.

7.2.1 Validation – Did You Design the Right Product?

Designing a product to meet specifications does not necessarily ensure that the product will make a customer happy. You have to make sure customers actually like what you have designed – this process is called design validation. Validation during product realization is typically done by giving users the product and getting feedback on what they like and don't like.

Validation testing can be done formally or informally. Informal validation testing can be as simple as giving the product to a user and asking "What do you think?" Formal testing can involve structured experiments. For example, you might videotape the user installing the product and do time and motion studies to identify aspects of the design that makes it challenging to install. Validation is used to answer the questions about the product such as:

- **Does the product performance match what customers expect?** For example, the specification document may call for a two-hour battery life. When the battery life is tested under a range of environmental conditions, it meets that requirement. However, when the product goes through open-ended user testing, it does not meet customer expectations: customers state that they want a 3–4 hour battery life.
- **Does it have the right the fit and feel?** It is very hard to quantify the right feel of a product, and there will always be issues that only become apparent when the physical product gets in the hands of the customers. For example, a certain IoT (internet of things) product did not have a weight requirement since the product was rarely handled while in use. It was assumed the lighter the better. However, when the EVT units were given to new users, they said that the product felt too light and tinny given the price they were being charged. Even though the weight had no bearing on the performance of the product, customers thought the lack of weight made it feel cheap, which was in stark contrast to the price point. The product team ended up beefing up the internal structure and adding weight so the product did not feel hollow. Other examples of fit and feel include button actuation, uneven gaps/seams, and texture issues.

- **Does it work in an intuitive way?** The user interface may seem evident to the designer who has lived with it for years. User testing checks to see whether the software, hardware, and other aspects of the user interface are intuitive and easy to use.

7.2.2 Verification – Did You Build the Right Product?

Verification, on the other hand, is used to evaluate whether your product matches your product definition. Verification typically comprises tests that are more formal and structured than those of validation testing or bench testing. Verification testing systematically evaluates each individual requirement. Calibrated equipment is used and a formal standard operating procedure (SOP) is defined for each test. The verification test plan answers questions such as:

- **Does it adhere to functional specifications?** Does the actual product, as built, match the target performance requirements? For example, does the battery life match what is promised?
- **Does it have the right aesthetic?** Customers do not like scratches, dings, or sink marks in their products.
- **Is it durable enough to withstand expected handling and wear?** We have all dropped a cell phone on the ground and spent a moment of panic wondering whether it survived. We know that these devices can't be made shatterproof without compromising size or aesthetics, but we still expect that they will be able to survive minor falls and daily wear without becoming unusable.
- **Does it turn on properly when unboxed?** There is nothing more frustrating than a DOA (dead on arrival) product that will not turn on or that breaks within a few uses. Teams must ensure that the manufacturing process will reliably produce the products, and that the packaging will protect the product so every unit will perform as expected when a customer purchases it.
- **Can it recover from failure?** Failures should not "brick" the product. If the product fails, is there a way to bring it back to life?
- **Does it work reliably over its expected lifetime?** Customers expect the product to work over and beyond the warranty period, but products can't be economically designed to last forever. Wear and tear can degrade performance or cause the product to stop working. Customers expect certain parts to degrade more over time (e.g., rechargeable batteries and brake pads on your car), but they expect others to reliably work over the expected life of the system (e.g., automotive timing belts must work for at least 100,000 miles).
- **Does it adhere to legal requirements?** Products must comply with regulations in every country and state where it is sold. There are electromagnetic field (EMF), safety, and environmental requirements that are imposed by countries and individual states (Chapter 17).

7.2.3 Developing and Executing the Quality Test Plan

Quality testing doesn't start by picking up the prototype and randomly running experiments on it. As with everything else in product realization, quality engineering starts with the team developing a careful quality testing strategy and a method for documenting the results. It should go without saying, but so many teams make mistakes during this process that we'll say it anyway: *every step of the quality testing process should be planned and formally documented. Ad hoc and disorganized testing is a recipe for overlooking critical product errors.*

Table 7.5 describes each step in the quality test process: what happens and why, which documents are created, and who is responsible. To help illustrate the process, we'll use a standard drop test – a test used on most consumer products – as an example.

Typically, the quality engineering team works with the engineering team to develop the first draft of the quality test plan, which outlines all of the tests that will be run during the pilot phase. The quality test plan (Section 7.3) is generated based on the specification document (Chapter 5) and additional risks identified in the risk management process (Section 4.3).

After the test plan is created, the testing engineering team will generate a SOP for each test in the quality test plan. The SOP details each step: the equipment needed, the test procedure, and the evaluation procedure to determine whether the product passes or fails the test. Developing the complete set of testing SOPs for a new product can be daunting; however, your quality team (whether internal or at your CM) will often have many of the standard test SOPs already specified, and many of the SOPs can be reused in subsequent product launches.

After test engineers complete a test, they generate a report of the results along with pictures and other supporting materials. For example, a drop test report will contain photos of any damage incurred from the drops and a summary of all functional tests done before and after the drops.

Even a simple product can have dozens of test reports, each of which can run to many pages. It is not feasible to expect all members of the design team to read and interpret all of the individual reports. At the end of each pilot run, the quality engineering team typically creates a summary test report highlighting which tests passed, which failed, and a root cause analysis and action plan for any failures. The complete pilot test report contains all of the individual test reports, the test summary, and other relevant results from the pilot run. For example, if a product had minor aesthetic damage during a drop test, the report may include pictures so the team can decide whether the failure is critical or not.

The rest of the chapter will focus on the generation of the quality test plan. If you are interested in learning more about test SOPs, test reporting, or summary documents, you can find numerous examples and templates online.

7.3 Pilot Quality Test Plan

The starting point of the pilot quality testing is the creation of the quality test plan. The test plan outlines the tests to be done, their timing, the number of samples needed, and links to related documents. Ideally, the quality test plan is created well ahead of product realization, because many teams across the company will use it in preparing for product realization. The finance team will use it when determining budgets for samples and calculating non-recurring engineering (NRE) costs (Chapter 8) for testing and equipment. The supplier management team will use it during the CM selection process to understand what capabilities are needed, and will ask the CM to quote on test costs. Most importantly, design engineers will reference the quality test plan to help ensure that they design a product that will pass all required tests.

The quality test plan evaluates many aspects of quality including:

- **Specifications and functionality**. Does the product meet the specifications and perform as expected?
- **Aesthetics**. Does the product look and feel as expected?
- **Durability**. Can the product survive one-time stresses?
- **Reliability**. Does the product perform as expected over the life of the product with no failures?
- **Response to accelerated stress**. What are the inherent structural weaknesses in the product?
- **User**. Does the product meet customer expectations?
- **Packaging**. Can the product withstand shipping and handling?

Table 7.6 provides an example of a quality test plan. Table 7.7 provides a summary of the key information for Sections 7.3.1–7.3.8.

The quality plan includes enough information for the testing engineering team to develop the testing SOPs, for the manufacturing team to plan how many samples need to be built at each pilot phase, and for the quality engineering team to verify and validate all of the specifications in the specifications document (Chapter 5). In addition to planning the tests, the quality

Table 7.5

Documents associated with the quality test plan

Document	Goal	Team	Timing	Drop test example
Specification document (Chapter 5)	The master list of all requirements. Outlines the product requirements and goals. Every requirement in the specification should have a verification and/or validation test associated with it.	Engineering and product teams.	Ideally it should pre-date the design concept phase. It will be updated throughout the development and product realization processes.	The specification may include the need for the product to be used while it is being carried, and the kind of drop it is expected to withstand: "It needs to survive a drop from 30 in. onto a carpeted floor."
Risks identified by the team (Section 4.3)	Throughout the design and product realization phases, new risks/questions may arise that need to be evaluated. Ideally, these should be captured in the specification document, but often they are kept in the risk list maintained by the project manager.	Engineering and product teams. The project manager maintains and updates the risk list.	The list is continuously updated throughout the product realization process.	The team may be concerned that the drop test is not sufficient for normal use and that customer expectations may be different from the specs. The test may be augmented with accelerated testing that stresses the product to understand the maximum height it can be dropped from and the likely failure modes.
Quality test plan	The list of all of the tests that are going to be executed in a single file. The plan typically contains only a high-level description of each test and its key parameters. In addition, it gives the timing of the test and the sample size requirements.	Quality engineering team with the product development team.	Ideally, the plan should be prepared in parallel with the specification document. The complete version should be finalized before EVT starts.	"The product will be dropped from a 32" height onto a carpeted floor and only have minimal aesthetic damage and be fully functional after testing." "Accelerated testing done during which the product is dropped from increasingly larger heights and tested for functional performance after each drop."

Document	Goal	Team	Timing	Drop test example
Test SOPs	These are the detailed test procedures that define exactly how the test is run and what equipment is used. Clear pass/fail criteria are specified. One document for each test listed in the quality test plan	Test engineering team.	The testing organization may already have some standard SOPs written. SOPs need to be complete before the pilot phase begins, and need to be given to all teams in sufficient time for review.	The SOP is a detailed description of fixturing, setup procedures, drop heights, surface on which to drop, which edges/surfaces/corners will strike the surface, and numbers of drops. The SOP specifies when inspections are to be done with clear criteria for aesthetic damage (including photos of maximum allowed damage). Also described are the functional tests to be run following each drop or set of drops.
Test reports	These detail the results of the tests. They will typically include the results, any deviations from the test protocol, and photos or measurements of any failed product.	Test technician or test engineer.	A report should be filed as soon as practical after completion of a test.	Includes pictures of the damaged product, functional test reports, and a pass/fail indication.
Quality test summary report	For each phase, all of the tests are summarized into one document with the test, deviations, pass/fail verdict, and root cause analysis, along with an action plan for each failure. It is used to identify which quality tests need to be rerun and should drive the updating of the risk analysis, the spec documents, and the pilot phase reports.	Quality engineering, engineering, and product teams.	The report should be completed after testing is finished but before the next pilot phase begins.	Summary of all tests with the pass/fail results and recommended design changes to address issues. The risk document is then updated with the needed design changes, and additional schedule risks are noted.

Table 7.5
(*Continued*)

Table 7.6
Example of a quality test plan

#	Category	Spec	Description	Pilot phase	Criticality	Sample size	Sample sharing	SOP link
1	Durability	4.34	Drop test: drop from 30″ onto a wood surface. Four times, once each face. Inspect functionality and aesthetics on all parts after test.	DVT	Minor to major depending on the damage	2	Drop tests must be run on pristine samples; these can be used later for surface chemical durability tests only.	Droptest_SOP_V1
2	Functionality	2.1	Battery switches to charge mode within 3 s of plugging in power source, and LED light changes color.	EVT, DVT, PVT	Major	100%	All functional test units can be used for other quality tests. No damage is caused by the test.	Battery_test_SOP_V1
3	Reliability	7.57	Button press test. Each of the five buttons are pressed 5000 times with a force of 5 N.	PVT	Minor	5	Samples can be used for other reliability testing including battery charging and camera function.	Button_test_SOP_v3
4	Aesthetics	1.01–1.50	Check all aesthetic requirements[a]	DVT, PVT	See aesthetic requirement	100%	All aesthetic samples can be used for other testing.	Aesthetic_inspection_v4

[a]Aesthetic requirements are typically documented in a separate table, with each aesthetic requirement listed with the minor/major/critical levels indicated.

engineering team needs to determine when during the piloting process the tests should be done. The following questions should be answered:

- **Which tests are done when?** Teams should run tests that provide useful results and drive improvements. Testing for testing sake drives cost, not quality. For example, early samples are often not representative of the final aesthetics because the tools have not had their final polish. Running aesthetics inspections or durability testing on early samples would therefore be unhelpful. Subsets of the test plan can be run as appropriate: for example, battery life on a works-like prototype with production-intent electronics.
- **How early can a test be started?** The earlier the team can learn about problems (or close out risks), the better. In addition, some reliability tests may take a long time to run and need to be started as early as possible in order to finish them before mass production starts. For example, safety-related fatigue testing requires several years of life to be simulated on a statistically significant number of samples. If these tests are started late in the piloting process, they may not have returned results before the product is launched.
- **Which tests are likely to be failed, and thus need to be repeated?** Not all tests are perfunctory – some tests can surprise you and find unexpected problems. Products will fail to meet specifications and need to be redesigned and retested. For example, if the design team had to trade off product weight against durability, they might know the product is at risk of failing the drop test. Additional ribs could be added if needed between pilot builds but the team didn't want to add them before they understood whether the design would work with less weight. Extra samples will be needed to re-run tests after design changes.

The quality test plan typically includes the following information:

- **A unique test number** is used to track the tests through the pilot process.
- **Category** (e.g., durability, functionality, aesthetic, reliability) keeps the list of tests organized and can help teams to identify where testing is not comprehensive enough.
- **Related specification** links each test to the unique specification numbers and ensures that the tests are consistent with the specification document.
- **Test summary** allows teams to quickly review the scope of the test without having to read multiple SOPs. It also enables the testing organization to estimate the time and cost for each test before developing comprehensive SOPs.
- **Pilot phase** indicates when the test is run. Some tests can't be done until late in the verification process because they require production-intent parts. Other tests can be done early in the process and once the product has passed, it does not need retesting unless there is a design change.

Table 7.7

Types of tests run during the pilot phase

Section and test type	Description	How and what are tested in each pilot phase		
		EVT	DVT	PVT
7.3.1 Functional	Ensures that the product meets the functional specifications. Typically, 100% of the products produced in each build are tested.	100% of units are tested with 100% of functional tests.		
7.3.2 Aesthetic inspection	Checks whether the product adheres to all aesthetic requirements. Can include color matching, surface defects, and FOD (foreign objects or debris). Aesthetic inspections are performed on all products during piloting. Also, it is typically done on 100% of units in full production.	Rough aesthetics are checked but the look of the product isn't typically representative.	100% of units are evaluated for most aesthetic criteria (other than those that require final tool polish).	100% of units are tested using all look and feel criteria.
7.3.3 Durability	Ensures that the product can withstand single-event damage. These tests include testing at extremes of expected environmental conditions. In consumer products, a majority of durability test SOPs are shared across most products. For example, most consumer electronics undergo electrostatic discharge (ESD) testing to confirm that the product can withstand static shocks.	Only those durability tests are run that do not rely on production-intent parts.	A limited number of samples is subjected to most of the durability tests. Most tests require one or two samples unless it is a safety related issue.	Follow-up tests are run to verify that failures found in DVT were corrected or that design changes did not introduce problems.
7.3.4 Reliability	Ensures that regular use does not result in loss of functionality or excessive aesthetic damage. These tests depend on the likely usage scenarios, user profiles, and warranty duration.	Early testing is done on critical components that require long reliability testing.	The majority of reliability testing is run during this phase. Results may not be available until PVT. Reliability testing is run on the number of samples needed to get statistically significant results.	Follow-up tests are run to verify that failures found in DVT were corrected or that design changes did not introduce problems.

(Continued)

Table 7.7
(*Continued*)

Section and test type	Description	How and what are tested in each pilot phase		
		EVT	DVT	PVT
7.3.5 Accelerated stress testing	Tests the product until it fails in order to identify weaknesses (e.g., raising temperature). These tests do not give an accurate estimate of the life of the product, but rather identify inherent weaknesses.	N/A	The majority of accelerated tests are run during this phase. Results may not be available until PVT. Typically tests only a couple of units because of the cost involved.	Follow-up tests are run to verify that failures found in DVT were corrected or that design changes did not introduce problems.
7.3.6 User testing	Tests whether the product as designed/built meets what the customers want. For example, the actuation force of a button might be reviewed by several customers to see how they like it.	Several units are tested for overall functionality and aesthetics.		
Certification (Chapter 17)	Checks whether the product meets legal requirements to ship or sell. Some certification testing is done by outside labs; others can be done in-house.	N/A	Samples from DVT can be used for certification. Certifications typically only require one or two samples.	Some certifications may require PVT units.
7.3.7 Packaging	Tests the ability of the packaging to protect the product during shipping, storage, and sale.	N/A	DVT samples are tested when packaging is complete.	Follow-up tests are run to verify that failures found in DVT were corrected or that design changes did not introduce problems.
7.3.8 Other verification tests	Tests non-functional specifications (e.g., weight, color) that need to be validated. These tests are based on the specifications not tested by any of the quality tests above.	A majority of the specs not covered by other categories are checked. Typically verified on a single unit.	Tests are done to verify unresolved specs and the effects of any design changes.	Tests are done to verify unresolved specs and the effects of any design changes.

Inset 7.3: Critical vs. Major vs. Minor Failures

 Not all tests have the same importance; having a small scratch on a surface is much less critical than having a shock or overheating hazard. Typically, failures are classified into three categories:

- **Critical.** These are safety-critical issues that cannot be accepted, such as risks of laceration or fire.
- **Major.** These are failures that prevent the product from operating but do not pose a safety risk to the consumer. These include DOA (dead on arrival) failures when a product won't turn on the first time the product is used.
- **Minor.** These are failures of aesthetics and minor annoyances. These could include scratches, slight color mismatches, and products that are not seated properly in the packaging.

- **Criticality** ranks the tests so the team focuses on the most critical. Some tests are more important than others. Any failure related to safety requires significant attention, whereas minor aesthetic defects can be put lower on the priority list (Inset 7.3).
- **Sample quantities** indicates how many products are used for each test. The total number of samples helps the teams plan how many samples have to be built at each phase. The first draft of the quality test plan will often require many more samples than can be economically produced. The plan typically is iterated to identify how to share samples or reduce the testing.
- **Sample sharing** enables the same unit to be used to run several tests (if the testing itself causes no damage). The more sharing of samples between tests, the lower the costs. However, sharing samples increases the overall test time because the tests have to be done serially.
- **SOP** links to the document that defines exactly how the test will be run.

After drafting the quality test plan, the team may find that they do not have enough time, samples, or resources to execute the ideal test plan. In that case, the project manager may decide to combine certain tests to reduce testing time or costs. The following sections describe the different test types.

7.3.1 Functional

Functional testing checks that every aspect of the product meets the functional requirements. For example, can the robotic arm reach the specified distance? Is the camera in focus? Or, is the right Bluetooth communication protocol triggered when a button is actuated?

Functional tests are typically done on 100% of the products produced during piloting to ensure both design and build quality. Two types of functional testing and examples of each are discussed below:

- **Comprehensive functional testing** (also called bench testing). Some functional testing done during product piloting requires specialized equipment for testing and data collection that will not be feasible on the production line; for example, using an oscilloscope to trouble-shoot a circuit. Bench tests provide a rich dataset for root cause analysis if there is a failure, and further the team's understanding of the product's performance. One example is:

– *Battery discharge tests.* These answer the questions of whether the battery's power and capacity match the specifications from the vendor. Repeated charge/discharge cycling of rechargeable batteries may be done to ensure that the run time of a product is acceptable. These tests can take significant time, but they give useful insights into the power control circuitry and can't be used during production.

• **Production-intent testing** (Chapter 13) are short-duration tests that can be incorporated into the production line. Typically, the tests are all implemented and checked during the PVT pilot. They include both built-in self-tests and custom test fixtures. These tests are designed to be consistent and reliable, to have short cycle times so they don't slow the production flow, and don't require the operators to take or interpret complex measurements. For example:

– *Image quality* needs to be quickly tested to check for the accuracy of the focus and any lens defects. Early in the pilot phases, camera experts systematically evaluate all aspects of the camera quality. During production, that isn't feasible. A custom test fixture is typically built that automatically records and analyzes an image, and gives a red or green light to the operators to tell them if the product passes or fails.

7.3.2 Aesthetic

As mentioned in Chapter 5, the aesthetic requirements for the product are usually defined in the specification document, though some teams choose to maintain them in a separate document. The aesthetic specification can include positive qualifications (color and surface matching requirements between parts) as well as negative restrictions (unacceptable scratches). It is critical during the pilot phase to fine-tune the aesthetic requirements and clearly define acceptable and unacceptable criteria. Having a comprehensive aesthetic test protocol can reduce the risk of your suppliers shipping parts to you that are "acceptable" according to the specifications on paper but do not meet your standards. Misalignment on expectations can lead to unexpected part shortages, poor quality, price increases, and conflicts with suppliers.

Aesthetic inspection is typically done on 100% of all units produced in the pilot phase. The aesthetic requirements usually contain the following:

• A definition of the A, B, C, and D surfaces (Table 7.8). Different surfaces of the product will be held to different standards.
• Drawings of the part with A, B, C, and D surfaces indicated.

Table 7.8
A/B/C/D
aesthetic surfaces
and examples

Surface	Description	Where on a laptop	Example of a flash spec	Spec for each level of criticality
A	Primary surfaces the customer sees and interacts with	Top of the computer and keyboard frame	Flash will not be visible	Minor: small deviations from spec Major: large deviations from spec Critical: any flash that creates a laceration hazard
B	Surfaces the user can see with deliberate effort	Underside of hinge	No flash larger than 0.10 mm	Minor: small deviations from spec Major: large deviations from spec Critical: any flash that creates a laceration hazard
C	Surfaces the user cannot see	Inside of the computer	No flash larger than 1 mm	Minor: any deviation from the spec Major: any deviation that interferes with function
D	Structural or functional surfaces that are called out in engineering drawings	Mounting surfaces for the hinge	Flash must not interfere with functional surfaces	Minor: any deviation from the spec Major: any deviation that interferes with function

- The types of defects allowed (e.g., flash or sink marks), and a quantified way of measuring the error and the allowable size (often based on the size of the surface). A list of typical aesthetic defects is given in Table 7.9.
- The severity of the defects (critical, major, or minor).
- The number of defects allowed.
- The testing conditions for each of the surfaces, including the amount and type of light and any golden samples used as comparators.

Table 7.9 provides a list of some of the aesthetic criteria typically applied to products. Some criteria apply to all aspects of a product and all processes. Others are unique either to fabricated parts (e.g., injection molding or casting), finishes (e.g., paints and decos), or packaging (e.g., gift boxes).

Table 7.9 Types of aesthetic requirements

Aesthetic requirement	Fabrication	Finishing	Packaging
Beading		●	
Belting/blend lines		●	
Bleeding of color		●	
Blistering/peeling		●	●
Burn	●		
Burrs	●		
Chip	●	●	
Cold shut	●		
Color matching		●	●
Color variation		●	●
Contamination/dirt	●	●	●
Corrosion	●		
Crack	●		
Delamination		●	●
Dents	●	●	
Depressions	●	●	
Die marks	●		
Discoloration	●	●	●
Ejector pin marks	●		
Finger prints	●	●	●
Fisheye		●	●
Flash	●		
Fractures	●		
Gate marks	●		
Gloss variation	●	●	●
Gouge	●	●	●
Irregular color	●	●	●
Knit lines	●		

Aesthetic requirement	Fabrication	Finishing	Packaging
Lump		●	●
Mismatch	●		
Nicks	●	●	●
Orange peel	●	●	●
Pinholes	●	●	●
Pits/void	●	●	
Porosity	●		
Protrusions	●		
Rainbow effect		●	●
Removable inorganic foreign material	●	●	●
Removable organic foreign material	●	●	●
Roughness	●	●	
Run/sagging		●	
Runs/drips		●	
Rust or corrosion	●	●	
Scratches or gouges	●	●	●
Scuffs, marks	●	●	●
Sharp edges	●		
Short shot	●		
Sink marks	●		
Stains	●	●	●
Stop marks		●	
Streaking		●	●
Surface finish	●	●	●
Tool marks	●		

7.3.3 Durability

All products will need to survive a range of abuses and environments. Products must survive being shaken, dropped, heated, and frozen; just imagine what a smartphone has to survive over the day in the life of the purse of a working mother. Some damage is acceptable (small scratches) and some is not (critical safety failures). While there is room for debate on the relative criticality of a dent or performance failure, there is no room for debate if that failure could result in harm to a user. The durability tests answers the question whether or not the product can survive its entire life safely.

Durability is the ability to withstand occasional stresses that are outside the normal usage, such as drops, vibration, and impact. It is often referred to as "shake and bake" testing. The durability tests should reflect the user scenarios and durability specifications outlined in the specification document. Ideally, the spec document is defined first; however, when the quality test plan is developed, additional durability requirements may become apparent, and the specification document is updated to reflect the addition of the durability tests. The setting of the durability requirements can be challenging. Teams should set standards that will meet the customer expectations without overdesigning the product (Inset 7.4).

When generating the list of durability tests, the quality engineering team needs to think through all of the likely stressors that the product will undergo in normal use. Figure 7.2 shows the types of questions the team can ask to produce a comprehensive durability test suite. First, the product must

---

Inset 7.4: How Extreme Should the Durability and Reliability Tests Be?

Teams are often tempted to over-test products and set extreme limits for what the product should be able to survive. "It might happen" becomes the mantra during quality test planning, leading to tests of extreme circumstances that the product almost certainly will never experience. The tests *should* be limited to what is considered reasonable use and in line with user expectations. The customer expects a phone to survive falling off a kitchen table but does not expect it to withstand being run over by a car.

When extreme limits are used, you risk missing critical failures. If a test fails at an extreme level, there is pressure to do the next pilot run, and the results might be dismissed because "of course that will never happen." Not only has the money to run the test been squandered, but you have also lost a valuable sample and probably missed information about how the product responds when it is dropped from a reasonable height.

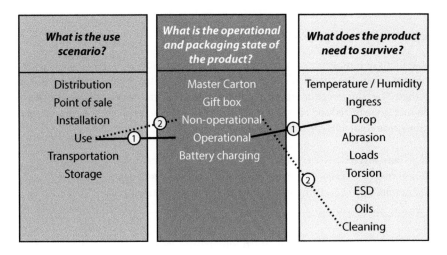

FIGURE 7.2
Questions answered
by durability testing

① Fully charged, product running, drop test from 36" height onto a concrete floor. It should survive all functional tests with only minor aesthetic damage.

② In non-operational state, the product should be cleanable with a range of cleaning fluids, including ammonia or bleach-based products by spraying and wiping with a soft cloth. It should pass all functional tests and no minor damage to surfaces or deco.

survive all of the normal use scenarios including transportation, storage in a warehouse, storage at home, and routine maintenance (Chapter 5). For each scenario, the operational state of the product should be described. For example, it matters whether a product is plugged in when it is knocked off the table, because that can stress the connector. In each scenario and operational state, the product can be subject to a range of stresses: thermal, electrical, mechanical, and chemical. Finally, the team needs to decide what damage is acceptable at each stage. If the damage creates a safety risk, it has to be addressed. In other cases, small aesthetic damage may be acceptable to the customer.

The following list gives the typical durability tests used on a consumer electronics product. However, your product may have some unique durability expectations. For example, a product that has wheels and is dragged to the beach may need to be tested for sand damaging the wheels and bearings. Products that sit on counters in a kitchen need to be tested for the effects of sitting in coffee spills for periods of time.

- **Temperature/humidity range**. The product needs to operate in a range of environments both hot and cold, dry and humid (Inset 7.5). Temperature and humidity can have detrimental impacts on surface finishes, electronics, and batteries. Figure 7.3 shows a picture of a typical thermal testing chamber.

Inset 7.5: Temperature and Humidity Durability

Combinations of temperature and humidity can have disastrous impacts on some products. For example, the chemical degradation that triggered the Takata airbag recall was exacerbated by high temperature and high humidity [70]. Extreme ranges of temperature and humidity are tested and are set to mimic the typical ranges in different environments. The three cases that are typically tested are:

- High temperature, high humidity (HTHH) – Left in a car in Florida on a summer's day
- High temperature, low humidity (HTLH) – Left in a car in Arizona on a summer's day
- Low temperature, low humidity (LTLH) – Left outside in North Dakota in January

- **Vibration**. Most products with electronic components will experience some vibration while being transported or used, potentially leading to mechanical or electrical failures.
- **Water and debris**. Naturally, products will be exposed to water, dust, and debris during normal use. However, it is prohibitively expensive – if not impossible – to design for full dust and water protection on every product. Products need to adhere to specifications (called IP or ingress protection requirements) within the likely use conditions as well as within the target price point (Inset 7.7).
- **Transfer of oils and other contaminants**. Some plastics can react badly to sweat, oil, food, or other chemicals; they can stain, delaminate, and degrade. Products that are handled routinely or exposed to chemicals need to be tested against these contaminants. Typical test protocols include sweat (yes, you can buy artificial sweat!), hand lotion, bug spray, perfumes, hairspray, oils, and mustard.
- **Cleaning materials**. Cleaning products can react with materials and adversely affect electronics as well as degrade the surface. Products should be tested against a range of common household cleaning agents. One medical device kept having seemingly random electronic and mechanical failures. Eventually it was determined that the hospitals were opening the case and spraying an antibacterial, and that spray was damaging the materials and electronics.

Inset 7.6: You Can't Always Know What You Are Designing For

The author's favorite engineering urban legend about product durability is the story of the phones at the front of an airplane. The servicers kept getting called for cracked handsets. It was a mystery why they kept breaking until a flight attendant was observed using the phone as a makeshift ice breaker to break up bags of ice. When they added a small mallet to the kitchen, the phones stopped breaking.

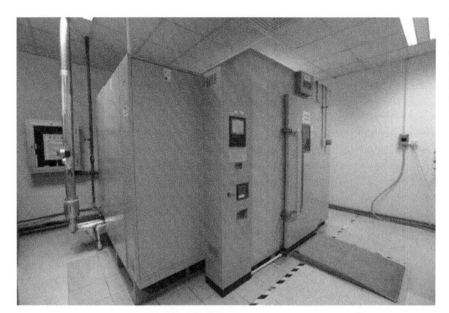

FIGURE 7.3
Thermal testing chamber. *Source*: Reproduced with permission from Shenzhen Kaifa Technology

- **Electrostatic discharge (ESD)**. Small shocks can damage the electronics. For example, the author's phone recently went on the fritz when she reset a fuse on her house's electrical circuit panel while calling an electrician. The failure was probably due to a low-level ESD which temporarily disabled the phone. ESD tests should be done at levels to check whether the device is affected and, if it is disabled, whether it can be restarted. In the case of the author's disabled phone, restarting it brought it back to life. However, some ESD shocks can brick a product (i.e., turn your product into the equivalent of a brick). ESD testing is standardized in International Electrotechnical Commission (IEC) standard IEC 61000-4-2 ESD Immunity and Transient Current Testing [71].
- **Ultraviolet (UV) exposure**. Ultraviolet exposure can degrade the mechanical and aesthetic properties of plastic. Anyone who has seen their white plastic appliances yellow and become brittle over time has seen UV degradation. UV is typically tested using an accelerated testing mode where the products are exposed to extreme UV light for a few hours to simulate many weeks of normal outdoor exposure to the sun.
- **Corrosive environments**. Salt air and water can have detrimental impacts: rusting, clogging, and shorting of electronics. The author's family had to buy corrosion-resistant locks for their house near the sea because, by the end of the season, a normal lock had to be removed with bolt cutters.
- **Abrasion**. It is expected that a product will not show scratches when exposed to normal use. The abrasion test typically uses pencils of varying hardness to check if the product will be scratched when rubbed by fingers, cloth, cleaners, or other expected sources of abrasion.

Inset 7.7: Ingress Protection (IP)

 IP defines the standard for how well a product can resist intrusion from debris, objects, or liquid. It is based on the IEC 60529 [72] standard set by the International Electrotechnical Commission. The IP rating is defined by three numbers described in Figure 7.4. The first number states the needed protection against solid objects, and the second against liquids. The third gives codes for other protection. For example, IP31C will be resistant only to small wires and tools and some dripping water, but can't be easily opened. On the other hand, an IP68X is entirely dust and immersion proof. Each rating has an associated testing protocol to ensure that it passes the rating used. Some products require IP testing as part of their certifications, and others use IP to communicate clearly to the consumer what the product is promising.

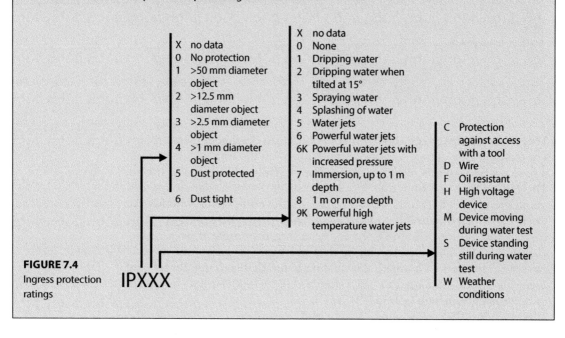

FIGURE 7.4
Ingress protection
ratings

- **Drop**. Almost all products will be dropped at some point. The standard test typically outlines what height it will fall from, onto what surface (concrete, wood, or carpet) and which surfaces, corners, and edges will strike the surface.
- **Impact**. Products will get hit and vibrated during use. For example, a device used to monitor fire hydrants was installed near the edge of the road and had to withstand being backed into by a car or hit with a weed-whacker. There was no standard weed-whacker test equipment so the company had to create their own test fixture.
- **Torsion and tension**. People who pick up their computer by the monitor reasonably assume that the hinge will not break. People will drag their tablet by the molding from between two books in their bag. They expect the plastic frame to stay attached to the LCD.
- **Compression**. Products are stacked and stored. For example, a computer will be left in a backpack with books on top.

7.3.4 Reliability

Reliability testing, in contrast to durability testing, ensures that a product will withstand regular use during the expected life of the product with an acceptable failure rate. If the product is less reliable than expected, warranty costs can surge and eat into profitability. Reliability is a function of three factors – the causes, the rate of failure, and the period over which reliability is measured.

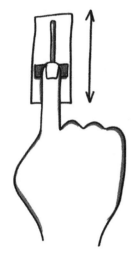

For example, an LCD touch panel should not fail when it is touched repeatedly, the hinges on the computer should not break, and all of the keys on the keyboard should stay functional for the life of the product. For example, through repeated use, the thin film over the switch has cracked. While the failure occurred after the warranty period, the author won't buy that radio with this kind of button again when the unit eventually fails. Most consumer electronic production facilities have automatic button testing equipment that will repeatedly press a button to see how it stands up to use.

Companies will also develop their own test equipment specifically for their product. For example, Figure 7.5 shows a picture of a test fixture used by SRAM to test the reliability of a bottom bracket of a bicycle. The bottom bracket connects the cranks/pedals on a bike to the front gear set (crankset), and comprises a shaft and bearings. It is necessary to understand the expected life of the bottom bracket under a range of loading conditions.

FIGURE 7.5 Bottom bracket reliability equipment
Source: Reproduced with permission from SRAM

Inset 7.8: Cuisinart Food Processor Recall

When it comes to safety, there can be no tolerance for safety-related failures. In addition, the period over which the product must perform safely far exceeds the warranty period and can last into decades. For example, in December 2016, Cuisinart issued a recall for the food processor blades produced from 1996 through 2015. Some of the eight million blades recalled were purchased 20 years prior. The blades broke near a rivet and fragmented into small shards which caused multiple mouth lacerations [73]. Even though the product was well outside the warranty period, Cuisinart still had to spend significant money to replace all affected blades and address the cases where people were injured.

One of the first highly published examples of reliability testing occurred around 1954 on the De Havilland Comet – the first commercial jet airplane with a pressurized cabin. Three Comets crashed after losing structural integrity and breaking up in flight, causing over 120 deaths. To understand the cause of the failure, the team at De Havilland used a large water tank to repeatedly and rapidly pressurize and depressurize the cabin over many cycles. After about 9,000 cycles, fatigue at a corner of a window caused the fuselage to rupture [74]. While De Havilland never gained commercial success, other aerospace companies learned that lesson and continue to rely heavily on reliability testing to ensure the safety of their products.

No one expects all of your products to work perfectly over a lifetime without repair (Inset 7.8). The only exception to this is for failures that cause injury or death. It would be cost-prohibitive to build that level of reliability into every non-safety aspect product, not to mention the weight and volume required to establish that level of redundancy. There will always be a small percentage of products that fail and need to be repaired or replaced. The acceptable rate of failure and length of warranty period depend on the product type, product expectations, and customer needs. For example, most electronics have a one-year warranty, but most cars come with at least a seven-year warranty. Any reliability failure over that warranty period results in a cost to the company. To emphasize this again, *there is no acceptable failure rate for safety-related failures.*

To verify reliability, testing needs to prove, with a high statistical confidence, that the actual failure rate will fall below the target rate over the warranty period. The target rate is set to ensure that you aren't going to bankrupt a company with product returns and aren't going to over-design the product and make it too expensive.

Proving that your product meets this rate involves budgeting an allowable failure rate to each of the failure modes, and testing each failure mode using enough samples and cycles to prove reliability. There is a vast body of work on reliability budgets and the statistics of testing and sampling sizes. Most large companies have statistical experts who will assist in the application of these techniques. However, for smaller companies, those resources may not be available. Small companies have one of three options: become experts, hire someone, or over-test. This section gives a basic introduction to designing a feasible reliability test plan [75]. The numbers that follow are based on the author's and other experts' experience (100+ years of combined experience on hundreds of products). The numbers can be used as a starting point but aren't hard fixed rules.

The reliability test plan is determined by answering the following questions, which are detailed below:

1. What maximum return rate are you targeting over what time?
2. What is the product reliability budget?
3. What are the likely reliability failure modes for product features, and what is the budget for each?
4. How many cycles is a given product feature likely to experience during the warranty period?
5. How many samples and cycles are needed to prove reliability?
6. How long and how much money will reliability testing take?
7. When can you start the reliability testing and how quickly does it have to be done?

1. What Return Rate Are You Targeting?

For non-safety-related failures, the first step in developing a reliability testing plan is to determine the target return rate for the product over the warranty period. The target return rate includes all of the potential failure modes combined, including those introduced by manufacturing variation and those that are due to non-quality related failures. During the financial planning phase, an allowable warranty rate should be selected to balance product cost, profitability, and cash flow. For example, many consumer products target a return rate of 5% for the warranty period. The only exception to this is for safety-related failures. For these, the expected failure rate needs to be essentially zero for the expected life of the product.

2. What is the Product Reliability Budget?

The product reliability budget is less than the target return rate. The target return rate, described above, includes both failures of reliability as well other factors. First, products may be returned for no reason other than "I did not like it." Second, unexpected failures will crop up because of a use scenario you didn't test for or because of unforeseen changes that happen during the manufacturing process. Third, your factory will be allowed to produce a small percentage of defective products without penalty (called the acceptable quality limit, or AQL). To get to the product reliability target, the return rate has to be adjusted by the expected number of dissatisfied customers, the acceptable defects introduced by the factory, and likely unexpected failures. Table 7.10 shows three example failure budget calculations for low-, medium-, and high-quality products. These numbers are representative of typical products that have done a good job of designing for quality through the product development process, and have done sufficient testing during the product realization process to shake out any problems. If you don't do the upfront work on quality, the rates will likely be much higher. The following describes the steps used to calculate an example reliability budget.

a. Reduce the target return rate by the number of dissatisfied customers. A good starting estimate is that around one third of the total returns might be from dissatisfied customers. Removing the non-quality related failures gives you the budgeted number of units that can be returned because of quality failures.

b. Reduce the adjusted return rate (from step 1) by the major AQL you have negotiated with your factory (typically around 0.6% for a consumer electronic device). If the defect rate falls below that number, you are responsible for the cost of the returns. For anything above that rate, the factory has to reimburse you if the failure is due to factory error. Removing the AQL rate gives you the allowable number of returns due to all product quality issues not caused by the factory.

c. Reduce the adjusted return rate (from step 2) by the number of returns caused by unexpected problems. Most products will experience a temporary spike in warranty claims due to manufacturing errors, unforeseen quality issues, and unknown use cases. These spikes are typically transient because companies act quickly to contain the failures. When annualized, these can drive the average in-warranty return rate up by 0.25–4%. Depending on the level of complexity, technology maturity, and the sensitivity of customers to quality issues, organizations need to try to predict the unknown unknowns. Unfortunately, no science can predict what you do not know, and significant failures (especially those that occur late in product life) can bankrupt companies.

Table 7.10
Failure budget

Product quality	Low (%)	Average (%)	High (%)
Target return rate (1)	8.0	6.0	4.0
Dissatisfied customer (2a)	3.0	2.0	1.0
AQL level (2b)	1.0	0.6	0.25
Unexpected (2c)	1.0	1.5	2.0
Product reliability budget	*3.0*	*1.9*	*0.75*

This reliability target is used to set the test parameters going forward. The reliability tests will evaluate whether the product has a higher or lower expected failure rate than this target.

3. *What are the Likely Reliability Failure Modes and What is the Budget for Each?*
The third step in reliability planning is to determine the list of potential causes of reliability failures that are included in the reliability budget. Theoretically, these should be listed in the specification document. Causes of reliability failures fall into one of two categories:

- **Safety-critical failures** are failures that could result in injury when the product is used normally. The allowable rate of these should be zero over the expected usable life of the product.
- **Non-safety-critical failures**. The team needs to list all of the failure modes of the product that can cause a return. These modes don't include parts that are typically replaced through regular maintenance such as filters or batteries. The reliability of critical components and functions can include:
  - Motors and actuators which fail over time.
  - Mechanical connections and joints (sliders, hinges) that can wear or break.
  - Input devices (touch screens, keyboards, touchpads) that break or stop functioning.
  - Sliding mechanical parts part that wear during use, setup, and storage.
  - Battery capacity which will degrade with multiple charges and discharges.
  - Cables that can fray with wear and vibration.
  - Sensors that can degrade through exposure to temperature and humidity variations.
  - Parts that can fracture due to cyclical fatigue (vibration, actuation, repeat loading) or normal impacts.

A simple way to get started is to split the allowable budget from the prior step and allocate a portion to each potential reliability issue. The allowable failure rate should add up to the product reliability budget.

Creating a reliability budget can seem like a daunting process given the number of possible failure modes. However, your suppliers can help – suppliers for both COTS and custom components should provide a spec sheet that contains data about the expected life of each component. If the service life of the component is less than the warranty period, the specs of the part may need to be revisited or mitigation efforts put in place (e.g., if a component's failure rate is a function of temperature, then fans may need to be added to the product). A good rule of thumb is to select components that significantly exceed the expected life of the product or that align with the regular maintenance cycle. If this is done, then these failure modes can be removed from the warranty reliability budget.

Some failures will trigger a known repair and replace process; for example, replacing the brakes on your car. These are typically tested separately from

Inset 7.9: Repair Expectation

Not all reliability failures result in a product return. Customers may expect that products need regular repair. For example, customers expect to have to replace brakes regularly but will be upset if the maintenance cycle is shorter than usual. Some of these expectations can have significant implications. For example, the timing belt on some cars is typically preventatively replaced at 10 years or 100,000 miles. The repair is expensive, but a failure can destroy the engine. Ensuring that the timing belt does not fail before 100,000 miles is critical to customer expectations. When setting the reliability budget for repaired items, the life-cycle should be set to the value of expected life before repair.

the reliability budget and are tested to the number of cycles or amount of time the customer expects between repairs (Inset 7.9).

4. How Many Cycles is the Failure Mode Likely to Experience During its Warranty Period?

The likelihood of failure within the warranty period is typically proportional to the number of times that feature is used. For example, the channel button on a television remote always wears out before the seldom-used buttons at the bottom of the remote. To prove reliability, the channel button would need to be tested over more cycles than the lesser-used ones.

For each failure point, the quality team needs to estimate how many cycles the product functions will undergo over the period of interest (typically the warranty period). For example, a smartphone may experience 50 clicks of the home button a day (more if it is a teenager using the product). If the warranty period is 365 days, then the total cycles over the warranty period are around 20,000 button presses.

Of course, the actual number of cycles that a product will undergo depends highly on the individual user. One person might use their food processor once a year at Christmas, while others might use it every day. For non-safety-related failures, the average user is typically selected. For example, based on customer surveys, the company making remote controls might determine that their average customer presses the button 20 times a day and set the reliability testing accordingly. For more critical failures, a more rigorous test may be required. The company would want to ensure that even a person using the remote several hours a day won't experience a critical failure.

5. How Many Samples and Cycles to Prove Reliability?

Once the target reliability rates for each subsystem are known, the team needs to set the testing requirements: how many times the feature needs to be tested on how many samples. Some teams take the naïve and incorrect approach of just testing one unit through one life-cycle and think that they have proved reliability. However, when verifying the reliability, you are looking for failure modes that may only occur once in ten thousand products (0.01%).

Because you are searching for small probability events, a considerable sample size and cycles are needed to prove that the product meets the target rates. Teams have the options of running a single unit for many cycles or many samples for fewer cycles. The number of cycles and samples is dependent on several factors:

- **What is the target reliability rate?** The lower the acceptable rate of failure for a given failure mode, the more cycles and samples you will need to test to prove the unit won't fail.
- **What is the target confidence?** If the failure is an annoyance, then you can test to a lower confidence level; if it is safety-critical, you are going to want more certainty. The higher the confidence you require, the more samples and cycles you need to test.
- **What is the type of failure?** The numbers of samples and cycles needed depend on the shape of the reliability curve (typically a Weibull distribution). If the fault is an early-life failure such as an incorrectly soldered antenna, then fewer cycles will be required; if the failure occurs late in the life of the product, more cycles are needed to approximate late-in-life product failures.

There are several software packages you can use to estimate the number of cycles required to test for a given failure mode based on the target reliability, confidence level needed, the Weibull distribution, and the number of samples. For example, an online calculator based on the parametric binomial reliability demonstration test (Weibull distribution) was used to generate Figure 7.6. The calculator was used to calculate the number of cycles needed for a failure rate of less than 0.1% for their default Weibull distribution for two confidence levels. It is assumed the product experiences 2,000 cycles during the warranty period. *NOTE: The data should be used for illustrative purposes only. It is for a specific use profile, Weibull distribution, and confidence level. Each product reliability will have unique reliability, confidence, and failure curves.*

In this example, to get high confidence with two samples, the product would need to survive 35,000 cycles without failure (or the equivalent of 17 years of use). To get low confidence with 10 samples, the samples would need to withstand 12,000 cycles (or six years of use).

6. *How Long and How Much Money Will it Take?*

After adding up the number of samples and testing time, it will be evident that teams cannot test all components and subsystems in the time required with the number of samples the team has available. The teams need to balance what is ideal against what is feasible.

FIGURE 7.6

Samples vs. test cycles

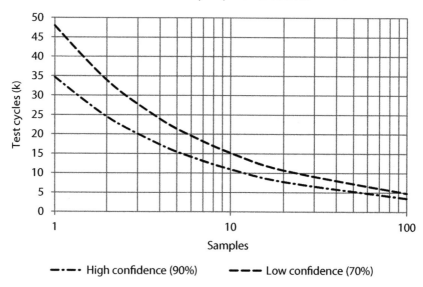

Test cycles vs. Samples to prove reliability
Mission time = 2,000 cycles
Reliability requirement 99.9%

Several strategies can be used to reduce the testing resource loads:

- **Combining tests**. Multiple systems can be combined into single tests to reduce sample usage and overall test time. For example, teams can test for speaker life while measuring battery charge and discharge cycles.
- **Re-budgeting reliability**. Failure modes that are expensive to test can be given bigger budgets (to reduce the testing time and samples), and those that are less expensive and time-consuming less (to increase the testing time and samples),
- **Creating performance curves**. For example, batteries show degradation in performance over time. Early performance degradation can be used to predict likely failures later in life without waiting months or years to see how the battery actually performs over time.
- **Using proxy data** such as the reliability data from other similar products or spec sheet data from your supplier. Bearing manufacturers hopefully have tested their product over thousands of cycles, extended periods of time, and under a wide range of environmental and loading conditions so they can provide reasonable estimates of lifetimes in your product.
- **Increasing the intensity** of some tests to decrease the testing time. For example, there are ways to accurately simulate the long-term impact of vibration by increasing the amplitude of the vibration. Teams can then use computation methods to approximate the reliability rates at lower intensities.
- **Reducing the confidence level** required for failures that result in maintenance or nuisance errors (though never for safety-critical failures).

- **Purchasing more reliable components** can increase the reliability of the product. Doing this may be expensive in the short term, but later in the product life, when the actual reliability performance is better understood, the teams can test cheaper components and replace them.
- **Simulation**. There are a number of computer software packages that can be used to simulate the performance of a product under different conditions. These simulations can be used to gain insights into where failures are likely to occur but need to be verified on actual hardware.

7. Starting the Reliability Testing

It is critical to start the reliability test planning and execution early enough in the product realization process that the results are completed before the launch of the product. As reliability failures are identified, the team should do thorough root cause analyses and corrective action.

In some cases, reliability testing may need to continue through initial production and into the mass production ramp. If there is a tradeoff between delaying the launch of a product and risking a non-safety critical failure, the team may opt to take the risk. However, this doesn't mean that you don't do the testing. As long as the failure is not safety-critical, production can start, and only if a problem found, a running design change can be implemented.

7.3.5 Accelerated Stress Testing (AST)

At times, it will not be possible to (i) test enough cycles and samples to get the confidence needed or (ii) predict all of the failure modes. In both of these cases, teams can use a class of techniques called accelerated stress testing (AST). AST puts products through increasingly stressful conditions to identify weaknesses. These stresses typically include increased temperatures, humidity, and vibrational stresses. AST is a broad class of testing strategies that include highly accelerated life testing (HALT), failure mode validation test (FMVT), multiple environment overload stress test (MEOST), and multi-step stress testing (MSST).

Products are operated at standard conditions for a short time period to establish a baseline performance, and then subjected to increasing stress. In the case of a temperature stress test, the temperature is increased in small increments as the product continues to operate. When the product fails, the failure is documented and then bypassed to allow the product to continue to function (unless the test poses a safety risk). The temperature increase is resumed, and the process repeated until either the product can't be fixed or the stress condition far exceeds the expected maximums (i.e., the housing will begin to deform).

The purpose of AST is not to prove the reliability within a warranty period but to identify what inherent weaknesses exist in a design. By increasing temperatures, vibrations, charging rates, and other stressors, the product will experience failures that would have typically taken a longer test or more testing cycles to appear under expected use conditions. The challenge of accelerated testing is that you may be solving problems that will never occur in real life, and it is relatively expensive.

7.3.6 User Testing

The tests described in Sections 7.3.1–7.3.5 are used to evaluate whether the product meets specifications. Testing should also be done to find issues that the specifications may have missed – that is, validation testing.

In general, the more thought put into designing the user test, the more the team will get out of it. Just giving the product to users without specific questions in mind will only identify the most glaring issues. Friends and family especially will not want to hurt your feelings and will not naturally have a framework in mind to critically evaluate your product. For this reason, it's important for teams to test the product with users who don't have a personal connection to the team and have a way of documenting their unstructured responses. In addition to open-ended testing, teams should define more formal test structures. The user testing should be tied both to the specification document and to any risk items identified.

- **Give the user specific tasks to perform with the product**. For example, you might ask them to set up the unit, or run a couple of functions.
- **Define specific questions that you want the user to answer**. For example, "Is it easy to install?" "Is the weight, right?" "Is the interface intuitive?" "Are users going to misuse it?"
- **Present the user with some open-ended questions** such as "what do you like most about the product?" or "is there anything about this product that you didn't expect?" You want to allow users to provide feedback you do not expect.
- **Ask for the unusual.** What else might the user try to do with the product other than the specified uses? What if someone used a smoothie blender to grind nuts? Would that use need to be warned against or might it work?
- **Define the population of people you are going to ask**. Which user profiles will you test? Make sure you have a range of ages, genders, and professions that represent your range of potential customers.

7.3.7 Packaging

During handling and shipment, the product will need to withstand a range of vertical loads (stacking), temperatures, humidities, drops, and vibrations. The International Safe Transit Association (ISTA) defines standard test parameters for ensuring that product can survive shipment without damage. Most package testing for consumer products is based on the ISTA-2A [48] standards. Typically, the tests are applied to a master carton filled with the inner packs and gift boxes (if that configuration is used for shipping). Testing of the master carton includes:

- **Drop test**. The master carton is dropped on each of the corners and edges.
- **Compression test**. The master cartons are stacked and compressed.
- **Rotary and random vibration tests**. The master carton is vibrated in several directions and vibration modes.
- **Thermal**. The packaging is cycled through different temperatures and humidities to simulation transport across the ocean and by road.

After testing, the product is inspected for damage (to the packing or the products), and all of the product is again tested for functionality and any aesthetic damage. The shipping testing can be more costly than many of the other tests because all of the units in a master carton (sometimes several hundred) have to be scrapped or used as samples, especially if there is damage.

Inset 7.10: Bio-compatibility

Bio-compatibility and allergy tests are not just limited to biomedical products. Fitbit Force had to recall about a million units after almost ten thousand complaints [76]. The Consumer Product Safety Commission (CPSC) stated that "Users can develop allergic reactions to the stainless steel casing, materials used in the strap, or adhesives used to assemble the product, resulting in redness, rashes, or blistering where the skin has been in contact with the tracker." The Fitbit CEO wrote to consumers stating that the allergies were due to "very small levels of methacrylate, which were part of the adhesives used to manufacture Force or, to a lesser degree, nickel in the stainless steel casing. Methacrylate is commonly and safely used in many consumer products [77]." McDonald's had the same problems two years later and was forced to recall almost 33 million give-away toys [78].

There are several methods for doing bio-compatibility skin irritation testing. For example, ISO has a standard ISO 10993-10, which tests for irritation and skin sensitization. While individual materials may be pre-qualified as bio-compatible, additional glues, contamination, and finishes combined with particular uses and designs can irritate the skin of some users, even when the constituent parts are deemed safe [79].

7.3.8 Other Verification Tests

Each of the specifications listed in the spec documents needs to be verified to ensure that the product meets the performance and aesthetic targets. Some specifications don't fall into any of the testing methods described above, so teams will have to decide how to test for them on a case-by-case basis. Often these specs can be checked during early pilot runs and are not rechecked unless there is a design change. Some are as simple as weighing a product to make sure it satisfies the weight limits; others may take more extensive testing such as bio-compatibility (Inset 7.10).

Summary and Key Takeaways

- ❑ The term quality is very broad and can apply to both quality as designed and quality as produced.
- ❑ All aspects of quality need to be verified and validated during the pilot phase. Validation tests ensure that you designed the right product and verification tests ensure that you built the product correctly.
- ❑ Quality testing covers a wide range of product characteristics including functionality, aesthetics, durability, reliability, and packaging quality.
- ❑ The quality testing plan needs to be consistent with the specification document.
- ❑ The quality testing must be carefully planned in preparation for piloting.
- ❑ Quality testing during piloting is a document-intensive process.
- ❑ Testing parameters need to align with the actual usage scenarios.
- ❑ Safety is always most important and failures in safety cannot be tolerated.

Chapter 8
Costs and Cash Flow

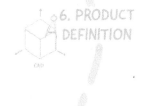

PRODUCT PLANNING

5. SPECIFICATIONS

6. PRODUCT DEFINITION

CAD

7. PILOT–PHASE QUALITY TESTING

8. COST & CASH FLOW

Managing costs throughout the product development and product realization process is critical to having a financially viable product. First, teams need to be able to predict the non-recurring engineering costs required to get a product to market. Second, they need to be able to estimate the final product COGS and distribution costs to ensure sufficient profit margin. Third, they need to understand how different revenue models will impact the timing of their revenue. Finally, they need to be able to model their cash flow to ensure that they raise sufficient funds and do not run out of cash at critical times.

Product Realization: Going from one to a million, First Edition. Anna C. Thornton.
© 2021 John Wiley & Sons, Inc. Published 2021 by John Wiley & Sons, Inc.

You can design the most appealing, high quality, and technologically sophisticated product; but unless you have the capital to pilot and produce it, and unless you can sell it at a price that's appealing to customers while covering your costs, you won't have a business.

Meeting cost targets is critical to ensuring good profit margins and managing cash flows. If the cost to produce and deliver a product is too close to or above the price the consumer is willing to pay, the product is not financially viable. Unfortunately, engineers often believe that cost is an output of the design process rather than a target to guide design decisions. Inevitably, many teams significantly underestimate the actual cost to build, ship, and support a product. For example, many organizations underestimate the costs to test and certify the product including the costs of the samples required. Additionally, teams often miscalculate the timing of payments and consequently find themselves short of cash at critical moments. Other typical cost potholes include:

- **"Just get it done, we have enough money."**  Throwing money at specific problems can solve them quickly, but indiscriminate spending can lead to cash shortfalls later. Teams rush to get prototypes, expedite parts, hire additional staff, and buy up materials to avoid downstream problems. They spend excessively in the short term only to run out of cash later on.
- **"Just pay now, we will cost-reduce later."** When prototyping, teams often assume that cheaper parts or suppliers will be available later in the process. For example, teams may buy a more expensive multi-cavity injection-molding tool to reduce the part cycle time and the per-part cost. However, this expense may drain cash when cash flow is tight.
- **"I thought samples and quality testing were provided by the contract manufacturer."** Teams often miscalculate their costs because they don't read the fine print in contracts or they make assumptions about what suppliers and partners will provide. Having a clear understanding of what is and isn't included in every contract can avoid unexpected costs.
- **"What do you mean, I have to do it again?"** A late design change can drive significant quality costs. Products may need additional pilot runs and, in the worst case, may need to be recertified.

The key to avoiding these potholes is to start cost estimations and cash flow analysis as early as possible. It is tempting for engineering types to abdicate responsibility for the money issues to the finance people and for the finance people to make assumptions without understanding the actual engineering costs and tradeoffs. Every company needs to answer the following four key questions using the skills and knowledge of all of the teams:

- How much is it going cost to get your product design ready, to design and build your production system, and to run multiple pilots?
- Once the product is ready for mass production, how much is it going to cost to produce your product and get it to the customer?
- When will revenue come in to balance the outflows of cash?
- What do you owe to suppliers and your CM, when do you owe it, and do you have enough cash on hand to make it through the process?

This chapter will outline the costs required to get a product designed, built, and delivered to the customer. Also, it will explain how to estimate costs early in the process and maintain the cost model throughout the life of the product. Finally, it will describe how to predict the cash flow based on the costs and the timing of payments.

8.1 Terminology

Before diving into how to estimate costs, several terms and concepts need to be defined.

- Price vs. cost.
- Financial metrics.
- Difference between non-recurring and recurring costs.
- Other terms that are used in cash flow calculations.

8.1.1 Price Vs. Cost

To quote from the *Sound of Music*, "Let's start at the very beginning, it is a very good place to start; when you read you begin with A–B–C." When you develop products, you need to begin by understanding the difference between price and cost. While it may be very obvious to some, a large number of students do not know the difference.

- **Price** is set by what value you create for your customers and how much they are willing to pay for that value.
- **Cost** is how much you have to spend to buy something.

Teams are often tempted to set price as a fixed mark-up on their total cost to produce the product or to choose a number that's just below a competitor's. However, either of these strategies can lead to oversetting the price or leaving money on the table. For start-up companies, crowd-funding campaigns can help determine whether a specific price point is compelling; varying price points are used to see where people stop signing on to the campaign. There are entire fields of study dedicated to setting the right price, whether through market analysis or large-data analytics, so this chapter will not address this topic. Most of the market analysis on price-setting, though, presumes that teams have an accurate understanding of their costs to get the product to market, but this is often not the case.

8.1.2 Financial Metrics vs. Cash Flow

Companies use several metrics to determine the economic success or failure of a product, including:

The total profit that can be made by launching a new product. The price (P) is the amount the customer pays; the landed costs (C_L) are the costs for making the product and shipping it to the customer, the warranty costs are the average cost of a warranty claim (C_w), the percent returned (r) is the number of units returned by the customers, and volume (V) is the total number of products sold. Finally, the product development costs (NRE) are the total costs including non-recurring engineering, marketing, and sales:

$$\text{Profit} = V\left(P - C_L - rC_w\right) - NRE \tag{8.1}$$

Profit margin is the percentage above the landed costs that you can sell the product for. Low-margin products are at risk of failing if a competitor can undercut your price. High-margin products give teams more room for potential price reductions later if needed.

$$\text{Margin} = \frac{(P - C_L)}{P} \tag{8.2}$$

Return on investment (ROI): This metric gives investors a sense of how much their investment will return as a percentage of the NRE:

$$\text{ROI} = \frac{\text{Profit}}{NRE} \tag{8.3}$$

A business will ultimately be judged by the profitability, margin, and ROI of a product development project (these metrics give investors a sense of whether the product is worth investing in); but before any profit accrues, teams need to have enough cash to get the product to market.

Cash flow is especially vital in start-up and smaller companies where capital is not as readily available. It is also crucial even for larger companies to understand and track cash flow because there are limited resources for new product development.

8.1.3 Non-recurring Engineering vs. Recurring Product Costs

The world of financial accounting has precise rules on how expenses and incomes are recorded and reported. These rules have both legal and economic impacts on how a company is valued and how the accounting organization reports performance of a company to the shareholders. For this book, we group costs into two broad categories – recurring and non-recurring costs – and leave the accounting details to the CPAs (certified public accountant).

- **Non-recurring engineering (**NRE**) costs** are all of those it takes to get to the point of producing the first saleable product. Before starting product realization, NRE includes all of the design and development costs. NRE costs associated with product realization include all of the piloting, testing, and tooling costs. Most often, the total NRE costs are much more significant than new engineers expect. Teams need to assess the NRE of the product to ensure that they have enough cash and resources to get through to production. No matter how profitable the product is projected to be in the future, if the company runs out of money before getting a saleable product, the product introduction will fail.
- **Recurring product costs** are the costs required to build and support each individual product produced. For example, the cost of a machined part that appears in the bill of materials (BOM) would be a recurring cost, whereas the cost to create the CNC path would be non-recurring. Quantifying the recurring costs is critical to understanding whether the product can be sold with sufficient margin to be profitable.

The terms used to describe costs sometimes aren't entirely standardized across industries, but the basic concepts are. These are the terms we will use for various costs in this text going forward:

- **Landed costs** are the total recurring costs to get the product into the hands of the customer (COGS + distribution costs).
- **Cost of goods sold** (COGS) is the cost of the product just off the factory line including packaging before delivery. The COGS includes materials, labor, and any factory overhead.
- **Distribution costs** are the costs of getting the finished product to the consumer. The distribution costs can vary widely depending on whether the product is shipped by air or by sea.

- **Tariffs and duties**. These are the costs imposed by countries when importing from overseas. At the time this book was authored, there was significant uncertainty about the tariffs and duties imposed on products imported to the United States from China and vice versa.
- **Support costs/warranty**. Some of your customer-purchased products will need to be serviced, repaired, or returned. The money to pay for warranty costs is held in reserve after the income is recognized (when the revenue is earned).
- **Overhead**. In addition to the cost to produce each additional unit, there are base costs to manage the operations, to keep the lights on, and pay people's salaries. These expenses can be separately accounted for or can be rolled into an overhead rate on the COGS.

8.1.4 Terms in Cash Flow

In addition to knowing what you need to spend, you need to know *when* you have to spend it, and whether you have enough cash on hand to cover your bills at the end of each day. Each of the recurring and non-recurring costs will have to be paid based on the contracts you have with your suppliers. Several of these concepts will be discussed in detail in the chapter on suppliers, but here are a few key concepts you need to understand for this chapter.

- **Lead time** can significantly impact cash flow because you have to purchase materials far ahead of when they are built into saleable units.
- **Purchase order** (PO) is the document you issue to tell your supplier and CM you want to purchase something. Typically, the PO requires an upfront deposit with the remainder due some time before or after delivery of the parts. The lead time on the purchase order (how far in advance you have to put in the order to make sure you get the parts on time) and the percentage of the total you have to pay upfront can have significant impacts on your cash flow.
- **Master services agreement** (MSA) is the document that legally defines the relationship with your supplier or contract manufacturer. In that agreement, all of the costs and timing payments will be laid out.
- **Minimum order quantity** (MOQ) is the smallest order you can place with your supplier. Most items you will purchase from suppliers will have an MOQ. The more you order, the lower the price per unit, but you also have to pay upfront for all of the materials and then hold the excess in your inventory until it is needed (and tie up capital).

8.2 Non-recurring Engineering Costs

The first of the four questions posed at the start of the chapter, "*How much is it going to cost to get your product design ready, to design and build your production system, and to run multiple pilots?*" is answered by estimating and managing the non-recurring engineering costs. The following sections list some of the typical non-recurring engineering costs experienced in product realization after the majority of the technical development is completed. The total NRE, as used in this book, includes all of the development costs up to product realization. NRE costs include those for development, capital equipment, marketing and branding, quality, and operations. The financial organization typically maintains this information in financial models and financial statements. Also, there are several SaaS products available that can help model the NRE expenses required to get to production.

8.2.1 Development Costs

Development costs are the costs to design the products, build the manufacturing system, and bring the product through the pilot process, but not those used to purchase capital equipment or tooling. Most of the development costs are service (non-material) costs.

- **Costs of services**. If other firms are engaged to provide design, engineering, marketing, or industrial design services, these costs can add up. Your company can outsource any of the roles described in Section 4.1. If you have the capability in-house, you still need to pay salaries or equity. For outsourced service providers, typically there is a deposit required to start work and a monthly retainer for an agreed-upon period. Contracts are typically set up such that the service provider only hands off the final deliverable when the final payment is made. Understanding these terms is critical. For example, in one case, the firm couldn't get revenues until the app came out, but the development firm wouldn't release the app until they got paid. Both companies lost their investments because the app and the product were never released.

  It is hard to estimate the costs of services without getting quotes from suppliers – the costs are dependent on too many factors. People who have outsourced similar services can give you a ballpark figure, but the range of possible costs will be extensive. Product design services can cost anywhere from tens of thousands to hundreds of thousands of dollars. App development costs also range widely.
- **Samples and prototype costs for development.** Samples and prototypes are usually expensive hand-built units using one-off manufacturing methods. The costs per unit are typically quite high and often an order

Vignette 5: The Importance of Terminology in Cost Modeling

Amanda Bligh, PhD, Principal Consulting Engineer

Anthony Giuffrida, Manager, North American Implementation Services

aPriori Technologies

Coming from a background of product development, marketing, and manufacturing, we are both painfully aware of the complexity of understanding cost as early as possible. As part of the aPriori team, we are working to help our customers better understand their product costs and drive manufacturability from early in the product development process. By having transparency in costs during concepting and detail design, engineers are able to make changes and modifications while the design space is still wide open, versus later in the process when fewer changes are possible due to constraints from elsewhere in the product.

When using a manufacturing cost model to perform early estimates, the difficulties facing teams are often not in learning the software or understanding the design to cost rules, but in a consistent understanding of terminology. Fundamentally, everyone involved in making and using a cost model needs to speak the same language. Models assume precise definitions to terms as encoded by a cost modeler. A model user may have a different interpretation of the same terms based on their manufacturing environment or background. While these vocabulary differences may seem small, the reality is that even small differences in assumptions can lead to vastly different manufacturing analyses and estimates. For example, if a customer assumes that setup time means part setup, but the model interprets it as the batch setup, the output will

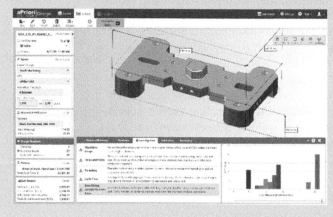

not match their assumptions. In the best case it will lead to a few hours of rework of the model by a consultant, but in the worst case it can lead to errors and a loss of confidence in the model.

The table below shows how the same term can be interpreted differently by the cost modeler responsible for building the model and the person who uses the model to drive engineering decisions:

Manufacturing term	Assumption embedded in cost model	Common assumption by user or SME
Cycle time for machining	Part load, tool movement, chip cutting time, part reorientation, part unload	Chip cutting time only
Setup time	Batch setup (once per batch run of parts)	Part setup (required for every part in the run)
Part cost/price	Part costs only: not including markups, logistics or strategic business factors	Price from a supplier including markups, packaging, logistics, expedite fees
Fully burdened part cost	Part cost + amortized capital investments (programming, fixtures, hard tooling)	Piece part costs only. Amortized capital investments captured separately
Process	A single process, like 3-axis milling	All processes required to make a part, such as band saw, 3-axis mill, CMM inspection
Machining operation	A feature made by a tool type, such as face milling, drilling, boring	All features created with one machine, for example 3-axis milling
3-axis mill	CNC milling center	Manual Bridgeport
Hourly machine rate	Annual cost to power and maintain machine including depreciated machine cost / # of hours machine is planned to run for the year	Annual cost to power and maintain machine including depreciated machine cost / # of hours machine is planned to run for the year + the hourly wage to operate the machine
Utilization	The percentage of the rough mass used to make a part	The percentage of time a machine is actively making parts in a year

We have learned the hard way that the downstream impact of vocabulary differences can be very dramatic. As such, our teams are deliberate when engaging in model customization with customers and use a selection of techniques to ensure both sides are speaking the same language. We often spend significant time defining terms and mapping the needs of the model and the data available from the client. In most cases we find cost elements and terms that do not align directly and require adjustments of assumptions either in the model or from the client's side.

If we would have one piece of advice for those building cost models, it is to clearly define your terms. This increases the confidence users have in your model, a key measure of its success.

Text source: Reproduced with permission from Amanda Bligh, Anthony Giuffrida, and aPriori.
Photo source: Reproduced with permission from aPriori Technologies Inc.
Logo source: Reproduced with permission from aPriori Technologies Inc.

of magnitude higher than those of the final product. The costs are also highly dependent on the accuracy and quality of the finishes. The samples and prototypes are needed to get feedback from user tests, to enable manufacturing analysis, to provide a platform for firmware and software testing, and to be used for marketing (Inset 3.2).

- **Piloting/setup costs**. Your factory may charge you one-time fees for running pilots to cover their labor and equipment utilization costs. These fees will depend on the factory and your MSA with them.

8.2.2 Equipment and Tooling Costs

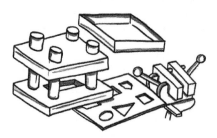

It is unlikely that you will be able to purchase "off-the-shelf" all of the parts needed to make and assemble your product without needing to create at least one tool or fixture. Fabricated parts will need custom tooling, automation will require custom grippers, and assembly will need fixtures (Chapter 12). Equipment and tooling costs can represent a large portion of the cash outlay for hardware products and can have long lead times. The equipment and tooling NRE costs can consist of:

- **Specialized dedicated equipment**. Many products may require the purchase of specialized manufacturing equipment. For example, robotic systems may be purchased for a specific production line. Depending on the contract with the factory or CM, the CM may or may not charge you for specialized equipment. You can get a sense of the cost of commonly used equipment by searching for equipment manufacturing costs online, but you will need to get quotes from manufacturers directly to accurately estimate the costs of the equipment, software, and accessories you need. Significant capital expenditures shared across multiple products are not included in NRE. For example, if a new CNC machine is purchased to support a new product, but that machine will be used on other product lines, it is not included in the NRE costs.
- **Tooling**. Tooling encompasses designed and fabricated items used to produce custom-fabricated parts. For example, a fabricated stamped metal part for the hood of your car will have a unique die explicitly made to form that part. Simple injection-molding tooling to produce simple parts can cost several thousand dollars, while complex tooling to make complex parts for medical devices can cost tens or even hundreds of thousands of dollars. Teams can pay for tooling in one of two ways. Some CMs will build the cost of the tool into the cost of the parts and own the tool going forward. In other cases, your company pays for the tool (typically a deposit at the start of tooling fabrication and full payment at the start of mass production).
- **Tooling changes**. During the pilot phases, teams may discover failures or want to make design changes that would require changes to the tooling: adding ribs to a part, changing radii, or tuning fit and finish. Tooling

suppliers typically charge for changes as additional time and material on top of the tooling charges, unless you have included changes in the initial contract. Tooling change costs are highly dependent on the scope of change, location of the supplier, and your agreement. When talking with tooling vendors, it is vital to get a quote for both the tool and the costs of changes.

- **Fixtures**. Fixtures are used to hold parts, provide guides to assembly workers, and locate parts relative to each other. These are typically charged as a separate line item on the PO by the CM and include both material and labor costs. The payment date will depend on the MSA, but fixtures are typically paid for during the pilot phase when they are installed in the production line. Fixtures can be very inexpensive if they are 3D printed, or they might cost hundreds of thousands of dollars if custom-made (e.g., custom aerospace fixtures).
- **Custom plates/stencils**. Some manufacturing processes may require the use of custom components that can also drive up costs. For example, the printed circuit board (PCB) process requires a custom stencil, and pad printing requires custom-etched plates. Stencils typically cost in the range of $100–$1,000 each depending on their complexity and size.
- **Material handling**. To ensure that the products are not damaged during assembly or transportation within the facility, manufacturers may need to create custom totes, fixtures, or carts. Naturally, these can range in price depending on the size, complexity, volumes, and manufacturing processes required to make them.

8.2.3 Quality Costs

Quality costs are typically an order of magnitude lower than the tooling costs, but they still represent a large cash outlay. Quality testing for regulated products (e.g., medical) is the exception; it can cost millions to get products approved for sale. The more complex the product, the more quality testing will typically cost. Quality costs include:

- **Testing fixtures and equipment**. During the factory build process, subassemblies and finished products will be tested to ensure that the products work correctly. If the factory has to check more than a sampling of subassemblies and final products, it is necessary to build testing equipment that can do the quality tests consistently and quickly with minimal operator intervention. As with fixtures, payment for testing equipment is typically due when the factory starts using it. Testing equipment can cost thousands to tens of thousands of dollars.
- **Certification costs**. Certification costs, which include testing and filing costs (Chapter 17), can range in the thousands to tens of thousands of dollars depending on the technology and the countries where the products are sold. Certification is typically paid for during the design verification testing (DVT) and production verification testing (PVT) stages.
- **Testing costs**. Durability and reliability testing for an electromechanical product can run into tens of thousands of dollars for equipment rental, analysis, labor, and samples (Chapter 7).

8.2.4 Business Operation Costs

In addition to the direct costs to design and produce the parts for your product, there is the team that has to be paid and have facilities to work from. The business operations costs depend on how the team is structured, how much is bootstrapped (e.g., not paying founders), and how frugal you are (it is incredible how much time and money is spent on designing custom T-shirts with cool logos before companies have a working product).

- **Costs to keep the lights on and other overhead** include rental spaces, leases on equipment, IT expenses, and software expenses. These are typically monthly costs that are known from the start of the project. In larger companies, these costs are included in the salary overhead; in smaller companies, overhead is calculated as a regular monthly cost.
- **Travel** can be costly, especially when the factory is overseas. During the pilot phase, you may be sending many team members to the factory for each pilot. The cost of last-minute airfares and hotel costs can be multiples of a trip planned several months in advance. Typically, the engineering and production teams schedule travel around pilot runs and ramp-up.
- **Salary** includes not just the wages paid to each employee, but the taxes, benefits, and overhead that go into an employment package. When a company is growing, teams need to hire in anticipation of the needed skills and frequently ahead of revenues. The team size typically increases significantly around the start of mass production when cash reserves are at their lowest.
- **Tradeshows** are a great way to get press on new products and thus can drive sales; but travel, booth rental, and swag can be costly.
- **Team/HR expenses** may include headhunting, recruiting fees, and team bonding activities.
- **Shipping** of products, parts, and materials between suppliers and the design team should be included in business operation costs. The costs of expedited international shipping can add up quickly and should be avoided if possible.

8.3 Recurring Costs

The second of the three questions posed at the start of the chapter was, *"Once the product is ready for mass production, how much is it going to cost to produce your product and get it to the customer?"* This is answered by accurately estimating and then managing the recurring costs.

Recurring costs are the expenses that occur each time a product is pro-duced. Cost estimates should begin early in the product development cycle and be updated and refined throughout the product realization process. Waiting to understand the cost of goods and the distribution costs until too late in the product realization process can lead to products that can't be sold at a high enough margin. By informing the team immediately of how a given design decision might impact costs, teams can re-estimate recurring costs before the choices become very difficult to unwind.

A word of warning: estimating recurring costs is often more of an art than a science. Once you have a supplier and a part drawing, it is easy to get accurate quotes. However, until then the estimates have a great deal of uncertainty.

There are many approaches to estimating the cost of goods, and each uses its own jargon [80]. For this text, we will use NASA's terminology and frame-work [81]. NASA categorizes estimation tools into:

- **Engineering build-up cost estimating (also called grassroots)** (Section 8.3.1). This involves getting accurate quotes for the parts and building up the cost estimate from those quotes. **Parametric** estima-tions can be used to predict the cost of materials when quotes are not available – they use mathematical models to model the part cost as a function of performance.
- **Analogous estimating** (Section 8.3.2). This uses the cost structure of existing products to assess the new product's cost. Analogous estimating identifies the differences between the existing and the new product and estimates the costs of those differences.

The BOM is an excellent place to store cost estimates. Also, several soft-ware systems provide methods to collect and maintain cost data. Costing estimation is not done once but is continued throughout the whole prod-uct development and product realization process. Because estimates will change over time, it is necessary to keep track of the estimates and the rationale behind them. It is too easy to put a placeholder number in an estimate and never go back and update it, only to be surprised late in the process with a significantly higher cost. As new, more accurate estimates and quotes are collected, the BOM and COGS numbers need to be updated. In addition to cost data, it is vital to keep track of the following information:

- The source and predicted accuracy of the quote, along with supporting documentation.
- The date of the last time the estimate was updated.
- What assumptions went into the quote including MOQs and process selection.
- Which BOM, drawing, and spec document version was used as the basis for the estimate.
- Whether any information is missing from the calculation.

Too often, a design change or an error in an assumption can lead to an unexpected cost increase later in the process. It is a rare exception when the change is in your favor.

8.3.1 Build-up Cost Estimation

Figure 8.1 shows the breakdown of what costs need to be estimated. Doing a built-up cost estimation starts with finding or estimating the cost of each part in the BOM. Once the BOM costs are calculated, estimates are made of labor, scrap, profit, and overhead. Next, the distribution cost is added to get the total landed cost. Warranty and customer support costs are included to determine the net margin on the product. Each of these will be described

FIGURE 8.1 Factors that go into the overall COGS, landed cost, and price

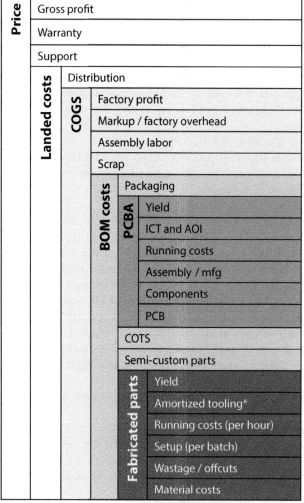

* Sometimes tooling is included in NRE and sometimes in COGs

in the sections below. Estimating these costs can seem like a daunting task early in the design process; however, it is understood that early estimates will be rough. As suppliers are identified, the cost model becomes more refined.

COGS: BOM Cost Estimation

Typically, the most significant contributor to COGS is the cost of the materials in the BOM. The other costs in COGS – scrap, labor, overhead, and profit – all typically scale with the cost of materials. For example, the more parts there are, the more labor will be needed to assemble those parts. Getting a reasonable estimate of the material costs provides a solid foundation for understanding the overall cost of the product.

Different types of parts – fabricated, purchased, and semi-custom – use different methods for estimation. The following sections provide an overview of how to start estimating each type.

Fabricated parts

Most electromechanical products will contain at least a few custom-fabricated parts. For example, a typical e-reader has three significant injection-molded parts that make up the housing and several smaller ones for buttons. A car, on the other hand, has thousands of fabricated parts. The cost of a fabricated part is a function of:

- **Raw material costs**. The part will never be less expensive than the raw material (C_R) that goes into the product. The raw material costs are a function of the cost of the material in dollars per weight C_m, the volume of the part (v) and the density of the material (ρ)

$$C_R = C_m v \rho \tag{8.4}$$

- **Wastage/offcuts**. Fabricating parts often involves removing and scrapping excess material. When purchasing the parts, you have to pay for all the material used, not just what ends up in the product. Some of the material can be recycled, but the recaptured value will always be less than the original material costs. Some examples of manufacturing processes and the waste they generate are:
  - *Injection molding*. The sprue, gates and risers can add 3–5% to the total material costs.
  - *CNC machining*. Depending on the amount of material removed through machining, the wastage can be significant. For example, Apple makes the Mac Airbook unibody from a single piece of aluminum. A majority of the material is machined away.
  - *Laser cutting*. The amount of wasted material will depend on how closely the part outlines can be packed on a sheet and what percentage of the blank is discarded.

- **Capital equipment costs**. Capital equipment is infrastructure used for specific fabrication processes, such as CNC mills and injection-molding machines. Some people include the amortized capital equipment and tooling costs in the part costs, and some don't (Inset 8.1). Dedicated manufacturing facilities typically charge for using the capital equipment as an hourly rate. The per-part operating cost of the equipment is a function of the hourly machine costs and the cycle times.
  - *The hourly rate* depends on several factors including location, equipment capability, and volumes. Some industry consortia annually publish average rates by region and tonnage. Often the consumables (e.g., cutting fluid, bits, and cutters) and labor are included in the hourly equipment cost.
  - *The cycle time* per part is determined by the part size, material, tooling, accuracy requirement, and equipment.
- **Labor costs**. Often the hourly rate for the equipment includes the labor charges to run the equipment, while at other times it is charged as a separate item.
- **Setup and breakdown costs**. Unless the equipment line runs 24/7 producing a single part, the fabricated parts are produced in batches. Between each batch, operators must change material, tooling, and settings. In addition, machines typically have a warm-up period which generates scrap. The cost of the setup and breakdown is often built into the MOQ prices – prices go down as order size goes up. Sometimes the costs of setup and breakdown are charged separately.
- **Yield**. A certain number of parts will fail inspection. The yield will be a function of the robustness of the designs of the parts and the tooling, the process control, and the quality targets. The rejection rate (which ideally is zero) can range between a fraction of a percentage to 50–80% depending on the design, quality requirements, and manufacturing processes.
- **Post-processing**. Parts may require post-processing steps to get the surface, durability, and aesthetics to the right standards. For example, the supplier will typically perform these steps and charge on a time-and-materials basis.

Inset 8.1: Whether to Amortize Tooling or Not

Unless your supplier is charging you extra per part to cover the cost of the tooling, amortized tooling costs should not be included in the COGS. The contribution of tooling to the COGS is highly dependent on the production forecast. For example, if your tool costs $10,000 and you estimate the volume at 100 units, the cost per unit will be $100. On the other hand, if you have a volume of 1 million units, the cost of the tooling per part is only $0.01. Besides, amortizing tooling does not help with predicting cash flows because the tooling costs have to be paid up front, while the COGS are typically paid as the parts are produced.

There are several ways to estimate fabricated part costs:

- **Expert estimation**. Talk to a manufacturing expert who understands these processes and can give you a rough estimate based on experience. Sometimes you are lucky enough to have these resources within a company, but outside cost estimation consultants can be hired.
- **Multiply the material costs** by a factor of between 1.2 and 3 depending on the complexity of the part. Often the first factor is a guess, and the factor is adjusted as the team learns.
- **Online estimation tools** will give a rough estimate for part costs based on US prices. If your product is made overseas, the costs will likely be lower by 15–33%.
- **Low-volume manufacturing suppliers**. Suppliers are starting to provide automatic quoting services online. These quotes are based on lower volumes and US manufacturing. The advantage of these services is that they will often give you feedback on the manufacturability of your parts quickly at no cost. However, they won't quote complex or challenging parts without going through a formal RFQ process.
- **Estimating software**. Some software can give detailed cost estimates based on CAD. The software uses proprietary models that predict machine time and labor costs from first-principle models. Cost estimation software tends to be expensive and is often used only in larger companies. However, many software companies are starting to offer their services in a SaaS model.
- **Get quotes from vendors**. The most accurate estimate will be an actual quote. However, vendors are hesitant to send quotes to companies they do not know or to customers they suspect might not be serious. It takes time to build a quote, and they are not paid for that time. Talking with vendors should start when the team is ready to begin the purchase process and are going out to competitively bid the product. If your CM has a relationship with a vendor, you may be able to leverage that relationship to get an earlier estimate.

Custom parts

Custom parts are parts or subsystems that are unique to your product, but are based on a supplier's expertise in a given technology. These can include items such as custom LCD panels, motors, custom bearings, and custom connectors. The designs are often derivatives of components that the supplier produces for other customers. When specifying a custom part, two factors will influence the costs:

How much customization is needed? If the supplier can easily reconfigure from an existing offering, the cost to you will naturally be lower. If the systems needs a significant custom design, the price will be higher. If customization includes making a product outside the typical bounds of their business (e.g., the supplier typically makes 500–3,000 mAh [milliampere-hour] batteries, and you want a 5,000 mAh one), the cost may jump as well.

Design customization can also drive cost if there are upcharges for adding specialized connectors, wires, mounting brackets, and housings.

If custom tooling is required, the price will be higher. The tooling may be charged as an upfront cost or can be built into the cost of each item. Typically, the tooling is owned by the supplier. Finally, for customized products, verification testing can also drive costs higher. For example, if a bearing supplier needs to run accelerated life (reliability) tests uniquely for you, they may charge you for it.

What are you asking for? Obviously, a 300 mAh (milliampere–hour) battery is going to be significantly cheaper than a 50 Ah (ampere–hour) one. Size, capacity, and performance will all impact the cost of a given component. However, custom part costs rarely scale linearly with performance or size. Instead, there are breakpoints in the cost curve due to differences in technology, packaging, and commoditization. For example, in the battery market, there are ranges of voltages and Ah ratings in which the extra cost of additional capacity is significantly higher than in others. For example, Figure 8.2 plots the average and ranges of quotes found on Alibaba for LiPo (lithium-ion polymer) 3.7 V batteries for a variety of Ah ratings. There is quite a wide range for the same battery attributable to differences in MOQ, volumes, connectors, and supplier quality. Even with the noise from these factors, several trends become apparent. For example, all batteries below 500 mAh (lower left graph) cost roughly the same. The material differences are minimal between them, so most of the cost is likely to be driven by assembly and handling costs. Between 500 mAh and 10 Ah (lower right graph), the cost increases at about $0.80 per Ah, whereas for larger batteries in the range of 100-300Ah (upper right graph) it costs $1.40 per additional Ah. Insights into these factors give teams the ability to estimate costs parametrically using market data.

How much you are buying? The price and how the fixed costs are charged is a function of the likely order quantities, total volumes, expected growth, and life of the product. If you need a small number of customized units, the price will be higher than if you can order in larger quantities. It is important to sell your company to your suppliers. If they think they can win a large long-term contract, they are more likely to give you a competitive bid. If they think you are only going to sell a handful or that you are likely to run out of money, they will price their product accordingly.

Which technology you are using? Cost is not a linear function of performance. As the required capability (e.g., size, power, accuracy) of the part increases, the supplier may need to change technology or manufacturing processes to achieve the marginal performance improvement you need. If you can keep your performance requirements below these transition points, the cost will be significantly lower.

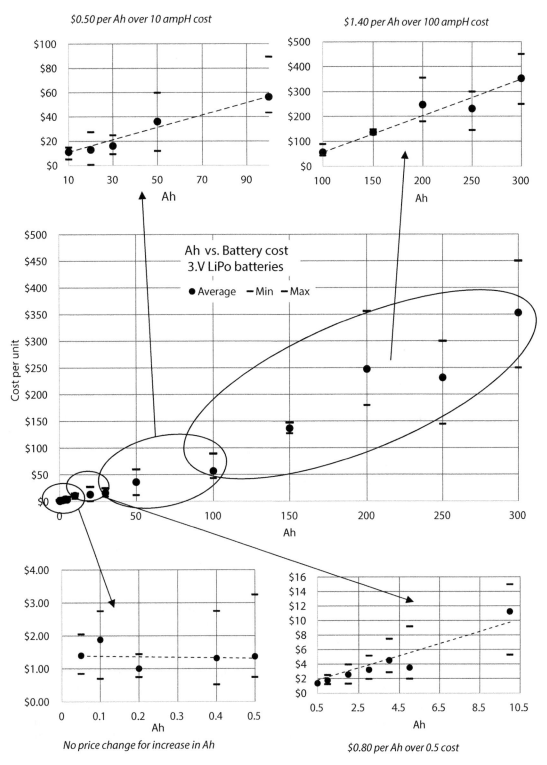

FIGURE 8.2 Costs for a battery as a function of capacity

There are several ways of getting estimates for custom parts:

- Look up similar parts online.
- Get estimates from several vendors when you are ready to get actual quotes.
- Build parametric models like those in Figure 8.2 to identify the break-points and the costs and the range within each performance class.

The cost of fully customized subsystems is often tough to quantify ahead of time. The costs are typically negotiated as part of a long-term partnership with the supplier.

COTS and consumables

Almost all product contains at least a few commercial off-the-shelf (COTS) parts. Most mechanical products will have COTS parts such as fasteners, mechanical joints (hinges), O-rings, and springs. While COTS are essentially commodities, quotes for the same nominal part may vary from supplier to supplier because of:

- **Quality of the part specified.** Inexpensive fasteners with low-quality finishes will have a lower cost than higher-quality ones.
- **Location of the supplier.** If you buy through a large US supplier, the costs will be higher than if you find the nominally equal part overseas. Keep in mind that working with overseas vendors might mean making compromises on lead time, quality, or flexibility in ordering.
- **Volume orders.** Order size is the biggest determiner of cost in COTS parts. If you can order parts in sufficiently large quantities, you can reduce your parts costs by as much as 50%.
- **Bulk purchases**. If you are purchasing several types of parts from a single vendor or distributor, it is possible to reduce the overall costs by buying a larger set of products.
- **Who is purchasing.** Your CM or factory may have agreements with certain vendors, and they can share volumes across multiple products (e.g., bulk purchases of screws).

Figure 8.3 shows how the cost of the same AA 1.5V alkaline battery from a US supplier's website varies between manufacturers and by volume. The price of the battery dropped somewhere between 30% and 50% over the range of MOQs. The lowest cost batteries are 60% less than the highest cost at comparable volumes. The costs for purchasing off-label from China at high volumes were as low as $0.08, but the quality is not guaranteed.

Sometimes selection of components involves a tradeoff between COGs and certification and development costs. For example, designing the Bluetooth transceiver to go on your board will save some money. However, you can save on certification costs and development costs by buying a preapproved Bluetooth module, even though the materials cost might be a bit higher.

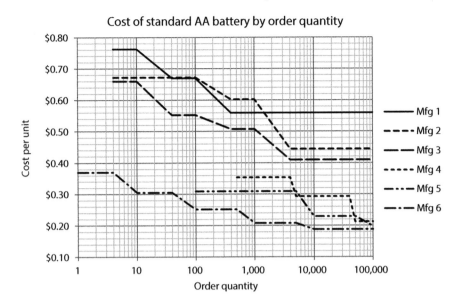

FIGURE 8.3 Cost of batteries by MOQ

You can get quotes for COTs parts from several locations:

- If you get quotes from Amazon, you will *significantly* overestimate the costs. You are highly unlikely to purchase in bulk through that distribution channel.
- Quotes from Alibaba or other marketplaces will give you the bottom end of the potential range of costs.
- Some distributors provide APIs (application programming interfaces) that enable you to upload a BOM and get quotes automatically.

Printed Circuit Board Assemblies (PCBAs)

Most of the cost of the PCBA will be in the PCB and the major electronic components such as microprocessors, Bluetooth modules, and Wi-Fi modules. A first basic estimate can be created by adding up the board cost, the most expensive components in your BOM (i.e., Wi-Fi or Bluetooth chip), and a small factor for assembly, testing, and packaging. Some low-volume PCBA vendors can provide automated quotes, which will typically be higher than the final high-volume overseas costs. PCBA costs are driven by the assembly design and will include the costs of:

- **Printed circuit boards (PCBs)**. These are the substrates that hold the components. The number of layers, trace widths, cutting methods, size, hole sizes, and other factors can have dramatic impacts on the cost of the boards. There are online quotation systems that can give you a ballpark figure for the parts, but if you want anything unusual, you need to get a quote from a vendor. Figure 8.4 shows how prices quoted online for PCBs can vary based on MOQ and number of layers.

FIGURE 8.4 PCB
costs by MOQ and
number of layers

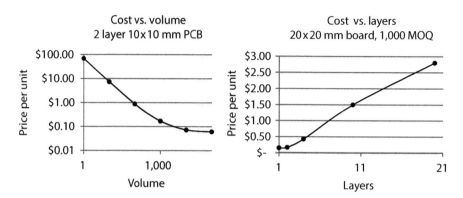

- **Components**. Components are individual chips, modules, connectors, and mechanical elements that are mounted on the PCBs. These can be easily quoted using a distributor's website. Many distributors can automatically quote the cost of individual parts by uploading your BOM. The prices will depend heavily on the volumes.
- **Reflow soldering cost**. This is the hourly cost to "rent" the equipment to mount and affix the components on the PCB and pay the staff to run it.
- **Secondary population process**. Some IC chips (also called modules) may need to be attached using methods other than reflow soldering. These other methods include wire bonding, manual insertion, and manual soldering.
- **Secondary PCB processes**. For example, if you need to cut your PCBA out of a larger board or if you need to conformal coat the product, you will be charged extra.
- **Assembly testing costs**. Each board needs to be inspected and tested before shipment. The cost of testing will depend on the cycle time of the test, which tests are done, and how much time they take.
- **Packaging materials**. A padded package or conductive envelope is often needed to protect each unit against physical and electrostatic discharge (ESD) damage. The padded packages and envelopes are usually not reusable.
- **Supplier**. US vs. offshore suppliers can have dramatically different prices. Within a given country, the equipment and labor costs can also range from low (e.g., old Chinese-manufactured equipment used to make through-hole boards) to very high (e.g., the newest Japanese equipment which can produce boards with pitches of less than 0.5 mm).
- **Batch size**. PCBA fabricators will often charge a setup fee for each run. The larger the MOQ of the batch, the smaller the overall setup costs will be per part.

Accessories and Packaging

Packaging and accessories can range from a $0.10 cardboard box to an elaborate unboxing experience that can cost $15. Packaging and accessories are highly dependent on the quality of the materials chosen and the amount of customization. The best method for managing the cost of packaging is to set a budget, then work with your packaging supplier to design a solution within that budget. You can get a sense of packaging costs by looking at the websites of major US-based packaging suppliers, where you'll find prices of many of the boxes listed. To get a complete estimate, including accessories and packing materials, you'll need to speak to a supplier to get a direct quote.

The packaging budget should include all of the parts described in Section 6.6 including:

- **Master carton**. Your package will need an outside master carton. It is cheaper to go with a standard size, but that may require you to make your inner packs and gift boxes standard as well.
- **Inner pack**. The inner pack needs to be of sufficient quality to enable direct shipping in this box.
- **Gift box**. The gift box cost can vary depending on the quality and detailing.
- **Inserts**. Insert costs will vary depending on the material, print quality, and shipping security.
- **Labels, tape or stickers**. The cost of items such as stickers, while negligible individually, can add up in the aggregate.
- **Dust and scratch protection**. Some products are shrink-wrapped to reduce damage and theft.
- **Manuals and quick-start guides**. These need to be printed and assembled. The cost will depend on the size, paper, print quality, and volumes.
- **Parts shipped with the product**. Also, the packaging will include several accessories such as cables, cases, adaptors, spares, wall warts, manuals, and templates. The cost of the accessories can typically be estimated using the processes described in the COTS and custom estimation sections.

COGS: Markup and Overhead Costs

Above and beyond the material costs, the CM will charge for their time, equipment, and labor as well as add a markup for profit. When working with a contract manufacturer, you will negotiate these markups as part of the terms in the master service agreement (MSA) (Section 14.5). These charges are typically applied cumulatively and are based on the COGS. If your factory is in-house, the markup will depend on the internal financial structures of your company. The markup and overhead include:

- **Labor costs**. The assembly, shipping, and other labor to make and distribute the product. A good starting point for the hardware labor is 10–15% of the material costs.
- **Yield/scrap**. A certain number of units will be scrapped because they don't meet quality targets or were damaged during assembly. Rather than tracking each failed unit, the factory will charge a flat percentage to cover these losses. This can range between 0.5% and 3% depending on

the quality of the design for manufacturing (DFM) and on the product requirements.

- **Testing losses**. Some units of your product will need to be destructively tested to ensure ongoing quality (Chapter 13). You may need to pay your factory for these as if you were having them delivered for sale.
- **Material handling charge**. Your factory will add a percentage to the cost of the materials that are purchased by them. This handling charge is used to cover the costs of ordering and inspecting all incoming materials. If parts are consigned, however, this charge is not applied. Some factories may discount the handling charge rate for particularly costly parts. For example, for parts under $10, they may charge 9%, whereas for parts over $25, only 3%. These handling charges can range between 5% and 20% depending on other contract terms and the supplier you have selected.
- **Rework**. Even after failing specific tests, some units of a product will be able to be reworked and sold. If the problem is caused by the factory, they will rework it for free; if the failure is due to a design error, you may be charged an additional labor cost.
- **Quality control/shipment audits**. The product will be inspected at the end of the production line. Sometimes this is done by the factory QA group, while in other cases it is done by a third party. In developing countries, the cost of a shipment audit can range between $1,000–$5,000 per shipment depending on how long a batch inspection takes. The cost in the US is significantly higher because of the higher labor costs.
- **Profit**. The factory can add a markup percentage to the COGS costs to ensure a profit. Typically, if the material handling charge is low, then the profit markup will be higher.

Distribution Costs

Needless to say, it costs money to get a product to the customer from the factory (Chapter 16). Depending on the shipping and distribution method, the landed costs can vary even for the same product (e.g., if the product is directly shipped to a customer by air, costs will be higher than if it is packed in a container and sent by boat and truck). When estimating the distribution costs, you need to account for the following:

- land transport from the factory to the main carrier (sea or air).
- main carriers who move the product most of the distance by either sea or air.
- land transport from the main carrier to the distribution site.
- order fulfillment (i.e., transport from the distribution site to the customer).

There is always a tradeoff between cost and speed. Shipping by sea can take between 12 and 45 days, whereas air freight is typically 3–7 days. Shipping a full container is significantly cheaper than shipping individual packages.

It is relatively easy to get cost and delivery time quotes from freight forwarders (intermediaries who coordinate shipment from the factory to its destination). Just remember, there are seasonal trends, and at certain times of year (e.g., just before Christmas shopping season) transport can be more expensive and slower.

Warranty and Customer Support

Your spending does not end when the product is in the hands of your customer. Almost every consumer product sold today will provide customers with a customer service number to call; these call centers must be staffed with trained operators. Customer support costs include the cost of fielding calls from customers, referring more challenging technical problems to the right engineers, and repairing or replacing faulty products. Also, users may need to return the product for repair, replacement, or credit. All of these costs need to be included in calculating gross profit. Finally, you will need to buy and maintain spare parts for your products. Chapter 18 discusses warranty and customer support in more detail. Customer support can be divided into several categories:

- **Installation or service costs**. The company may have promised installation and support as part of the sales contract.
- **Customer service support**. Rarely can a product go into the hands of customers without a few users needing help and dialing a toll-free number. The cost of customer support may be calculated as a percentage of products requiring a call, and the average cost of each call, or it might be included in the overhead costs. The rates for calls typically are estimated in the range of $1 per minute, but rates can be higher if call service people with greater expertise are needed, or can be less if the call center is outsourced overseas.
- **Warranty and repair costs**. In Section 7.3.4 we discussed testing whether the product is reliable enough to meet a target return rate. That return rate will be the dominant factor in the warranty and repair costs. Inevitably, some products will be returned by customers, and this cost needs to be estimated. The warranty costs will be a function of the percentage of products returned and the average cost of that return (including replacement, repair, reimbursements, and the reverse logistics required). Warranty and repair costs are typically paid out of a warranty reserve (Inset 8.2).

Other Ongoing Support Costs

Once the product is in production, people and resources will need to continue to support it. Teams of people still need to sell the product, ensure that it is manufactured at the needed rates, that the product is improved and cost-reduced, and that the quality is sufficient. These costs are typically rolled into an overhead cost.

Inset 8.2: Warranty Reserve

 Companies must hold a warranty reserve to cover any warranty expenses during the promised life of the product. When a product is sold, a percentage of the revenue is held in a separate warranty reserve or warranty accrual account to cover future warranty expenses. The warranty reserve is recalculated every month and adjusted to reflect current sales and the predicted return rates for each product. If the return rate turns out to be higher than expected, then the company has to add money to the reserve. When warranty expenses are incurred, they are paid out of the reserve. There are several reasons for maintaining a warranty reserve. First, it ensures that the recognized revenues accurately reflect the actual revenue; and secondly, it ensures that there is enough cash on hand to provide customer service late in the product's life.

8.3.2 Analogous Estimating

The prior section on build-up cost estimation focused on getting accurate quotes for the costs of parts, production, and delivery and then adding them up to get the total costs. For many teams, especially in smaller start-ups, a grass-roots estimate can be time-consuming and, depending on the skill of the cost estimator, fraught with errors. Teams can also use analogous estimation to get a second view of costs. Using both estimation methods will enable organizations to triangulate a better estimated cost of the final product. Analogous estimating identifies similar products – either within the company or externally – and builds an estimate by looking at the differences. In general, the steps for analogous estimating start with the overall selling price of the existing product. Then these steps are followed:

- **Estimate the COGS** for the analogous product by removing the likely margins and cost of distribution from the selling price.
  - *Estimate the likely margin of the product to predict the landed costs.* For consumer electronics, margins are 30–60%, and for medical products are 80–200%. Some of this information can come from company financial reporting and industry reports.
  - *Estimate the likely COGS assuming their distribution method.* For larger companies, you can assume the product is probably sent by sea to reduce costs.
- **Estimate your adjusted material costs** by removing the overhead and labor costs and scale for volume.
  - *Estimate the product's likely scrap rate*, factory overhead, and labor costs and remove them from the above COGS estimate to get the BOM material costs. Larger companies are usually going to have lower overhead rates than smaller companies.
  - *Estimate the likely impact of scale.* Scale the cost by the volume and supplier leverage discount likely achieved if the product is produced by a larger firm (e.g., Apple and Amazon can order parts in much larger MOQs than smaller companies can). Scale can impact costs by between 5% and 20%.

- **Allocate the adjusted material costs** to the major components in the proxy product. You can get estimates of low-volume costs for the major components and then allocate the rest to the structural, mechanical, and other parts. By also doing bottom-up estimation, teams can get a sense of the relative costs of each of the major parts.
- **Identify the differences between the proxy and your product**, then add these differences to the material costs. For example, you can ask:
  - *How many additional PCBA or flex circuits?* Each PCBA will add $2–$5 depending on the cost and the types of connectors.
  - *What additional high-cost electronic components are being added* (e.g., Wi-Fi or Bluetooth)?
  - *Do you have the same set of expensive components?* Adjust the cost of expensive components (e.g., LCD or cameras) up or down based on the differences in capacity, accuracy, and quality.
  - *How many more fabricated parts do you have?* You can use cost estimation software to estimate the relative difference between your product and theirs.
  - *Are you using a different assembly method?* For example, if their product uses screws and your product uses snaps, take out the cost of the screws.
  - *Are you using a higher or lower grade of materials?* Adjust by the percentage increase or decrease in material costs.
  - *Do you have a higher or lower finish quality?* Adjust for the cost of additional paint or more expensive tool finishes.
- **Once you have your estimates for the material costs, rebuild the overall cost** by adding your markups, overhead, labor charges, and shipping.

8.4 Revenue and Order Fulfillment

The third question, *"When is revenue coming in to balance the outflows of cash?"* is answered by estimating the revenue streams from each customer order fulfillment model.

Companies can generate revenue in several ways: they can get paid for the product they produce, they can charge for software, or they can use a razor/razorblade model (e.g., Keurig coffee-maker/coffee pods). Each type of revenue model has different timelines and margins, and each can have a dramatic impact on cash flow. Typical revenue models in consumer products include:

- **Single, one and done**. In this case, the customer purchases a product, and that is the end of the revenue stream from that customer. A simple example of this is buying a notebook at an office supply store. The customer has options for sets of colors, sizes, and features, but the purchase of that notebook is a one-time event and is not linked to other purchases.

In this case, getting the price and cost aligned is critical because the single sale is the only chance to generate revenue.

- **Product family model**. In this case, the purchase of a product will encourage the purchase of a second or related product. For example, with a wireless speaker system, the consumer can easily add additional speakers with minimal setup, and the whole system can be driven off of a single app. Other companies' speakers won't be compatible, so customers are incentivized to buy a second product from your company.

- **Data collection model**. Software-based businesses have been using data collection as a revenue model for a long time. Facebook is "free" to its user, but makes money by selling user data to advertisers. At the time this book was written, there was significant controversy about data collection in consumer hardware products (Inset 8.3).

- **A device as a conduit for a service**. The Ring doorbell company sells its smart doorbell for several hundred dollars, but then sells the accompanying cloud support for a monthly fee. The product is probably profitable in itself, but the cloud services provide an ongoing income stream.

- **Razor/razorblade model**. In this standard revenue model, the primary device is sold close to cost, and the company makes profits on the consumables that customers must purchase to use the device – for example, ink for inkjet printers, and disposable Keurig K-cups for Keurig's coffee machines. This model can be very useful as long as the company can maintain a captive market for its proprietary product. Keurig struggled with maintaining margin on the razor part of their model (K-cup) when the product went off-patent.

- **Service contract model**. Some products are sold for close to cost, but then the service contract ensures a regular revenue stream. Medical equipment often follows this pattern. The customer is interested in a low upfront cost and ensuring that the product is always running. They are willing to pay a higher recurring cost for this service. In this case, the company may want to spend more during development to optimize design for ease of maintenance and for reliability.

Inset 8.3: Cautionary Tale about Selling User Data

We all like "free" services. How many people use Google's free email services? However, nothing is free. Many of these companies sell your data as their business model. While many companies successfully create products where customers are not upset by their data being sold, some companies have created negative press for themselves. In 2017, iRobot, the robotic vacuum company, announced its intention to sell mapping data from its vacuums to third parties and got considerable pushback from the public who weren't happy about people mapping the insides of their houses [82]. Strava, a training tracking company, inadvertently identified military bases through heat maps of typical run patterns of military personnel who were using their products while overseas, thus exposing a potential vulnerability for national security [83]. Ring doorbell is running into surveillance concerns as it partners with police forces to supply video taken by their cameras [84].

Depending on the way the product is purchased by the customer, the timing of the cash flow may be very different. When planning cash flow, teams need to understand the timing of when they will get paid. In some cases, companies get the revenue ahead of the sale, while in others (service contract or razor/razorblade), they will only receive revenue much later. The following is a list of different sales models and how they impact cash flow:

- **Pre-orders**. For start-ups, pre-orders can provide a necessary influx of cash to allow them to order materials and begin production. Pre-orders can be taken through a crowdsourcing platform or through company platforms. This revenue can be used to pre-fund the NRE and production costs. For example, Tesla used the revenue-ahead-of-sale model. They used the customer's deposit on their lower-end model to fund the build-out of the factory. They could do this because customers were willing to wait months, if not years, to get their cars.
- **Company-owned websites**. Selling directly to consumers captures a more significant margin on the product because they don't have to share the profits with another channel (e.g., Amazon). However, the company has to have enough inventory on hand to fulfill orders quickly.
- **Online (Amazon) or big-box retailer (Best Buy)**. A retailer will either purchase your inventory or take it on consignment and sell the product through its platform (fees range widely but average around 13%). Depending on the lead time demanded, you may be able to build to their orders and ship directly from your factory, or you may need to fulfill the orders from inventory already in your warehouse. Depending on the contract, you may get paid before delivery or may have terms that pay you 30, 60 or 90 days after delivery.
- **B2B with large companies**. If you are in a business-to-business (B2B) model, your customer may take as many as 120 days (yes...almost four months) to pay you if you are supplying some larger companies [85]. Large businesses use this model to improve their cash flow, and you have little negotiating power because they are so large.

8.5 Cash Flow

The last question, "*What do you owe to suppliers and your factory, when do you owe it, and do you have enough cash on hand to make it through the process?*" is answered by predicting the cash flow during product realization and into mass production. Companies often underestimate the cash required for development costs, tooling, and inventory. Also, they tend to overestimate how quickly they can get through the piloting phases. As a result, they do not see profit as soon as they expect to. Many start-ups and small companies fail because just before they are set to start ramping production, they run out of money and cannot pay for their purchase orders. Unfortunately, investors can often sense desperation and might refuse to invest more cash, or do so only on terms that border on extortion.

Most start-ups operate on a cash flow knife-edge. They are continually balancing spending too much and spending too little. If you spend more than you have on hand, you either have to beg for extensions on payments or cannot continue as a business. In the case of start-ups, it can be hard to raise money when you are desperate for cash. On the other hand, having excess cash is also risky. If you hold money that you traded for equity without using it to advance the product, you aren't getting any return on it. Investors want to know their capital is being deployed (but not squandered).

The best finance managers check cash flow plans many times a week, if not daily, to ensure that the company is leveraging all its resources without going into the red. To operate on the knife-edge, you need to have an accurate cash flow model that contains all cash influxes and outflows. You must also understand how every decision made might impact cash flow, and you should continuously update the model with any new information.

Cash flow analyses can predict the net amount of cash you have on hand at any moment. No matter how substantial the ROI is projected to be on a given product, if you do not have the money to pay your suppliers, you cannot build the product. Teams need to plan when money will come in and go out to safeguard against overextending. Understanding the cash flow requirement is also critical to fundraising the right amount.

8.5.1 Tradeoffs in Cash Flow

In many cases, organizations must choose whether to pay more upfront to save on downstream costs, but using up precious cash. For example, teams will need to decide between:

- **MOQ and cost**. The higher the order quantity, the lower the cost, but the team will trade off between tying up cash in inventory and downstream savings.
- **Tooling costs vs. part costs**. Less expensive tooling (fewer cavities, soft tooling) is cheaper in the short term but will drive a higher overall per-part cost because fewer parts are made per mold cycle, and you need to replace the tooling more often.
- **Tooling payments and COGS**. Tooling can be paid for upfront, or you can pay a higher per-part cost, and your supplier will own the tool and will amortize the tooling costs into the part costs.
- **PO terms vs. COGS**. Some CMs will give better purchase order terms (lower costs upfront) but will charge higher margins.
- **Factoring**. You can take out the equivalent of a pay-day loan for invoices that your customers owe you from a number of credit institutions.

Instead of having to wait for payment, which as mentioned can be as long as 120 days for B2B, credit institutions will pay you the amount your customers owe you and take a small percentage of the invoice when it is paid.

- **Sea shipment vs. lower inventory**. Shipping by sea can reduce costs of transportation, but requires six weeks of inventory to be in transit at all times. This inventory can significantly impact cash flow because you have to pay for the inventory, but you don't get paid for it for several months.

8.5.2 Building a Cash Flow Model

The cash flow model documents the timing of the payments and how those payments are linked to both the sales and production forecasts (Chapter 15). To model the cash flow, the project manager will need information from multiple functional groups within the product realization process:

- **NRE costs**. The timing of each payment needs to be documented; and if changes occur, they need to be reflected in the cash flow model. For example, the model depends on when the tools are built (Chapter 12), the scope and timing of the pilot phases (Chapter 3), and the scope of the quality test plan (Chapters 7 and 13).
- **Recurring costs**. The timing and amounts of the costs will be highly dependent on the pilot schedule, production schedule, the lead time on parts, and the payment terms you have agreed to with your suppliers. A couple of things to keep in mind about the timing of when you owe money for each item are:
  - *Material authorization (MA) for long-lead items*. Sometimes there are parts required for PO that take longer to order than the PO lead time. For example, you might have an 8-week PO lead time but a critical part may have a 24-week lead. Because the factory doesn't have enough time to buy the materials when the PO is placed, they may need to pre-order those parts. You have to pay in full for those parts using a process called a material authorization.
  - *Initial PO payments*. When you place a PO, you need to pay then for a percentage of the total bill.
  - *Final payment*. The final payment to the factory for the order is made around the time of delivery.
- **Revenue.** You need to predict how much revenue is coming in when. Each sales channel will have a different forecast and, depending on your terms with each channel, different timing. For example, you may sell through an online marketplace as well as direct to customers. Each of these will have different forecasts and timing as well as margin.
- **Warranty**. The estimated warranty costs are added to the warranty reserve account at the time the revenue is recognized (Chapter 18). Unless subsequent warranty costs exceed what was predicted, they are paid out of the accrual account.

To create a cash flow model, you need to identify all of the expenditures and the revenues as well as their timing and relation to the sales forecasts and production schedule.

The following is a simple example of a cash flow model (shown graphically in the top graph in Figure 8.5) for a simple consumer goods product, assuming the following:

- $25 landed cost of goods per unit.
- 75% due to the CM at PO and 25% due on delivery.
- A $30,000 monthly office expense cost which increases to $45,000 at PVT.
- Sale price of $40 each.
- 8-week lead time on PO.
- A 4-week delay between product shipment (free on board or FOB) and delivery to the customer.
- 30-day terms from the customer (i.e., they pay you 30 days after delivery).
- two tooling purchases of $50,000 each: 16 weeks and 8 weeks before the first run.
- $25,000 per pilot for three pilots.

The chart shows, based on the cumulative cash flow calculation, that the team needs to have close to $1 million in the bank to get through the process. The product will break even at 18 months. However, relatively small changes – 40% due to the CM at PO and decreasing the COGS to $22 – can decrease the cash needed by $200,000 and move up break-even by four months (bottom graph in Figure 8.5).

8.5.3 How to Maintain a Cash Flow Model

There are several software tools that can help with building a cash flow analysis. Many of these are designed to help you manage existing cash flows and known POs, rather than looking out 18–24 months into the future. Other SaaS platforms have built-in intelligence specific to launching hardware start-ups and can help identify costs and cash flow impacts of which a novice might not be aware. Many organizations use custom-built Excel spreadsheets that are difficult to maintain and are often not integrated with other financial tools such as forecasting models and financial management systems. However, the Excel sheets allow a significant amount of control and customization of the model.

Building and maintaining a cash flow model can almost be a full-time job. The model should be set up to enable teams to easily update the model and see the impact on the cash requirements. Creating and updating cash-flow models on a regular basis can help companies navigate the tradeoffs between cash flow and product costs. Besides, it can help identify, early, when cash-flow needs may exceed the available cash from investors.

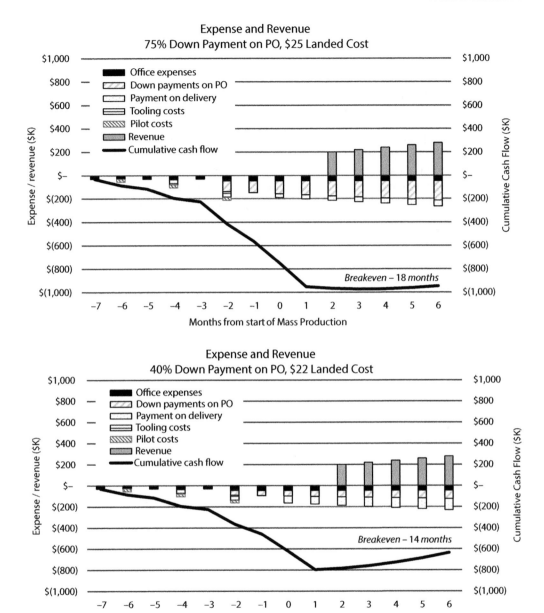

FIGURE 8.5 Two different cash flow models

Summary and Key Takeaways

- ❏ Cost estimation, early and often, is critical to ensuring a profitable product and having sufficient cash to get the product to market.
- ❏ Cost estimation is time-consuming and requires teams to use a variety of tools and methods.
- ❏ Continually updating and maintaining the estimates (along with assumptions) ensures that teams are not surprised by late changes in cost estimations.
- ❏ Teams should use more than one method to estimate costs (build-up and analogous) to check assumptions and triangulate the final cost of the product.
- ❏ You need to create, manage, and maintain your cash flow model throughout the life of the project.

Chapter 9
Manufacturing Systems

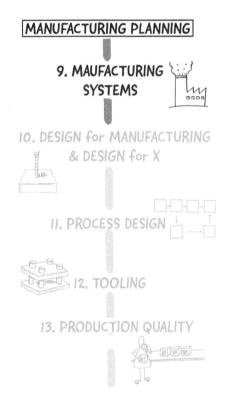

Manufacturing is more than just getting parts built and assembling them into a product. Significant work is required to design the supply chain, optimize the facility, plan the production, manage the materials, and ensure quality. This chapter introduces some of the basic concepts in designing a manufacturing system and sets the stage for the following chapters on manufacturing and supply chains.

Getting a product to market involves both designing the product as well as designing the production system to build the product. Even if all of the manufacturing processes are outsourced, it is critical for the core team to understand how the product will be made. For those readers whose companies own their own facilities, you will have more control over the design of the production system. For those using a CM, you have less control, but you will often have a strong hand in the design of the layout and manufacturing steps, as well as driving improvements to reduce cost. You will also need to review and audit the facility to ensure that their production system meets your needs.

Understanding how the production systems will operate enables teams to design their parts and products so they can be more easily made with quality and reliability, while keeping costs down. For example, the Samsung Galaxy Note 7 battery fire recall ultimately cost Samsung over $5.3 billion. After an extensive effort (which included building a large-scale automated testing facility), it was determined that the failures were caused by manufacturing defects [86–88]. However, the design of the phone battery created an opportunity for the failures to occur: insufficient space to robustly assemble the product and insufficient materials to both insulate and assemble the electrodes.

Designing the production system includes answering:

- How parts will be made and assembled (Chapter 10)?
- Is the product ready for manufacturing (Chapter 11)?
- What tooling and equipment is needed to support production (Chapter 12)?
- How quality will be ensured during production (Chapter 13)?
- Who will make your parts and provide any other services needed (Chapter 14)?
- How material will be ordered and how much finished goods inventories will be maintained and when (Chapter 15)?
- How products will be transported to the customer (Chapter 16) and how the product will be returned and repaired if there are problems (Chapter 18)?

To begin answering the above questions, teams must have a working knowledge of how manufacturing facilities are organized, what types of manufacturing facilities exist, and how product and process design decisions can impact the flexibility and efficiency of the manufacturing facility. This chapter provides an overview of different types of production systems and how facilities are organized. It will also give a basic introduction to Lean principles as they relate to manufacturing. In addition, Inset 9.1 gives a quick tutorial on the metrics used to measure the efficiency of facilities.

Inset 9.1: Manufacturing Terms

Cycle time. Cycle time is defined as the total time from the beginning of a given manufacturing process to the end. It is essential that you know the cycle times of your production facility because they will determine both your production rate and lead time.

The definitions of the "start" and "end" of a given process will depend on how the process itself is defined. For example, some people talk about the product's cycle time as the time between the factory receiving the order and materials and the time when the process/assembly/product is ready for delivery. Cycle time can also be used to describe the time for a single subprocess to be completed; e.g., the cycle time of an injection-molding machine may be about 20–30 seconds. Also, as with many aspects of the product realization process, the actual cycle time will usually vary from the planned cycle time. Finally, teams often confuse cycle time and labor hours.

Takt time. "Takt" is the German word for a conductor's baton. It is the rate or timing for how frequently a product order needs to be released to the factory (triggering the factory to start making a product) to satisfy the customer demand. It is the drumbeat of the factory. Takt time is important because it defines the amount of capital equipment needed and if the production line can meet demand.

Lead time. This is the time between getting an order and getting a product out. It may include the time to purchase or fabricate all of the materials necessary. It can include the delays due to high utilization rates (e.g., waiting for the factory to have the capacity to fulfill an order). Lead time determines how much inventory is in the system at a given time. The longer lead time, the more work in progress (WIP).

Yield. A certain percentage of units will fail, require rework, or need to be scrapped at each station or process. The percent that passes is the yield. For a series of processes, the first pass yield is typically calculated by multiplying the yields of each station in a process. This overall number gives a measure of the factory's ability to get a single unit through without any defects at any station. Rework is not included in this measure.

WIP (work in progress) is the amount of material in the factory at a given time. The longer the wait times in inventory buffers, the longer the processing time; and the higher the number of processes, the more material will be in process at any given time. Reducing WIP reduces inventory costs. In addition, it is hard to keep track of inventory in a system, and the less inventory in the system, the easier it is to manage it.

Bottleneck. The "bottleneck" in any process is the step whose effective rate in parts per hour is the smallest. The line cannot move faster than the bottleneck process. Adding capacity in other locations will have no impact on the production rate. Understanding where your bottlenecks are will help to focus resources and capital on the issues that will maximize production speed.

Downtime. A process will never be "up" 100% of the time. Glitches in the production process, variabilities in material, and machine failures can cause certain steps or even the entire process to shut down for certain periods of time. The downtime rate is the calculation of what percentage of time the machine is not available for use.

9.1 Production System Types

If you ask 100 different engineers to describe "production facilities" you will probably get 100 different descriptions. Some are full of loud and massive equipment such as a forging plant. Some are quiet, small, and kept in clean room spaces requiring the workers and visitors to suit up in a "bunny-suit" with a respirator. Some factories are dedicated to a single product and produce the same thing over and over, as in an automotive factory, while others flexibly build a variety of custom products in very small quantities. Which one is right for your product will vary depending on production volumes, production variability, and product mix.

Despite all of these differences, any person who has visited more than one factory will begin to see some commonalities. For example, every factory will have a place where incoming material is processed and an area where product quality is evaluated. The dominant difference between factories is whether a factory just produces a single product (dedicated production), or can flexibly produce a huge range of products (job shop), or falls somewhere in between (flexible production).

A single facility may include both dedicated and flexible production. For example, an automotive line might have flexible production for sheet metal stamping. Stamping equipment is changed over to produce different parts of the automotive body (it might stamp the hoods in the morning and the trunk in the afternoon) and send the parts to a dedicated production line for assembly. Terminology for and descriptions of the range of production facilities are as follows:

- **Job shop** is the term for a facility that makes small production batches using multifunctional manufacturing equipment. In a job shop, the sequence of steps and equipment used will typically vary significantly from product to product. For this reason, job shops are often used to make things you only need one of, such as prototypes or fixtures. They may contain a mix of machining

JOB SHOP

equipment, cutting technologies (laser cutters and turret presses), forming equipment (such as vacuum forming), and flexible assembly stations.

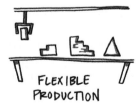

- **Flexible production** is the term for facilities that can produce a range of similar products in medium quantities. Typically they have dedicated equipment that supports the products they specialize in. For example, some CMs specialize in consumer electronics and will have their own injection molding and printed circuit board

FLEXIBLE PRODUCTION

assembly (PCBA) facilities. Others may focus on electronic medical equipment and will have certified quality testing facilities. Others may focus on single manufacturing processes such as electroplating or forging. Products are typically built in batches with the equipment set up with custom tooling and then reconfigured for the next product. For example, a flexible assembly facility will be able to place a number of manual workstations in different configurations to accommodate short or long assembly sequences. The flexible stations have places for electronic equipment and materials and typically have electrostatic discharge (ESD) protection to prevent damage to sensitive electronics.

- **Dedicated production** is the term for a facility that only produces a single product or family of products (e.g., cars). The equipment is fixed and does not change. For example, an automotive line is a dedicated production system. Each line can only produce a single car (or a range of related cars). All of the material handling equipment is fixed in place and can only be reconfigured with significant work.

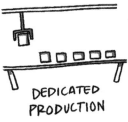

Table 9.1 lists the characteristics of each of these types of production systems and gives examples.

9.2 Dedicated Manufacturing Facilities

A single factory or company may have many manufacturing centers under one roof or may specialize in just one type of production. Here are several common types of centers which are described in more detail in the following sections:

- **Fabrication facilities** are dedicated to a single manufacturing process to make fabricated parts. For example, a facility may specialize in forging.
- **Secondary material and finishing facilities** specialize in coatings and heat treatments such as anodizing.
- **PCBA facilities** specialize in assembling and testing printed circuit board assemblies (PCBAs). These may range from low-tech through-hole assembly (in which wire leads are fed through metalized holes in the PCBA and soldered on the back) to highly sophisticated SMT (surface mount technology) processes.
- **Assembly facilities** specialize in assembling the parts into a working product. Some facilities are highly manual, while others depend heavily on automation.
- **Pick-and-pack facilities** are used to assemble finished goods into the final packaging. Typically, they also do the order fulfillment either direct to consumers or to other distribution centers.

Each of these are described in detail below.

Table 9.1
Types of production systems

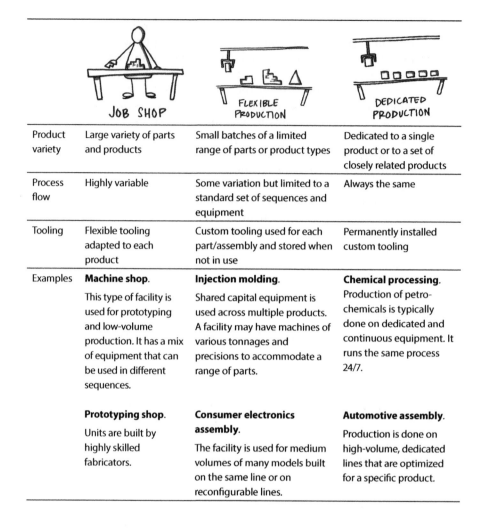

	JOB SHOP	FLEXIBLE PRODUCTION	DEDICATED PRODUCTION
Product variety	Large variety of parts and products	Small batches of a limited range of parts or product types	Dedicated to a single product or to a set of closely related products
Process flow	Highly variable	Some variation but limited to a standard set of sequences and equipment	Always the same
Tooling	Flexible tooling adapted to each product	Custom tooling used for each part/assembly and stored when not in use	Permanently installed custom tooling
Examples	**Machine shop.** This type of facility is used for prototyping and low-volume production. It has a mix of equipment that can be used in different sequences.	**Injection molding.** Shared capital equipment is used across multiple products. A facility may have machines of various tonnages and precisions to accommodate a range of parts.	**Chemical processing.** Production of petro-chemicals is typically done on dedicated and continuous equipment. It runs the same process 24/7.
	Prototyping shop. Units are built by highly skilled fabricators.	**Consumer electronics assembly.** The facility is used for medium volumes of many models built on the same line or on reconfigurable lines.	**Automotive assembly.** Production is done on high-volume, dedicated lines that are optimized for a specific product.

9.2.1 Fabrication Facilities

Fabrication facilities produce parts unique to a single product using specialized equipment. Examples of fabrication facilities include injection molding, casting, forging, machining, and blow-molding. Not all fabrication facilities are the same. Some facilities will produce materials at a lower quality and a lower cost than others. Others with more precise equipment and more in-depth expertise will be able to provide more accurate and complex parts but at an additional cost. All fabrication facilities typically have the following in common:

- **Specialization** in a single or interrelated set of manufacturing processes. For example, a supplier may have deep expertise in forgings but may also have machining capabilities to provide customers with a completely finished product.

- **Very capital equipment intensive** necessitating that the facility run at full capacity to cover the amortization of the equipment. In addition, these facilities will often have a range of machines for the same process to accommodate a variety of part sizes and product quality needs.
- **Deep expertise** in specific manufacturing processes, including the ability to evaluate design for manufacturability.
- **Tool-making capabilities**, or at least close relationships with tool manufacturing suppliers.
- **Production in batches** of varying sizes for a large number of customers or a large variety of products.
- **Close partnerships** with assembly or contract manufacturing companies and other fabrication facilities to provide "one-stop-shopping."

9.2.2 Secondary Finish or Material Treatment Facilities

Finish and material treatment facilities perform secondary processes such as electroplating, heat-treating, anodizing, and painting. Secondary processes are used to either improve the mechanical performance or add aesthetic details, or both. For example, anodizing is used to create an oxide – and often colored – layer on aluminum parts. The material is immersed in a number of different baths with different chemical and electrochemical steps to build up the oxide layer, color, and seal it. These facilities are typically:

- **Very capital equipment intensive** because the equipment for most secondary processes tends to be expensive and large. In addition, these types of equipment require significant maintenance and process knowledge to keep them functioning efficiently.
- **Large-batch processers** because the equipment is designed to coat or treat many parts simultaneously. It isn't efficient to run the process for just one part.
- **Faced with significant environmental regulations** because the chemicals they use are often toxic and can't be allowed to contaminate the water supply. Their processes are also typically energy intensive.
- **Expert in the properties of materials** which enables them to accurately control and assess their processes and the outcomes. For example, heat-treatment facilities will have testing equipment that enables them to evaluate the microstructure of the treated material and the ultimate material properties created.
- **Part of a complicated supply chain** because large batches of materials need to be shipped off to the facilities after fabrication and then shipped back to the assembly facilities.

9.2.3 Electronics Assembly Facilities

Electronics assembly facilities specialize in the manufacture of PCBAs and other parts related to the electronic components of a product. Many electronics assembly facilities also have engineers on staff who can help with PCBA design and firmware development, making the facility a more of a one-stop shop.

Electronic assembly facilities typically have a number of work areas. The main part of the facility usually houses the SMT lines (surface mount technology lines) (Figure 9.1). This automated equipment applies the solder paste to the raw PCB boards using a stencil, robotically places the individual components on the board, and then permanently attaches the components to the PCB through a process called reflow soldering. These facilities also will have areas which inspect and test the boards, add other components using dedicated processes such as wire bonding, and manual assembly stations where through-hole components such as connectors are manually soldered to the board. They also may provide processes such as potting and conformal coating, which reduce a board's sensitivity to environmental stresses. They share the following characteristics:

- **Printed circuit boards (PCBs)** are usually sourced from another supplier, typically a close partner of the PCBA facility.
- **Components (such as integrated circuits, memory modules, and resistors) are sourced from distributors**. Generally, the facility will work with a select set of suppliers to get preferential discounts on these components and to streamline the ordering processes.
- **The facility will typically assemble most SMT components using specialized pick-and-place machinery**. They will also have secondary operations such as wire-bonding, inspection, and testing areas. Almost all electronic products today use SMT technology; that said, it is not uncommon for boards to have a few through-hole components (such as connectors), and these are generally soldered by hand. SMT and other equipment can vary between low-cost older Chinese equipment to newer Japanese equipment that can produce a more densely populated and higher-quality board.
- **PCBAs are produced in batches to reduce setup costs**. Setting up the reels of components and the stencils is very time-consuming, and that cost has to be amortized over a large batch.

FIGURE 9.1
SMT line.
Source: Reproduced with permission from Shenzhen Kaifa Technology Ltd

- **Assess the quality of the board** using a variety of techniques including AOI (automated optical inspection), X-ray inspection stations, and ICT (integrated circuit testing) to ensure that the boards are manufactured correctly and have no defects (Section 13.3.3).

9.2.4 Assembly Centers

Assembly facilities assemble, test, and package the product. The facilities can range from highly automated to fully manual. Most assembly facilities share the following characteristics:

- Facilities produce a **variety of products** (often for various customers). Assembly lines for different customers are segregated to protect IP.
- Manual assembly lines are typically easily **reconfigurable** so the line can be changed over quickly to a different product. The line may also have automated equipment to assemble and move product.
- Product assembly is broken down into several **stations**, each of which has a workspace, an area for materials, a set of standard operating procedures (SOPs), and quality control checks and tests.
- **Labor** can range from low skill to high skill depending on the processes. For example, if the product includes a delicate soldering process to attach an antenna, higher skilled workers will be employed. Facilities **train their workers** to ensure they have the right skills to execute the assembly and test processes. For example, some workers will be qualified to solder, and others only to pack boxes.
- The line includes **in-line testing** as well as end-of-the-line functional tests.
- The facilities have large spaces dedicated to **incoming material** from suppliers, and significant effort is spent tracking orders to ensure that all of the parts are available on time to meet customer orders.
- The facility includes **rework lines** where defective product can be repaired. Defective product can come from the line or from customer returns.
- Some facilities may have **dedicated fabrication facilities** attached to them (e.g., injection molding or PCBA).

9.2.5 Pick-and-Pack and Order Fulfillment

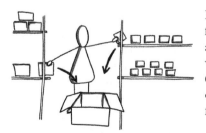

Pick-and-pack facilities and order fulfillment facilities take finished goods and repackage or co-package them with other goods to ship to a customer (who may be a consumer, distributor, or retailer). Pick-and-pack and order fulfillment facilities:

- **Are close to the final customer** and support the "last-mile" delivery.
- **Utilize low-skilled workers** who only need to be trained on relatively simple manual tasks.

- **Actively manage inventory** to ensure accountability for what materials are arriving and what is being shipped.
- **Have no dedicated equipment** other than material handling to move large and heavy packages and pallets.
- **Provide order fulfillment** services and ensure that the right products get to the right customers. In addition, many facilities will manage incoming orders and provide the production schedule to the factory to meet those orders.
- **Do simple product testing**, upgrading firmware and software, or other final product checks which may be among their services.
- **Customize products** and SKUs for different distribution channels. For example, the facility may package products in gift boxes specifically designed for one distribution channel.

9.2.6 Other

The types of facilities described above cover a majority of those used in consumer goods, but are far from representative of the entire range of manufacturing facility types. Other types can include:

- Continuous processing for chemicals.
- Fabrics and textiles.
- Food production.
- Biotech and pharma.
- Technology suppliers such as battery manufacturers.

9.3 Areas in a Manufacturing Facility

Each manufacturing facility will have unique processes depending on the company, technology, and products. For example, facilities that make carbon fiber composite parts will have large refrigeration systems to keep the pre-preg tape and fabric (carbon fiber with resin) cold before production, and medical device assemblies may have a dedicated cleanroom. However, most facilities share the following common functional areas (Figure 9.2).

- **Receiving**. Here is where materials are received from outside vendors. The incoming orders are logged, unboxed, labeled, and sent to incoming quality control (IQC). Often receiving uses bar scanners to track incoming materials. However, some companies require RFID (radio frequency identification) tags to reduce paperwork and errors as well as reduce the time between material coming in the door and being sent to production. It doesn't matter if you express-ship materials to the factory if they sit in receiving for three days waiting to be processed. In addition, some materials may require temperature and humidity control. Unloading docks are notoriously hot and humid in the summer, and product can be damaged

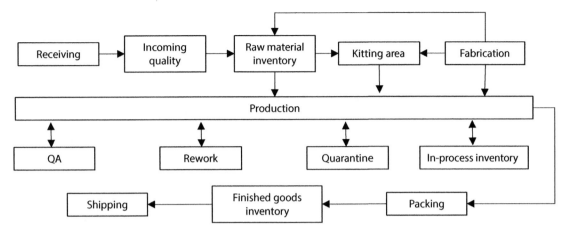

FIGURE 9.2 Typical factory areas

sitting on the docks. Specific instructions are needed for handling any perishable or temperature-sensitive materials.

- **Incoming quality control**. Some or all incoming materials will be inspected based on instructions from the engineering team. Incoming quality can become a bottleneck in some facilities. Most products will just be checked to see if the materials match the purchase order, but certain materials (e.g., those whose performance is safety-critical) may require specialized testing before use (Section 13.3.1).

- **Raw material inventory**. This is a location where materials are stored prior to use in fabrication and assembly. Depending on the sensitivity of the material, this storage may be in an environmentally controlled area. It may be also be set up to segregate or secure high-value materials (precious metals) or material with IP content such as custom firmware that may leak to the gray and black market (Section 17.2.4). Finally, it is easy to lose material in large storage areas; accurate and efficient tracing methods are necessary to make sure the materials can get to the factory floor when they are needed

- **Kitting**. Some assembly or fabrication lines will kit materials (bundle or group them) together from the raw material inventory so all materials can be shipped together as a group to each station on the factory floor. There is typically an area between raw material storage and the assembly line where the kits are assembled and stored for use. Kitting ensures that all of the correct materials are delivered to production in a way that streamlines material handling. It supports both Lean production and the use of Kanban (pull-based systems) to drive production planning.

- **Fabrication**. Some of the parts and assemblies will be produced outside of the main production line. These fabrication areas work like mini-factories that build parts or subassemblies that are later used in production. Typically, fabrication centers produce parts in batches and can make a wide range of parts using a single process (e.g., an injection molding center).

- **Assembly and production line**. The outcome of the production line, naturally, is the finished product. The line can include manufacturing processes (e.g., ultrasonic welding), material handling equipment (e.g., conveyors), assembly stations (manual or robotic), prep stations (e.g., cleaning), and testing stations (e.g., camera quality checks), among other processes. Significant literature exists on how to lay out the facility, how to assign work to stations, and how to optimize material flow between stations to reduce cycle time and WIP.
- **In-process inventory**. Depending on the process, materials may need to be stored waiting for the next step (typically called inventory buffers). If there isn't sufficient room on the line, the inventory may be moved to another location until the line is ready for it.
- **Packaging**. A separate area may be dedicated to final product inspection and packing. The package design – e.g., how complicated the boxing, and how easily stackable the master cartons – will significantly impact the layout of the packaging area.
- **Finished goods inventory**. Before shipping, finished goods might be stored in inventory and orders fulfilled out of this area (Figure 9.3). Companies that can accurately forecast their product sales and have level demand will typically have smaller finished goods inventory.
- **Shipping**. This is where the purchase orders are packaged and shipped. Depending on the distribution plan, orders may go to a separate distribution site or orders may be fulfilled directly from the factory.

In addition to the standard sequence of areas that most products pass through, products that do not meet quality requirements may pass through other parts of the facility, including:

FIGURE 9.3
Finished goods
inventory
Source: Reproduced
with permission
from SRAM

- **Rework area**. Within the facility a specific area is typically dedicated to reworking faulty products. This area will contain tools for disassembly, testing equipment, and spare parts. This area, along with quality assurance, does root cause analysis of the failures.
- **Quarantine area**. For products or material that have failed QA and cannot be sold or used in production, a separate quarantine area is typically demarcated (sometimes in a locked area). The RMA (return material authorization) process will return the defective materials to suppliers for refunds. Labeling and tracking of materials in the quarantine areas is critical to ensure that the non-conforming product or material does not get sold or reintroduced into the facility. The quality management system and testing documents will determine when a product should be shunted to quarantine and how the non-conforming product should be managed.
- **QA (quality assurance) labs**. The QA labs are typically separate rooms where product samples are tested (often destructively). These labs can include equipment for environmental, chemical, and material testing. The QA lab executes some of the tests defined in the quality test plan that require special expertise and equipment that can't be placed on the production line (Chapter 13).

9.4 Lean Principles

"Lean manufacturing" refers to a set of guiding principles derived from the Toyota production system [89], to improve quality, responsiveness, and cost-effectiveness in manufacturing. Lean practices reduce the waste in a system by identifying what is valuable to the customer and removing all other activities that don't support generating customer value. Lean manufacturing focuses on reducing waste which can include scrap, time spent hunting for tools, rework, inventory, and excess time in assembly. Second, Lean works to increase the responsiveness of the system, enabling it to meet customer needs more quickly. Third, it works to improve quality, which helps with both waste and responsiveness while improving customer satisfaction.

All of the methods and tools described in this book are consistent with Lean principles. For example applying DFM, DFA (design for assembly), and quality management practices early in design will help to reduce waste by ensuring easy assembly, reducing part count, and reducing scrap. In addition, having complete and up-to-date documentation is the foundation of continual improvement. Finally, a complete quality test plan ensures that the product is produced to the right standard. The following sections describe some of the more well-known Lean principles and how they relate to product realization practices. These guiding principles are highly interrelated and all work to drive down cost and waste. While each Lean guidebook has a slightly different list, they all emphasize the following six principles as critical. These are discussed in detail in sections 9.4.1 through 9.4.6.

- Minimizing inventory (Section 9.4.1).
- Reducing lead time (Section 9.4.2).

- Improving and leveling production (Section 9.4.3).
- Kaizen (Section 9.4.4).
- Quality (Section 9.4.5).
- Respecting people (Section 9.4.6).

9.4.1 Minimizing Inventory

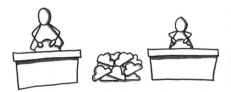

Excess inventory, whether in incoming materials, WIP, or finished goods, increases cost throughout the system. You must buy inventory which ties up capital and you must keep track of it. While it is on the floor, it can get damaged, it gets in the way, it can become obsolete, and there is a cost associated with the space it occupies. Ideally, the factory should work on a one-piece flow approach where no inventory builds up before, between, and after stations. One-piece flow (moving away from batch production) helps to smooth production rates as well as reduce inventory. However, you cannot always predict the exact demand for your product or whether demand will stay even. You also need to plan for potential disruptions in manufacturing cycle time or for quality variability. For these reasons, it will be necessary to keep some extra products, in various states of assembly, on hand. Strategies in product and process design aimed at minimizing inventory throughout the product realization process can include:

- **Choosing parts and subassemblies with shorter lead times**. The longer the lead time, the more inventory the facility will need to carry. Because long lead items are often purchased far in advance of the customer orders, it is often necessary to order extra materials (before knowing about customer demand) to hedge against order increases.
- **Selecting parts with lower MOQs**. Lower minimum order quantities (MOQs) enable more frequent orders to be placed closer to the production date, thus reducing the inventory that the manufacturing facility will need to hold.
- **Part commonality**. Reducing part count can help reduce the inventory carrying requirements because it reduces the effective MOQ and the impact of demand variability. As an example, if you can change a design to need one type of screw instead of four types, you can reduce the number you need to order. If the MOQ of a screw is 100, rather than buying 100 each of four different screws (total of 400 or 100 units worth of inventory), you only need to buy one type (total of 100 parts that will make 25 units).
- **Balancing the line through using DFA** will reduce WIP. Because it is not possible to precisely equalize the production rates of all of the steps, there will almost always be WIP between some of the steps in the production line. Reducing the number of stations (by reducing the labor required

to assemble it) and the variability in the cycle times (caused by difficult assembly steps) will reduce the need for WIP.

- **Late product differentiation**. If the product has multiple SKUs, the later the SKUs are differentiated in the production line, the better. For example, products that come in several colors can have the differently colored parts added at the start of the assembly sequence or at the end. If the latter, a common core assembly can be built in the same way for all of the SKUs. The undifferentiated subassembly can be held in semi-finished goods inventory and only differentiated when the orders need to be fulfilled. For example, if there are ten colors offered, the colored parts can be assembled when the order comes in. Otherwise, you have to keep finished goods inventory of all ten colors, some of which may never be purchased.
- **Mixed-model production**. If a variety of SKUs are being produced, the ability to change from one version to the next without changing the layout of workstations allows for flexibility in the assembly system and smaller batch sizes. For example, a company may provide several versions of the same basic coffee-maker. If the assembly sequence is the same and if there is sharing of common subsystems, all versions can be made on the same line by varying the materials and by adding different parts.
- **Multiple versions shipped as one SKU**. For example, an electronic product can be shipped to both the US and Europe using the same SKU by including a universal charger. While this marginally increases the cost of the unit, it means that teams only have to contend with the total worldwide demand forecast rather than having to predict how many go to each country. Also, if specific demand forecasts are not correct, it is easy to redirect inventory from other markets without having to repackage the product.

9.4.2 Reducing Lead Time/Just-in-Time Production

Just-in-time production reduces the time between when a customer demands a product and when production is started. Ideally, a company would get an order and be able to make and ship that product the same day. Lead time can be reduced by reducing the lead time of the components, reducing cycle times, and reducing the WIP in the system. However, just-in-time is rarely possible because there are lead times to order materials and produce the product.

Lead time reduction supports the Lean principle of reducing inventory. The relationship between the two is defined by Little's Law:

$$L = \lambda W \tag{9.1}$$

L is the average WIP in the system, λ the production rate, and W is the lead time to produce from the start of production to the end. The inventory in the system is directly proportional to the lead time.

Shorter lead times have other benefits including the ability to quickly roll out design changes and easier forecasting. If it takes you 100 days to purchase and build a product, you have a 4-month forecast to plan. If you have a 10-day lead time, you only have to forecast a month ahead. Lead time and inventory are correlated through a second mechanism: the more uncertainty you have about a forecast, the more finished goods you need to carry to account for variable demand.

9.4.3 Improving and Leveling Flow

Creating a balanced line – one in which the cycle time of each station is nearly identical – will dramatically improve overall cycle times by reducing bottlenecks, reducing the time material spends moving around a factory, and removing labor content. A balanced line helps to ensure one-piece flow between stations, thus reducing inventory and reducing lead times. Improving and leveling flow can also be achieved by:

- **Ensuring a constant process time** per part (low variability) to help to stabilize the manufacturing processes.
- **Reducing secondary processing steps** (e.g., coatings and cleanings) to reduce the number of overall steps, reduce batch sizes, and lower cycle times. Secondary processes typically have to be run in batches, so you have to build up enough inventory to create a batch large enough to make the secondary process cost-effective.
- **Applying automation** can significantly improve flow because the equipment is much more predictable than human assembly and material handling.
- **Optimizing factory** layout to reduce the distance material has to move between stations, to reduce the chance of damage, the need for material handling equipment, and inventory.
- **Leveling production rates** to keep the factory running at the same volumes every day, rather than producing in fits and starts. Leveling the rates reduces the overhead of setting up and stopping the line, stabilizes the labor required, improves material planning, and helps improve quality. Leveled daily production is enabled by accurate long-term forecasting and planning. The factory hates nothing worse than being called up and told "No, wait stop production, we have too much inventory" and then being called up five days later and told in a panic "We got a huge order, you need to double production tomorrow."

9.4.4 Kaizen – Continual Improvement

Kaizen is the Japanese word for improvement. Manufacturing processes can always be made more efficient and more reliable (Section 19.2 describes how to make continual improvement during mass production). Continually improving the product design and the production system can have dramatic benefits to the bottom line, and the lessons learned can be rolled over into the next generation of this product or the next product your company creates. Having well-defined processes is a critical starting point for driving improvement, and subsequent chapters on process documentation will describe these documents.

9.4.5 Quality

Ensuring good quality at all stages will always improve production flow because additional rework lines won't be needed, and the product will be less likely to fail at final testing. Quality should be both designed in through DFM and robust design, and managed through the manufacturing process. Poke-a-yoke (Japanese for error-proofing) reduces variability in cycle times and reduces errors. Errors can be reduced by designing better fixtures and tooling and laying out materials in a way that ensures that an assembler grabs the right part. Lean principles also promote Jikoka – the reduction of errors through automation.

9.4.6 Respecting People

Core to the idea of Lean principles is respecting people, including respecting and engaging with workers, with the community around you, and with the broader world. Respecting people also includes creating diverse teams whose members all have a voice. An engaged workforce has lower turnover, takes pride in the quality of the work, and is more likely to come up with improvement ideas. Not respecting people can have dire financial consequences. Bad press about poor working conditions can impact your brand (e.g., Nike) [90]. Also, a number of studies have shown that ethical companies tend to outperform their less-ethically focused companies [91]. Finally, ethics and sustainability are becoming differentiators for customer purchases.

Summary and Key Takeaways

❑ While there are many different types of production facilities, they have a large number of commonalities including managing incoming material, quality control centers, and process control.

❑ Decisions in product design can have dramatic impacts on the ability of the production line to operate efficiently with a minimum of waste.

❑ Teams should use Lean principles to drive improvements in their facilities.

Chapter 10
Design for Manufacturability and Design for X

MANUFACTURING PLANNING

9. MAUFACTURING SYSTEMS

10. DESIGN for MANUFACTURING & DESIGN for X

11. PROCESS DESIGN

12. TOOLING

13. PRODUCTION QUALITY

Designing parts and assemblies without considering how they will be made is the best way to make product realization long and painful. Teams need to balance how the product will made, assembled, and tested against the functional and reliability requirements of the product. This chapter introduces how to think about manufacturing, engineering, sustainability, and testing early in design and continue that evaluation throughout the product realization process.

Product Realization: Going from one to a million, First Edition. Anna C. Thornton.
© 2021 John Wiley & Sons, Inc. Published 2021 by John Wiley & Sons, Inc.

To build products, teams need to define both product design *and* how they will reliably fabricate the parts and assemble them while keeping costs down. As has been emphasized repeatedly in this book, many of the problems and inefficiencies identified during the pilot phases are avoidable by designing the product with manufacturability, testability, and assemblability in mind. It is much cheaper and easier to design products right the first time than it is to make design changes after cutting tooling. Prior chapters have described multiple examples of struggles and recalls. Many had, at their root, the misalignment between what the designers intended and how it was produced on the factory floor.

Prior to the Industrial Revolution, most products were both designed and made by craftspeople who simultaneously designed the product and figured out how to make it. As design professionals became more separated from the workers that built the products, it became necessary to train engineers on the manufacturing processes that would be used to build their products. Until the massive outsourcing of the last several decades, most engineers had the luxury of walking down to the plant floor and seeing the manufacturing process in action. Unfortunately, access to the equipment and the people that operate them is more difficult, as they are often halfway around the globe and speak a different language. Over the last 50 years, the field of design for manufacturing (DFM) has grown to address this challenge. As companies realized that DFM could have dramatic impacts on cost, quality, and schedule, other design for X (DFX) tool sets have evolved. Boothroyd and colleagues came out with their textbook on *Product Design for Manufacture and Assembly* in the mid-1990s [92], and the other tools and frameworks continue to emerge. DfX tools include:

- Design for aesthetics
- Design for corrosion
- Design for disassembly
- Design for the environment
- Design for ergonomics
- Design for logistics
- Design for maintainability
- Design for recycling
- Design for reliability
- Design for safety
- Design for serviceability
- Design for testability
- Design for utilization
- Design for variety

This chapter will give a brief introduction to six of the DfX methods that have direct impacts on product realization (Sections 10.2–10.7).

- **Design for manufacturing (DFM)** ensures that each part can be built cost-effectively (in both tooling and part costs) with consistent quality.
- **Design for assembly (DFA)** assesses the assemblability of a product, including both the labor content and the potential for errors and damage.
- **Design for sustainability** evaluates the environmental and health impacts of various materials and processes.
- **Design for maintenance** evaluates how easily the user or technicians can diagnose, disassemble, repair, and revalidate a defective product.

- **Design for testing** ensures that the factory can cost-effectively test the product during production.
- **Design for SKU complexity** ensures that multiple versions of a product can be built with minimal risk of assembly errors while reducing costs.

The last section of the chapter (Section 10.8) lists 11 fundamental DFX rules that can be used as a starting point for teams who have never done a DFX study before.

10.1 Selecting Manufacturing Processes

Many of the problems with manufacturability are caused by not thinking about the manufacturing process early enough. The author often meets with business school students and start-up teams who have ideas for new products. They bring in a computer-aided design (CAD) model and a 3D printed part and say, "here's everything we want our product to do. How do we make it?" Often the features they want can't be made or will be too expensive. Defining geometry before knowing what materials and processes are likely to be used can lead to poorly designed products that don't meet customers' needs at the right price point.

Most engineers learn to develop products using Figure 10.1 as their framework. They create geometry to meet the functional specifications, then think about what material will work, then finally decide on the manufacturing process. For example, the author recently had a sports device start-up team in her office. They had a rough design and a barely-working 3D-printed prototype, and they wanted to know "how do I talk with contract manufacturers (CMs) about getting this made?" When asked what material, they said, "either plastic or metal," and when asked what processes they wanted to use, they gave a blank stare and said: "I thought the CM would figure that out." With further discussions, it became clear the design details and manufacturing strategy would be very different for the two materials classes (and that's not even taking into account variations in different types of plastic and different types of metal). The team was sent back to the drawing board to understand the relative limitations and benefits of deep drawing (forming deep vessels out of a ductile metal) and blow molding (creating plastic bottle by forming softened plastics against a mold using air pressure). They had to consider a wide range of factors including weight, complexity, robustness, and cost.

Another larger company used computer numerically controlled (CNC) machining to produce their prototype and early production runs of their load-bearing mechanical parts. They assumed, erroneously as it turned out, that they could send the parts to a casting supplier and get the same parts

HOW do I SOLVE my CUSTOMER'S PROBLEMS?   WHAT am I GOING to MAKE out of THIS?   HOW am I GOING to MAKE this?

SPECIFICATIONS ➡ GEOMETRY ➡ MATERIAL ➡ PROCESS

FIGURE 10.1 Linear process that locks design into non-optimal solutions

cheaper and faster. However, without doing a major redesign, the parts couldn't be cast because there were too many sharp corners and uneven wall thicknesses. In addition, the parts required significant post-machining to make them function like the original parts. Had the parts been designed to facilitate casting in the first place, the team would have saved considerable money and time.

Ideally, the iterative discussion (Figure 10.2) of what combination of geometry, material, and process best satisfies all of the product specifications should start as early as possible. A designer may propose a geometry and material to satisfy a need, and a process is selected to satisfy most of the requirements. However, changes are needed to make it manufacturable. The designers may also change the material to one that better suits  the manufacturing process, but then need to change the design to accommodate the new material properties, which then impacts the manufacturing process again. The team continues to iterate until the optimal combination of geometry, process, and material are selected that best satisfies the specification. Only once that is done is the CAD model detailed and parts prototyped. By understanding your processes early, you can avoid baking-in features that are not manufacturable and take advantage of the unique characteristics of a process. For example, if you start thinking about injection molding early, you can add surface textures and features and avoid expensive painting processes and deco.

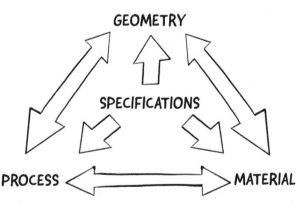

FIGURE 10.2
Iterative design process that balances geometry, process, and material

Generate initial list			
What material properties are important?	Are there standard processes your customer/industry typically uses?	What is your estimated production volume?	What is the rough size, shape, and complexity of the part?

Refine the options			
How critical is cost?	How quickly do you need the parts, and can you wait for tooling?	How will the part be assembled?	Does the part need secondary processes / operations?

Align with capabilities			
What capabilities exist already in a company?	What supply chain (if any) currently exists at your company?	Do you have internal resources that understand the process?	How comfortable is your manufacturing organization with learning new processes?

FIGURE 10.3 Questions used to refine the possible manufacturing and material selections

For many teams, the hardest step in this iterative process is figuring out which options they have for manufacturing processes. Many teams will have a rudimentary understanding of the most frequently used methods for fabricating parts: injection molding and machining. However, those two methods represent only a small fraction of the available processes. The expression, "if you only have a hammer, everything looks like a nail," is very appropriate here. The more processes teams can draw from, the better the designs will become. Unfortunately, globalization means that young engineers don't have the luxury of wandering down to the production floor and seeing the processes in action. To replace the on-the-shop-floor experience, engineers need to double down on reading, learning, asking questions, and taking every factory tour offered to them.

When deciding what methods to use, engineers need to educate themselves about what processes are available and what may or may not work. Initial ideas can be generated by looking at competitive products. Competitor's manufacturing processes can be determined through teardowns (pulling apart competitive products to learn how they make it), literature, and how-it-is-made videos.

The scope of possible processes and materials is so broad that it can be overwhelming. Teams can quickly down select both by asking the questions listed in Figure 10.3.

The first set of questions identifies what broad classes of manufacturing processes are possible candidates. Depending on the answers to the above questions, you may get pointed in different directions. For example:

- **What material properties are important?** You may know you likely need structural metal parts because of temperature and size limitations; you can use any number of processes including: metal-injection molding, die-casting, machining, deep drawing, or sheet metal forming. You may also know the maximum operating temperature. Machining can be used on any kind of metals. However, die-casting is limited to relatively low melting temperature materials. If the product has to be held at very high temperatures, then die casting may not be appropriate.

- **Are there standard processes that are typically used?** A prior company may have gone through this exercise already and you can learn from them. This doesn't mean that their choice is right for you, but you can ask yourself why they chose it and use that information to inform your decisions.
- **What are your volumes?** If low volumes, then machining may be the right way to go. If the volumes justify paying for tooling, then die-casting or metal injection-molding tooling might be the right choice. For example, laser cutting is used frequently in low-volume production for thin, large pieces, because it is a very flexible way of cutting sheet stock. However, designing a progressive stamping die is more cost-effective if you are going to make tens of thousands of units.
- **What is the rough size, shape, and complexity of the part?** For example, if you want metal boxes made, then a press brake is a very cost-effective method. However, you are limited to box-like shapes but there is no lead time for tooling. If you want a thin-walled symmetrical-shaped deep part (metal trash can or water bottle) with no seams, then deep drawing may be an option. However, there is a significant time and cost for the tooling. Die casting and metal-injection molding can produce very detailed parts less expensively than a machined part at high volume.

Figure 10.4 shows a simple scatter-plot of various metal manufacturing processes plotted on their relative tooling costs and the complexity of the geometry they can achieve. (A quick online search can explain all of the terminology here and describe processes you don't know.) By thinking through how each process satisfies the cost, complexity, and performance requirements, you can quickly down select to a few different options.

Once a set of options is selected, the second set of questions will help you to further refine your decisions. They can include issues such as:

- **How critical is cost?** If your margins are large, and your part costs relatively small compared with the overall cost of goods sold (COGS), you may opt for a more expensive process that ensures a high quality.
- **How quickly do you need the parts, and can you wait for tooling?** Lead times for die-cast molding tooling can be many weeks to several months, depending on the complexity of the tool. If you need the parts in a week, then machining them may be required.
- **What kind of secondary processing do you need to achieve the functional specs?** Referring back to the example of the forged part: forging the part may seem, on first glance, to be a more cost-effective process because of material savings. However, since the product required very

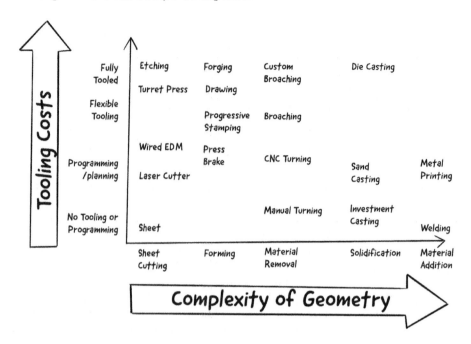

FIGURE 10.4
Possible ways to manufacture a metal part

accurately machined features to function, the cost increased dramatically. The cost and lead time of parts can increase if they need to be coated or heat-treated.

- **How will the part be assembled?** Metal parts need to be either fastened or welded. Material compatibility is critical. Some metals don't weld well, and others should never be fastened together. For example, steel and aluminum will lead to a weakening of the aluminum because of their chemical interactions.

Finally, you need to take into consideration the "devil you know" factor. While a new process such as superplastic forming of titanium may seem perfect to create a complex thin-walled geometry, your company may have not experience with it. Depending on what capabilities you have, the capabilities of your suppliers, and the strength of your supply chain and manufacturing organization, you may decide to go with a more conservative assembly of bent metal parts rather than risk a completely new process.

10.2 Design for Manufacture

Once you have the process selected, you have to make sure your design fits the limitations of that process (and takes advantage of its unique features). The central concept of DFM is understanding the fundamentals of the manufacturing methods and ensuring that the geometry, material properties, and features can be achieved reliably and inexpensively. DFM identifies where the design will:

- **Fundamentally prevent the manufacturing process from being used**. If your team has decided to use a deep-drawing forming process, they will be limited to roughly symmetric parts with constant wall thicknesses.
- **Result in costly processes or long cycle times**. Having an undercut in an injection molded part requires lifters or subsequent actions, both of which can dramatically increase the tooling costs and cycle time.
- **Result in poor quality**. Having uneven wall thicknesses in a cast part can cause warping or poor filling. Sharp corners in a thermoforming can cause tearing or wall thinning.

The best engineers apply DFM principles continuously through the design process. They understand the capabilities and limitations of a process and optimize the design for both function and process. Also, even the most knowledgeable engineering teams will typically do several formal DFM reviews in which peers, experts, and outside resources evaluate the plans and give recommendations (Inset 10.1). Formal DFM reviews should occur at least twice: during initial design and before tooling. The first review is intended to catch fundamental flaws that will drive cost and affect quality. Referring back to the example of the machined part that couldn't be cast, had the teams designed the part to be cast first and then figured out how to

Inset 10.1: How to Do a DFM Review

Most fabricated parts will need to be reviewed and modified to ensure the parts can be reliably built. Stand-offs may need to be added to support large surfaces, thicknesses made even to improve flow, and features changed to reduce tooling complexity.

There are a number of ways to get feedback on manufacturability

- **Hold a formal internal DFM review**. In this approach, you can bring in people from outside your team to review the designs and give you feedback. Typically, the reviewers have not seen the product before and provide a fresh set of eyes.
- **Online DFM tools**. There are a number of tooling and part suppliers who have online tools that will evaluate your designs and identify areas of risk.
- **DFM software**. A number of software packages can be purchased that will give you DFM feedback on the feasibility and cost of your parts.
- **Hire an outside firm to do preliminary design reviews**. There are a number of engineering firms who will evaluate your DFM. While these can be relatively expensive, they can save significant time and expense downstream.
- **Send your part out for quote**. Fabricators may give you feedback on producibility, but are often unwilling to invest too much time if you are (i) too early or (ii) have obviously not done your homework.

machine it in the short term, the total cost of the product would have been significantly lower. Examples of issues raised in the first review may include:

- **CNC machining**. Determine whether a part will require multiple re-fixturing (machined, repositioned, and then machined again) or expensive 5-axis equipment to create critical functional features. Teams may decide to change the location of essential features so they can all be machined from a single setup using a less expensive machine.
- **Injection molding**. Break up large unsupported high-gloss surfaces with either a texture or different configuration to reduce the chance of defects.
- **Press brake**. Ensure complex parts will have a feasible bend sequence. Teams may decide to make the housings in two parts and assemble them later to reduce the overall cost.

The second phase of DFM occurs after the parts are fully detailed and ready for production. The changes recommended by this second DFM review are hopefully minor and the changes can be made without dramatically impacting adjacent parts. These reviews are typically done in conjunction with the fabrication suppliers, who ultimately will build your product.

Design changes that arise from the second DFM process can include:

- Adding ribs to the backside of large injection molding surfaces to reduce sink marks.
- Adding radiuses and draft angles to features such as bosses.
- Making all wall thicknesses identical to improve the flow of molten material in the mold.
- Adding locating features (such as shoulders) to allow the cast part to be located in a CNC machine or lathe to post-machine critical assembly features.
- Adjusting snap geometry to reduce the need for lifters or actions in an injection-molding machine.
- Adding extra material for removal during secondary processes such as reaming or lapping to achieve the tolerances on the critical surfaces.

For engineers who don't have first-hand experience with manufacturing processes, knowing all of the manufacturability rules can be a daunting task. Teams would need to read and apply the contents of 20–30 books to optimize all of the processes that are used to make the typical product. Without an understanding of the processes, many heavily depend on the formal DFM reviews with experts. Some teams use DFM software tools to evaluate the manufacturability of parts automatically; however, the software tends to be expensive and very time-consuming to use. None of these are easy options. Designers need to ask questions of manufacturing engineers and do the hard work to learn about the relevant processes as early as possible.

Vignette 6: How to Talk with a Supplier   **toner**plastics

Steve Graham, Founder & CEO, The Toner Plastics Group

Toner Plastics make all kinds of molded and extruded products. We build tooling and provide extrusion and injection molding parts to a variety of companies. Imagine the different conversations we have with the customer who wants parts molded in a clean room for use with intravenous tubing versus the customer who wants Hula Hoops. We have to talk to customers ordering 1,000 special containers used for hydroponic growing of vegetables on the International Space Station, and then the customer who buys millions of the exact same bottle cap a month.

When we decide if we want to work with a customer, and we don't take all customers who come through the door, we use a number of factors. First, we want to quickly determine the dollar volume of the order, the gross margin potential, and the likelihood of repeat business. In order to do this, we need to have a number of questions answered. The better prepared you are when you come to meet with a supplier, the better we can determine how to help you. You should always ask your supplier to sign an NDA (non-disclosure agreement) prior to conveying any information.

Second, we need a drawing of the product, the type of material required, the tolerances, and the surface finish you need. You need to indicate which surfaces are critical and where the tooling design team can put gates and ejector pins. In addition, we need to know the labeling, pack-out, and certifications needed. You also need to have a believable estimate of your total annual volumes and the order frequency and size.

The thing you will be most interested in is "how much will it cost." The answer is "it depends."

When producing plastic parts, the cost to build the mold is the largest cost. The cost depends on the number of cavities, the complexity of the mold, and the amount of parts you want to produce before the mold wears out. If the supplier knows you are just getting started they may offer a less expensive mold made from aluminum rather than steel. They may also offer to work the mold cost into the part price rather than charge you for the mold up front. Make sure you understand that in this case the mold is not owned by you but rather the supplier. If you need mold design, ask whether mold flow simulations are needed to be run and the cost to do so. Also discuss if you have to pay for multiple trials and how much tool modifications will cost.

You also need to ask questions about lead time for the tooling, lead time for each batch, and payment terms (typically run 30 days). Make sure you ask about price breaks based on volume, so that you know what your cost will be as you order more. You can also ask them for an hourly cost to run your product on their equipment if you are in the development stage (you need to build small batches to produce parts for pilots). This is done all the time.

Most suppliers devote some of their time to good ideas not yet in the market. Be up front with your suppliers and sell your vision. If you do so, you will have a better chance of finding someone that will try to help you rather than look at you as just another transaction and charge you for every minute of engineering and manufacturing time.

Text source: Reproduced with permission from Steve Graham
Photo source: Reproduced with permission from Toner Plastics

10.3 Design for Assembly

Most products have to be assembled from a collection of parts using either hand assembly or automated assembly (or a combination of both). Optimizing the design for assembly (DFA) can reduce labor costs, reduce cycle time, and improve quality. As with DFM, engineers should take assembly into consideration throughout the design process, and do periodic formal DFA reviews to include outside perspectives.

There are two ways to think about DFA. The first is how the overall architecture of the product helps or hinders assembly, and the second is the design of the actual assembly methods to improve quality and reduce cost. As with DFM, the best time to apply DFA rules is during the early phases of design before changes become expensive. DFA heuristics include:

- **Minimize part count**. If two parts can be combined or the design simplified to reduce part count, then do so. First, minimizing part count reduces the fixed cost of tooling. While the combined part may require a more complicated tool, the combined tool is unlikely to be as expensive as two individual tools. Second, it reduces assembly time. Third, it reduces variation and quality failures that could be introduced during assembly. Finally, fewer parts simplify production management and ordering.
- **Poke-a-yoke (Japanese for error-proof) the assembly**. Design the pieces such that incorrect assembly is not possible. For example, use symmetrical parts so there's no such thing as "backward" to prevent incorrect assembly. Symmetrical parts also have the benefit of being easier to assemble robotically.
- **Standardize parts**. Having many different commodity part (e.g., ten different screws that only vary slightly) increases the material management overhead, inventory carrying costs, and the risk of the wrong part being assembled. Furthermore, part-type proliferation increases the design overhead. For example, if the design team standardizes screw types, then the self-taping holes or threaded inserts will also be designed identically, reducing the chance of a design error.
- **Design for a top-down assembly**. Ideally, the product should not be reoriented or turned over during assembly. Each time a product is picked up, reoriented, and put back down, it adds labor hours, increases the chance of a part becoming dislodged, and increases the likelihood of damage (structural or aesthetic).

Several methods can be used to run a DFA analysis. First, teams can apply the heuristics above. Second, more companies are relying on virtual reality (VR) or augmented reality (AR) simulations to identify issues. For example, Boeing has been using VR systems to simulate the assembly process with operators and engineers from as early as the late 1990s [93]. Third, Boothroyd and Dewhurst, the fathers of DFA, have several books and software tools to assess assemblability [92]. Finally, teams can walk through the assembly steps with 3D printed parts.

Once the overall approach to assembling the product is determined, the method for physically connecting the parts has to be designed. Assembling two parts can be accomplished in many ways, each of which has benefits and shortcomings. The following lists several common categories of methods (these are ordered from most permanent to least):

- **Make it a single part**, because, generally speaking, it is better to have fewer parts than more. If you can avoid an assembly process by combining two parts into one (e.g., by making a housing out of a single clamshell rather than assembled from 20 parts), you reduce assembly costs.
- **Material joining** can bond the parts at a molecular level using solid-state or fusion welding methods such as ultrasonic welding or arc welding. Welding is typically chosen when the pieces are never going to be disassembled, when a watertight seal is needed, or when the parts are too small to add screw bosses or snaps. These methods typically require capital equipment and unique skills to operate the processes.
- **Glue and tape** can be used to join parts that are too small for a mechanical joint. The selection and application of adhesives can be more of a science experiment than an exact science. In addition to being sensitive to material, the consistency of application, and material, handling is critical to the long-term durability of the joint. Finally, glue is messy. It can get everywhere and, without controlled dispensing, can leak onto other parts, causing interference or surface quality issues. In one product, the adhesive used for a polished nickel and a polished bronze electroplated part worked well, but when a dark bronze version was introduced, the glue joint started to fail. Significant research was done to find an adhesive that would work for all three coatings. Meanwhile the company had several hundred complaints about buttons falling off their products.
- **Permanent mechanical fasteners** such as rivets are useful when you don't have room for a removable fastener and you never want to disassemble the parts.
- **Snaps** are integral features in injection molded parts that semi-permanently connect two parts. Think about the top of your ketchup bottle and how the top and bottom can snap together and stay closed. Integral snaps enable simple assembly of parts without additional equipment or fasteners. Snaps have to be carefully designed to allow enough compliance to deflect during assembly but not so much that the parts will disassemble in transit or use.
- **Self-tapping screws** have the benefit of not requiring threaded inserts but cannot be removed and reinserted. In addition, they can be easily over-torqued and stripped.
- **Threaded inserts** can be co-molded or press fit into a plastic part. The metal thread can be used repeatedly without risking damage to the part. While inserts allow for reassembly, they increase cost and require more material for the bosses.

- **Others**. The list of other fastening methods runs into the hundreds: from fasteners as simple as Velcro and cable ties to ones as complex as Cleco fasteners.

No one fastening approach is inherently better than another, and the decision of one over another will be based on several factors:

- **Product specifications**:
  - *What materials are being used?* For example, silicone will be difficult to adhere with glue or tape.
  - *What are the forces and loads?* Some assembly methods do better when the parts are in compression; some are better in shear.
  - *What are the environmental conditions?* For example, if a part is subjected to significant vibrational loading, then screws may back out.
- **Capital costs**. Some methods will have a higher capital cost for equipment and or tooling.
- **Part costs**. Threaded fasteners can add significantly to the cost of the unit over self-tapping.
- **Manual vs. robotic**. Some processes are better suited for automation. For example, ultrasonic welding is well suited to robotic assembly where tape is not.
- **Need to disassemble**. If the parts need to be repaired, then using a threaded insert with a fastener is a better choice than using glue.

10.4 Design for Sustainability

In this era, it should go without saying that you should consider the social and environmental impacts of any products you bring to market. If you remain unconvinced, though, research has shown that products that are environmentally and socially responsible can give you a competitive edge. Every year Nielsen's *Annual Global Survey on Corporate Social Responsibility* [94] reports on the preference of customers for sustainable products, customers' willingness to pay extra for sustainability, and how sustainable products have faster sales growth. As with many of the DFX topics, there are many books and articles worth reading on how to design for sustainability. The following is a shortlist of items that can be addressed in product development and product realization processes:

- **Reduce waste during product realization**. Materials used in prototypes, air travel to manufacturing facilities, and scrapped material all have an impact on the environment. Additive parts and low-volume urethane cast parts are typically not recyclable. A single flight from Boston to Hong Kong has a footprint of 1.9 metric tons of carbon: the equivalent of driving an SUV 5,000 miles. The more efficient teams are with resources, the less impact they will have on the world's limited resources and the lower the carbon footprint of the product (and they will spend less).

- **Environmental regulations**. Chapter 17 reviews the legal certifications required to sell products in individual states, regions, and countries. Many of these regulations are related to environmental and safety concerns.
- **Material choices**. When faced with material choices, teams should look for materials that have the lowest environmental impact over the product life. Choosing materials that do not require painting (and hence reducing the need for volatile organic compounds), adhere to directives such as RoHS (Restriction of Hazardous Substances) [95], and reducing packaging materials will all have dramatic impacts on the environmental footprint of the product.
- **Supply chain decisions**. First, choosing partners who treat their workers and the environment well is both the right thing to do and can avoid publicity issues [90]. Second, some countries of manufacture hold their facilities to higher energy use and waste disposal requirements. For example, the EU is leading in environmental regulations. In addition, some countries depend on dirtier sources of energy, such as coal. Third, fair-trade certifications for materials and labor ensure that the material originators are paid appropriately and have safe and humane working conditions [96].
- **Waste prevention**. During the manufacturing process, reducing the amount of waste produced will have a direct impact on your environmental footprint. There are many sources of waste, including:
  - Sprues from injection molding
  - Offcuts from stamping or chips from CNC machining
  - Wasted energy
  - Wastewater from washing processes
  - Scrapped material due to product rejects
- **Reuse/repurpose/recycle plans**. You should design your product so that it can easily be reused, repurposed, or recycled. To do this the product has to be easily disassembled and the different materials easily segregated. In addition, teams need to think about creating follow-on uses for the product. For example, the smartphone market has a very large secondary market in developing countries where people are willing to use product that is older.
- **Use of hazardous materials**. Some paints, solvents, cleaners, and coatings require the use of toxic materials which are harmful to both workers and the environment. You can reduce the need to use damaging materials by reducing the amount of paint used, using in-mold deco, and simplifying color schemes.
- **Reduce packaging**. It is tempting to want an Apple-like unboxing experience. However, high end packaging typically contains inks, dies, and papers that cannot easily be recycled. In addition, the packaging is typically thrown away and requires energy to make and ship. You don't necessarily need a complex unboxing experience with expensive packaging to create a good customer experience. For example, an internet provider sent the author an upgraded modem/router. The color packaging was unnecessarily three times larger than the modem, and all of the printed materials were in color, none of which added any value to the installation

process, but did fill her recycling bin. The cables had beautiful color fabric and Velcro cable ties with the company's logo on them. Having a great unboxing experience was not part of the buying experience: the decision to stay with the internet service provider had already been made. It is valid to assume most customers would have been satisfied with much simpler packaging and the ~$5 savings taken off their cable bill.

10.5 Design for Maintenance

For many products, consumers expect that they will need to maintain or repair their products to keep them running. We replace brake pads on our cars, replace electronics boards on aging ovens, and annually replace filters on our air-conditioning and heating systems. Maintenance and repair services can be delivered through a number of channels, including:

- Dedicated maintenance functions (AppleCare) or service centers (automotive service centers).
- Third-party service providers (refrigerator retailers supporting the units they sell).
- Maintenance providers (electronics repair shops).
- Maintenance functions at the customer's location (aircraft maintenance crews).
- User doing minor maintenance.

Maintenance can come in the form of routine care (replacing the filter in a hot tub), replacement of parts that you expect to wear out (replacing a belt on a lathe), or repair (rewiring a broken switch on a table saw). When designing your product, you should think about how easily and how often it will be maintained and repaired. There are several reasons for this:

- If the repair occurs within the warranty period, the company has to pay for it. The easier and cheaper a repair is, the less the warranty costs will eat into your profit margin.
- Extended warranties – in which the customer purchases not only the product but the promise of it always working – can be a significant revenue stream. By reducing the cost of maintenance or repair, the profitability of extended warranty programs can increase significantly.
- Customers become very unhappy when what appears to be a minor issue with a product costs a significant amount to repair. Ice recently got into the rails in the author's freezer drawer because of a fault with the humidity sensor. While the flaw wasn't critical, the repair ended up costing a frustrating $700. Products that are easier and cheaper to repair will lead to happier customers.

Design for maintenance encourages design teams to think through how the product will inevitably fail, and how it can be easily repaired or maintained when that happens. If you're designing a product complex enough to expect repairs, you should keep in mind how easy it will be to:

- **Diagnose the failure and the necessary repair**. You don't want consumers to have to send you the product before you know what's wrong with it. Self-diagnosis (e.g., error messages) and standardized diagnostic tests (e.g., running engine diagnostics to indicate which sensors tripped the engine light in a car) can reduce the time and labor required to identify the root cause. Customers value self-diagnosing smart systems in many products: the author's fully electric car emails her monthly with the status of the car and any maintenance or repairs required. It also reminds her when it is time to rotate the tires.
- **Disassemble the product to gain access to the repair location**. The product architecture should both provide access to the critical parts (with easy entry and clear signage) and prevent damage when removing the defective part. In the example of the frozen drawer rails above, the maintenance person had to disassemble then reassemble the entire bottom half of the refrigerator. Repairing the broken rail took only a few minutes, but it took an hour to get the product apart and another hour to put it back together. This problem could have been easily solved if the designers thought about what maintenance each piece might require.
- **Replace the part and reorient the parts correctly**. Following the same DFA rules as above can help speed up and improve the quality of the repair process. The filters on the author's robotic vacuum don't identify the orientation or which side should be up, and it always takes several tries to get it to snap in correctly. A simple "this side up →" sticker would have made for a smoother customer experience.
- **Support the product with spare parts**. Products can often last far beyond the time the company stops making them, and companies generally stop making spare parts when they stop making the product. People who restore old cars often have to scrounge through junkyards to find the right replacement parts. Ideally, critical replacement parts should be backwardly compatible with products previously released. This is especially important in expensive products with a long life where parts are expected to wear and break – for example, cars, bikes, and washing machines. Part commonality (or backward compatibility) both reduces inventory management and avoids end-of-life issues. You might also think about designing your product to be packaged with spares (e.g., a blender that comes with a replacement blade) or one that can predict when spares are needed. For example, ink jet printers can email a warning when the printer is close to running out of ink. The email automatically brings up the re-order page with the right ink cartridge selected.
- **Isolate expensive parts from inexpensive maintenance parts**. For example, if a simple sensor is likely to fail and need replacement, don't put it on the same board as an expensive microprocessor chip.

10.6 Design for Testing

Products will need extensive testing during the pilot process to verify that the product works as expected (Chapter 7). During production, subsystems and the final product will also be tested (Chapter 13). Testing equipment itself can be expensive, but more importantly, testing increases cycle time and lead time, contributing to overall costs. Thinking about testing during design can reduce the time and cost of testing the product in production. For example, the team can:

- **Leave real-estate for sufficient test pads** on the PCBAs. Test pads are locations on a board where it is easy to place a probe to measure electrical properties. Test pads allow for engineers to comprehensively evaluate the boards before assembly to make sure they are working correctly.
- **Design products with the ability to connect subsystems to test fixtures so they can be tested separately**. For example, you probably want to test a speaker before you embed it in a product. The speaker subassembly should be able to be tested in a custom fixture that simulates the rest of the product. This reduces the cost of finding a problem after the whole product is assembled.
- **Include built-in testing (also called self-test and self-diagnostic) capabilities to check functions automatically**. Having the product be able to test itself reduces labor and increases test accuracy in production. The same self-test functions can be used for diagnosis once the product is in use. For example, the backup generator at the author's house turns itself on automatically every week and runs a complete test of itself. It calls the maintenance center if there is a failure. The septic system does the same process.
- **Avoid the need for burn-in**. Some products need to be run for a period to minimize the risk of early-life failures and to calibrate the systems. If possible, burn-in should occur in subsystems before they are assembled to reduce the cost of rework, reduce the space required to test the units, and minimize work in progress (WIP).

10.7 Design for SKU Complexity

Most products come in different colors, configurations, price points, part counts (2-pack vs. 10-pack), internal specifications (1Tb vs. 500GB hard drives), or models for different countries of sale (different wall wart plugs). Running multiple SKUs through the same line in mixed or batch production can result in errors, increased WIP, and excessive finished goods inventory. The following heuristics and recommendations can reduce the inventory and the potential for errors when producing multiple SKUs:

- **Keep the number of SKUs to a minimum at the initial product launch**. While it is tempting to launch ten colors in the first production run, it is easier to get the line stabilized (and assess market demand) by building only one or two colors at first. Henry Ford, when launching the world's first mass-produced automobile, kept complexity to a minimum by declaring that customers could get the Model T in any color they wanted, "so long as it is black."

- **Differentiate the product only at one point in the assembly process**. If there are different colors or options, they should be added to the product at only one assembly step, avoiding the need to match colors downstream. For example, cars typically have their doors and frames painted at the same time. The doors go down one line to be assembled and the body the other. And then, at the end of the line, they are mated up.
- **Differentiate products as late as possible**. Late differentiation allows the factory to pre-build a single base product and only differentiate just before shipping. It also enables faster changes to SKU mix, reduces WIP, and reduces the risk of assembly errors. For example, if products are built for different countries and languages, the stickers with the specific language are added at the end.

It will not always be possible to reduce the number of SKUs to one or two. It may be necessary to build multiple ones. In this case, the quality system can be designed to manage the SKU complexity by:

- **Clearly indicating during production when product SKUs are being switched** to prevent incorrect assembly. People need visual cues to see that they are changing from the low-end to the high-end version of a product.
- **Reducing the lead times and costs for differentiated materials**. Because of demand uncertainty, enough material for each SKU variation needs to be kept on hand. There is a risk that if one color or version is not sold, scrapping unused material can be very expensive. A shorter lead time means you have less of a risk of having to write off inventory, because you order material close to the date you need it when you have firmer orders.
- **Ensuring color matches before assembly**. If parts are coming from different vendors, each color may individually be in tolerance, but when two parts at extreme ends of the color tolerance are assembled, the color match between them can fail inspection (Section 6.3). Color match should be checked at incoming material or made by the same supplier.
- **Employing a final quality check and shipment audit** to verify that the SKUs are assembled, packaged, and labeled correctly.

10.8 Eleven Basic Rules of DFX

Most products will be assembled from parts made from tens of different manufacturing processes. For any engineer – new or experienced – it can be hard to keep all of the relevant rules front-of-mind. However, looking across the DFX rules for multiple manufacturing processes, some common themes emerge. The following list synthesizes these universal themes that apply across multiple processes. The list is designed to be easy to remember, so while it isn't completely comprehensive, it can help teams avoid some common pitfalls.

Rule 1: If there is an opportunity not to follow directions, someone will

Teams have to assume that anyone who builds, uses, or fixes a product will not read the directions and will "wing it." Designers should do whatever they can to make the assembly, use, and repair error-proof. For example, try to:

- Optimize the design so the processes have a large operating window (any small errors in setup won't impact quality).
- Avoid parts that have to be assembled in a specific order or orientation to be correct.
- Avoid complex wire routings or the need to solder in small spaces.
- Avoid requiring operators to use their judgment on qualitative quality factors (e.g., is the camera focused? Do the colors match? Is the sound good enough?).
- Avoid intricate bolt-hole patterns with parts that are hard to align and require a complex tightening sequence.
- Avoid parts that look the same but are not interchangeable (i.e., two screws with the same length, head, finish, and diameter but different pitches).

Rule 2: Things are always easier if you have fewer parts to manage

Reducing part count reduces the potential for reduces the overhead associated with managing the operations, including inventory control, ordering, and tracking. Finally, the fewer parts you have, the fewer tools you need to make them. For example, try to:

- Enforce part commonality (especially in fasteners: screws, bolts, etc.), which can reduce inventory and complexity (as well as supporting Rule 1 above). Also, common parts increase the net minimum order quantity (MOQ) and can reduce COGS.
- Combine parts where possible, to reduce assembly time, errors, tooling costs, and inventory costs.
- Group parts into subassemblies to reduce complexity. Subsystems can be built up and held in inventory to meet peak demand.
- Have suppliers provide pre-assembled subassemblies rather than boxes of parts.
- Stick to your MVP (minimum viable product) for the first generation of your product. Keeping to just the critical functions (and their associated parts) will reduce complexity and opportunities for defects.

Rule 3: The more you handle a product, the more defects get introduced

Each time a product is passed between stations, turned over to facilitate assembly, or stored in a warehouse, opportunities for damage arise.

- Whenever possible, make sure all steps assembly can be accomplished without having to flip or rotate the product.
- Protect critical surfaces and components if you do have to move them.
- The fewer assembly steps, the fewer opportunities the product has to get damaged.
- Design custom fixtures to hold the product safely during production.
- Store and move materials carefully. It is amazing how much product is damaged by forklifts.

Rule 4: Screws are bad, but glue and tape are worse

The best fastening method is one you do not need. If parts have to be assembled, the most error-resistant methods are snaps or self-locating joints that are fastened using automated equipment. These methods avoid installing the wrong fastener or stripping a screw; they also reduce the labor content and training required for assembly. If snaps are not a possibility, then screws may be needed. Tape and glue are also a possibility, but they can be the source of many quality problems. Both require proper selection and experimentation. Also, tape and glue adhesion is sensitive to surface preparation and application technique. If the surface has oil or isn't roughened, the glue might not stick. Finally, glue has a habit of getting on other parts of the product, causing aesthetic failures.

- Think through your assembly fastener strategy early in the design to ensure you have left room for the fastening method.
- Add mechanical backup to any glued or taped joint. This will reduce the stresses on the joint and keep things in place.
- If you have to use glue, use a glue dispenser and control the environment and surface preparation carefully.
- Avoid the need to assemble in the first place (Rule 2). Combining parts or getting rid of unnecessary parts means you don't have to think about fastening at all (and you save on tooling costs and assembly labor).

Rule 5: Material will never stay where you put it and tries to go back where it started

When teams create a CAD model, they often assume incorrectly that plastic and metal parts will be completely rigid. Spring-back, in-built stresses, and compliance all conspire to move and deform structures. Even metal, when it is machined, will warp as the material is removed and the internal stresses in the part are released. If anyone has tried to rip a large piece of lumber on a table saw, they know a hazardous situation can develop as the wood can bow and pinch the blade when a cut releases stresses. This rule encourages designers to:

- Avoid large floppy parts that aren't stiffened with ribs.
- Avoid deep bends. Parts that are bent or formed are likely to have spring-back.
- Hold parts in fixtures while they cool so they don't warp.
- Assume every sheet metal part will be slightly out of dimension and may need to be held in place during assembly.
- Create critical dimensions after the stresses have all worked themselves out.

Rule 6: You need to get the tool into and out of the part
Corollary: Parts like to be in 2½D and symmetric

When designing a part, the engineering teams typically think about what stresses are imposed on the part, what space it needs to fill, and how it will interface with other parts. When designing for manufacturability, engineers need to think from the point of view of the tool. Specifically, how is the tool going to get into the part and out of the part? For example, if you have a thin, deep cavity, is a small diameter tool going to break? If you have an undercut in the side of a machined part, do you have to re-fixture the part to cut it or use expensive 5-axis CNC? If you have an undercut in an injection molded part, do you need lifters or side-actions in the tool to create those features? If you have straight edges, can you get the part out of the tool without scratching the part or do you need a draft angle?

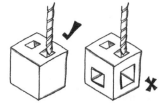

Unless the team is planning on using additive manufacturing in mass production, the simpler the geometry, the better. Simpler geometry will drive lower-cost tooling, lower cycle times, and reduce quality errors. Most traditional manufacturing processes do best when the geometry looks like an extrusion and where there is symmetry in the parts. Undercuts, features in the sides of parts, all require either re-fixturing in the case of CNC machining or multi-action tools in the case of molds. This rule encourages designers to:

- Avoid undercuts that make accessing the cavity with a tool difficult or requires tooling with secondary actions.
- Design features to match the tools you have. For example, if you have a ¼″ diameter end mill, don't make your fillet radii 3/8″.
- Try to make all features oriented in one direction to avoid re-fixturing.

- Always think like the tool rather than the part. Think about how the tool will get into of the part. How will be tool be stressed? How the tool will get out? What will it hit as you remove it?
- When in doubt, add draft angles to parts made from molten or plastically deformed materials.

Rule 7: Unpredictable things happen when the material goes from liquid to solid

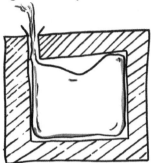

A variety of manufacturing technologies use the molten form of a material to facilitate the creation of the desired shape: such as injection molding, casting, and extrusions. When the molten material cools it typically contracts, warps, and moves. Depending on the design of the structure, the part can warp, twist, or deform. Teams can address these issue by designing wall thicknesses that are all the same, and minimizing large unsupported surfaces. Also, they should

- Avoid large masses of material.
- Ensure that large surfaces are supported correctly.
- Avoid thin wall sections where material can cool too quickly.
- Think about where the material will flow from and to, and leave room for gates and risers.

Rule 8: Material does not like sharp corners

When parts are bent, forged, or injected in a molding process, sharp edges are challenging to create. Creating sharp corners can embrittle material and provide locations for cracks to propagate. This lesson was learned long ago on the De Havilland Comet failure described in Chapter 7.

- Sharp corners are hard to manufacture when machining parts or molds.
- Tight bend radii in formed metals increase the work hardening, which can lead to stress concentrations or tearing.
- It is hard for molten plastic to flow into and hold a sharp-cornered shape. Internal corners create stress concentrations that can cause warping and buckling. External corners are hard to keep sharp because, as the material shrinks, it will pull away from the tool and round the edge.
- Sharp corners typically have more defects than rounded surfaces.

Rule 9: Vibration is your enemy

Vibration will cause connectors to disassemble, screws to back out, metal to fatigue, and wires to rub against structures. Products should be designed to reduce the effect of vibration, and they should be vibration-tested to ensure no long-term reliability issues persist. This is related to Rule 5.

- Avoid screws if possible, and use snap fits or use welding technologies that are more robust (e.g., ultrasonic welding).
- If possible, combine parts to avoid the need for fasteners (see Rule 4).
- Don't skimp on wiring connectors. If the product is subject to vibration, use locking connectors rather than friction ones, because locking connectors use a mechanical lock to hold the wires together rather than just sliding on.
- Ensure proper cable management in your product, so when wires vibrate they aren't putting stresses on connectors, and wires should be routed so they aren't pinched or can rub against sharp edges.
- If you're using screws, use Loctite or other adhesive to reduce the chance that screws will back out.
- Run accelerated and extended reliability testing for parts that may be subject to fatigue failure (Inset 7.7).

Rule 10: The best DFM will fail if the process is not controlled

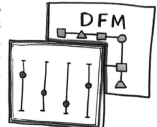

Even when the best DFM analysis is applied to the product design, if the process parameters are not correctly standardized and controlled, the quality will be problematic. The process control should be carefully designed and managed. These topics are discussed in Chapter 11.

- Ensure that all critical processes are tuned to optimize cost and cycle time. The facility should maintain documentation of the process settings in a controlled document.
- Factories should use statistical process control (SPC) – a statistical method to track trends in quality described in more detail in Chapter 13 – to track and manage the quality of the processes and identify any disruptions.
- Standard operating procedures (SOPs) and process plans should document in detail all of the fabrication and assembly operations.
- Adherence to SOPs and process plans should be audited to ensure compliance.
- Regular preventative maintenance (PM) and calibration of equipment are needed to keep the tools and equipment at peak performance.

Rule 11: Know when to break the rules

Most products would be big, blocky, and overweight if all of the DFX guidelines were followed to the letter. To achieve more innovative design goals, it may be necessary to push the edges of the DFX limits. Pushing the limits means that you are ahead of the competition, creating new markets and providing customers with new and exciting products. Manufacturing experts should be consulted to learn which rules can be broken, how to reduce risk, and where you can't violate DFX rules. Just remember, it is one thing to push a technology to its edge; it is another to violate the fundamental laws of physics. The question that engineers need to answer is: where is the line between the possible and the impossible?

- Apple broke the rules of "no large, shiny, white plastic surfaces." Large, smooth surfaces are hard to achieve, but by pushing the limits, they set the stage for a revolution in product aesthetics. Everyone else had to scramble to learn how to do what they did.
- Processor manufacturers, such as Intel, are continually pushing capabilities such as line width and clock speed. The new chips they design often violate the limits of the existing processes.
- Dell changed the mode of computer manufacturing by enabling a rapid turnaround on custom computer configurations.
- Ikea changed the model of furniture sales by selling the products unassembled in flat packs. By putting the assembly in the hands of the customers, they gave the ability for someone to walk out of the door with a living room's worth of furniture in the back of their car. In addition, they removed all of the potential damage, handling, and assembly errors that would have been introduced if they assembled the furniture before sale.

Summary and Key Takeaways

- ❑ The earlier you think about processes and DFX, the better. Changes early in the design are less expensive than modifying tools.
- ❑ Teams need to understand the processes they will be using so that they can create designs that have a chance of being producible.
- ❑ DFX is not just about making the product, but thinking about how design decisions will impact downstream costs.
- ❑ Always talk with experts in the fields to get their input. They will be able to both identify problems and make recommendations for easy fixes.
- ❑ Use the basic 11 rules of DFX if you don't know where to start:

  1. If there is an opportunity not to follow directions, someone will
  2. Things are always easier if you have fewer parts to manage
  3. The more you handle a product, the more defects get introduced
  4. Screws are bad, but glue and tape are worse
  5. Material will never stay where you put it and tries to go back where it started
  6. The tool has to get into and out of the part
  7. Unpredictable things happen when the material goes from liquid to solid
  8. Material does not like sharp corners
  9. Vibration is your enemy
  10. Best DFM will fail if the process is not controlled
  11. Know when to break the rules.

Chapter 11

Process Design

MANUFACTURING PLANNING

9. MAUFACTURING SYSTEMS

10. DESIGN for MANUFACTURING & DESIGN for X

11. PROCESS DESIGN

12. TOOLING

13. PRODUCTION QUALITY

It's not just your product that needs to be carefully designed – the *process* for making the product also needs to be designed, documented, and tested. This chapter reviews the major activities in defining your process, including planning, allocating work, handling materials, and writing standard operating procedures.

Product Realization: Going from one to a million, First Edition. Anna C. Thornton.
© 2021 John Wiley & Sons, Inc. Published 2021 by John Wiley & Sons, Inc.

Many first-time product designers imagine that once they've designed their bench prototype and picked the basic manufacturing processes, they can simply send the drawings off to the manufacturer and wait for the perfect product to be sent back to them. The reality is not that simple, though. The CAD drawings, bill of material (BOM), and other design information tells the factory what to make but not *how* to make it. In preparation for piloting and ultimately for mass production, manufacturing engineering teams need to define the process operations to fabricate parts and assemble finished goods. The design team needs to be involved to ensure the process meets the product requirements.

If fabricated parts are outsourced, development of the process plans for those parts will be managed by the vendor. If assembly is outsourced to a contract manufacturer, they will take responsibility for designing the process to get the final product built. Even if the core engineering and management teams are outsourcing all of the manufacturing processes planning, it is essential that they know and understand the documentation used to define the manufacturing process. For example, one telecommunications company thought that a failure in an antenna was due to a part defect. They spent significant time trying to trace the fault with the component. After watching the assembly process, they discovered there was no standard operating procedure (SOP) for assembly, and the CM was mishandling the part and damaging it. The CM should have had an SOP; but the engineering teams should have known to check for the SOP during piloting.

There are numerous terms used in the industry for developing this set of procedures: process engineering, manufacturing planning, operation planning, factory planning, and process planning. This chapter introduces the basic concepts of defining the manufacturing process and uses some of the more common terms. To define the overall process to produce the final product, manufacturing engineers and design engineers typically work together to:

- **Determine the overall process flow** (Section 11.1). The process flow lays out all of the steps to go from incoming materials and parts to finished goods.
- **Define the work that happens at each workstation**. Once the overall flow is understood, then the specific operation done at each workstation (also called a station or work center) must be defined.
  - *Decide between manual and automated processes* for each operation (Section 11.2). It may be cost-effective to automate some aspects of the process including material handling, assembly, and testing. Manufacturing engineering needs to define what operations will be manual and which require the development of automated equipment.
  - *Allocate work to the workstations* (Section 11.3). The operation required to build the parts, assemble the products, and test them has to be allocated to a set of workstations, some in series and some in

parallel. Work allocation is made to balance the throughputs to minimize overall cycle time, minimize material handling, and minimize work in progress (WIP).

- **Create a process plan** (Section 11.4). The process plan is a formal document that defines operations to make and assemble the parts from incoming through finished goods. It contains all of the resources needed for each operation, including time, labor, space, equipment, and materials. The information in the plan drives resource planning, and is the primary source for manufacturing resource planning (MRP), and enterprise resource planning (ERP) systems (Chapter 4.4).
- **Create SOPs or work instructions** (Section 11.5). For each station a detailed description of how exactly each operation should be executed along with the materials and parts required is created.
- **Create a plan for material handling and transport** (Section 11.6). Material will need to be moved between suppliers, areas of the factory, and stations. In addition, individual stations may require fixtures to hold units during assembly.

The process definition and documentation serve several purposes:

- **Ensure that the product is built the same way each time**. The documentation will prevent personnel from incorrectly making assumptions about what is critical and what is not. Also, the plans can be used to audit the process to ensure that the facility is executing everything correctly (Section 19.2) and drive continual improvement.
- **Support certification and regulatory requirements**. For example, quality management systems (QMS) such as ISO9001 (ISO) [34] require comprehensive documentation of processes. Regulated products such as medical devices require documentation as part of CGMP processes (Chapter 17).
- **Support the design and layout of the factory**. The process definition is used to estimate tooling and labor requirements and can also guide factory layout and material flow. If the production flow is too complex to design manually, factory simulation and optimization software can be used to optimize the process flow (Chapter 9).
- **Ensure that the right materials are purchased**. The process plan includes lists of materials and parts needed. The M-BOM will be updated as necessary to make sure the materials and parts are on hand.
- **Support MRP and ERP systems**. The data from process planning feeds the material planning processes. The process plan defines all of the cycle times, materials, and labor. MRP and ERP use the data from the process plan to ensure that materials are ordered on time and that product orders are released to the factory with enough time to meet demand.

During the product and process design phases, decisions about product architecture, testing strategies, and design details can help to make the production system run more smoothly. Ideally, the manufacturing teams will start to design the process flow in parallel with the product design itself. By simultaneously developing a product and its process sequence, teams can dramatically reduce costs and cycle times. For example:

- Your product architecture and assembly sequence will have a dramatic effect on how work is allocated to stations. Following DFA (design for assembly) rules (Chapter 10) will reduce the amount of labor, the material handling requirements, and the variability in the assembly times. For example, if the product requires each part to be put together in sequence (i.e., part A cannot be assembled until part B is put in, and so forth), then the overall cycle time is the sum time of all of the assembly steps. If the product can be built up in subassemblies, these subassemblies can be made in parallel. The cycle time is the longest time to assemble one of the subassemblies plus the final assembly.
- If robotic assembly is an option, parts should be designed with locator features and grip surfaces to enable robotic actuators to locate and pick up the product.
- Moving material between stations is a significant source of damage and contamination. Designing subassemblies that can be safely moved will reduce in-process damage and cost.
- Product differentiation late in the assembly will reduce WIP and potential errors (Section 10.7).

The process definition is validated and tested during both design verification testing (DVT) (does the process ensure the product performance?) and production verification testing (PVT) (does the production system work as planned?), as well as throughout production ramp up (speeding up the line during mass production to final production rates). Each of these verification stages will find and fix problems with the process design.

11.1 Process Flow

The first step in defining a process is to create a high-level picture of the overall flow of materials from incoming to finished goods. The process flow diagram provides a quick way for the team to visualize the whole process without having to wade through pages of SOPs and process documents. Figure 11.1 shows a simple flow chart for urethane casting. Urethane casting uses a silicone mold made from a pattern. A two-part thermoset material is mixed and poured into the mold to create a part. The process flow helps the team to identify where:

- **Operations can be done in parallel** (e.g., weighing the urethane and cleaning the mold). Parallel processes can be done on separate lines to reduce the overall cycle time and in-process inventory. The mixing of the urethane doesn't have to wait for the mold cleaning, and the overall time is just the longer of the two processes.

FIGURE 11.1

Example of flow chart for urethane casting

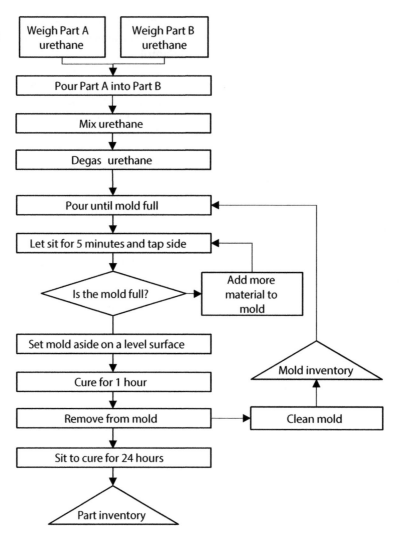

- **Critical checkpoints and iteration happen in the process**. In Figure 11.1, the mold has to be checked for a complete fill. If the cavity isn't full, it has to be topped up. Operations with iteration have higher variability in cycle time; they typically have inventory buffers ahead of and behind the process to prevent starvation and blockage of the upstream and downstream tasks.
- **Inventory is needed to hold in-process goods**. For example, in a process that uses molds, the facility will need a dedicated location to store the clean molds.

The information can be documented in a simple flowchart (Figure 11.1), swim lane charts, or value stream mapping. Each of these formats has benefits and shortcomings.

- **Simple flow charts** are the easiest to create but are limited in the information that can be represented. Flow charts are diagrams that show a workflow or process by connecting boxes and shapes with arrows.
- **Swim lane charts** are useful where work flows between multiple functions and the work allocation is an issue. They are the same as flow charts but create columns that distinguish who is responsible for which tasks.
- **Value stream mapping** is an excellent way of getting all information about a factory in one location. Value stream mapping is a complex version of a flow chart with more symbols to represent both work flow and information flow. However, the data is often not maintained or kept consistent with the information in other representations of the process. For example, the value stream map often documents the cycle times and labor requirement, but reconciling the data with the process plans often is ignored. As a result, the value stream map can quickly become out-of-date and inaccurate.

11.2 Manual vs. Automation

For any operation, the material handling and processing can be done either manually or by automation/robotics. Automation, as you might imagine, has several obvious benefits, but it also has certain shortcomings. Automation reduces labor costs, lowers the probability of quality issues due to human error, and typically improves throughput. However, automation is expensive – it requires significantly more debugging and optimization than manual assembly at the start of production. Elon Musk was quoted on Twitter in April 2018, saying, "Yes, excessive automation at Tesla was a mistake. To be precise, I made a mistake. Humans are underrated." [97]

If you're wondering whether to use manual assembly or automation for each of the operations in your production process, these are the factors you should consider as outlined in Table 11.1.

The term "automation" describes a vast range of possible solutions for assembly and material handling. Automation solutions will fall somewhere in the spectrum between wholly fixed automation and flexible automation.

- **Fixed automation** is explicitly designed for the product and factory, and it cannot be repurposed for other activities. For example, in a casting facility, there will be dedicated material-handling equipment for large castings. In most cases, fixed automation is reserved for dedicated production lines for which the volumes and speed of production justify the design and purchase of custom equipment.

- **Flexible automation** is generic equipment that can be used for many different products or processes. There are generic robotic systems which can be adapted by changing custom grippers and by modifying programming. Generic robots are sold to a wide variety of companies ranging from consumer electronics to medical devices. In another example, computer numerically controlled (CNC) lathes can automate the turning of a part and the material handling. Depending on what tools are put in the turret and how it is programmed, the lathe can be used to make a huge range of products (Figure 11.2).

11.3 Work Allocation to Stations

Most products cannot be fully assembled at a single work cell or station. There are too many operations and pieces of equipment involved, and a single operator will have difficulty mastering all of the processes. Typically, work is broken down and executed in a series of operations at individual workstations with different tasks assigned to each. There will be specialized equipment for each operation, and the right materials will be delivered to each station in time to meet demand. Each station will also have its own set of SOPs and quality checks. When determining how many stations are needed and how the work will be balanced between them, teams will

Table 11.1
Manual vs. automatic assembly

Metric	Manual vs. automated
Size of the parts	Manual assembly is usually useful when parts fall outside the range that is easily handled by an automated system. Automatic processes are often better than manual assembly at handling small delicate parts. Automation may also be required when parts are too large or heavy for humans to easily handle (e.g., automotive components).
Delicate parts	If parts cannot be touched and are easily damaged (e.g., small wires), automation may be better.
Orientations	Operations that require multiple part re-orientations are more suited to manual handling. Humans can more easily flip and rotate pieces than machines can.
Production volume	Small volumes may not justify the programming and equipment costs of automation. For example, a company producing a small number of bespoke products would opt for manual assembly.
Dedicated vs. flexible lines	If the process is set up and broken down repeatedly (as in seasonal holiday decorations), a manual assembly line is more flexible.
Labor costs	Where labor costs are low, and equipment costs are high, manual assembly may be better. This will depend on the country you are producing in, as well as the degree of skill required for manual assembly.
Contamination	Where human contamination is not an issue, manual assembly is okay. Where parts need to be kept sterile (e.g., medical devices), it may be better to automate the process.
Error sensitivity	If the process requires a repetitive but highly accurate process, automation can remove sources of variation.

FIGURE 11.2
CNC lathe.
Source: Reproduced
with permission
from SRAM

need to weigh having too much work at each station against increasing the
amount of WIP (the more stations, the more WIP in the system).

Each station will typically have some common characteristics:

- An area for incoming work – the materials needed to execute the assembly.
- Equipment required to perform the operations at that station.
- SOPs.
- Testing equipment.
- A space to work.
- A method for documenting the work completed.

In the case of electronics assembly, the station will probably be a work-
bench; in the case of aircraft assembly, the workstation will be the size of
the aircraft. Figures 11.3 and 11.4 show the assembly stations in an electronics facility and at Boeing.

11.4 Process Plans

After (i) the overall process is understood, (ii) the amount of automation
determined, and (iii) work divided between stations, the process plans need
to be defined. Although these first three steps, described in the sections
above, define what needs to be done, they don't define individual operations in sufficient detail to be able to design the facility, purchase equipment, and hire enough labor.

Process plans – also called route sheets, operations sheets, and operations
plan sequence – are structured documents, typically in a table format.

FIGURE 11.3
Electronics
workstation.
Source: Reproduced
with permission
from Artushfoto,
Dreamstime.com

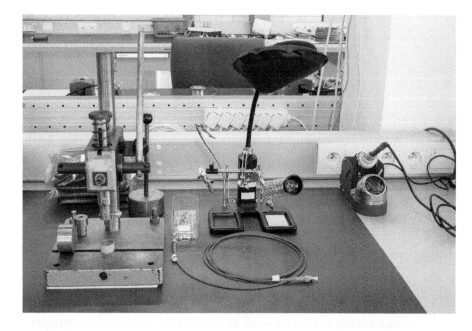

FIGURE 11.4
Boeing 787
assembly station.
Source: Reproduced
with permission
from Boeing

They define the operations in the process, the equipment and materials needed, and the location in the factory where the operations will occur. The process plans are used to comprehensively document all of the resources needed as well as the assignment of work to individual workstations. They are formally structured to ensure all of the relevant data is collected about each step. Mostly process plans are used to ensure the right resources are available when production starts and the cycle times are understood.

Vignette 7: Listen to your Factory

Sam Shames, Chief Operating Officer and Co-founder, Embr Labs

Matt, David, and I, the founders of what became Embr Labs, came up with the idea for Embr Wave as part of a student project as Materials Science and Engineering students at MIT. While brainstorming ideas in an over-air-conditioned lab in June 2013, we became so cold we had to put on sweatshirts to stay comfortable. We realized how little sense it made to air condition a huge lab space to the point where people were cold – especially one that sat empty most of the day. Upon winning that contest in October 2013, the idea went viral. After more than six prototyping rounds, with funding, we went into production in 2016. Over the 15-month period, we learned a lot about what would work theoretically and the practicality of building our product in volume.

Our biggest mistake during manufacturing was not listening to our CM about the feasibility of hand-soldering multiple lead wires on a printed circuit board that was 15 x 35mm, and the mistake cost us five months and tens of thousands of dollars. During the DFM stage, our CM told us that hand-soldering would not be scalable, but our US industrial design firm said they were wrong. We made the mistake of listening to the designers, and in our first prototype sample run, we got 2 working units out of 10. Worse still, after this low yield, we still didn't listen to our CM and made design modifications to the housing rather than switch to a connector. It wasn't until after the second round of prototypes where we got 2 out of 5 working units that we finally listened. Our CM then sourced a low profile connector, modified our PCBA design, and solved the issue. The first build we did with the connector in place had a 96% yield. Now more than two years later, we've shipped tens of thousands of units and consistently get 98–99% yield in our manufacturing. The lesson is to listen to your CM about whether or not a production process is feasible and to bring them into the decision process as early as possible. Had we done that, then we would have designed in the connector from the beginning and been able to get to market faster and cheaper.

Text source: Reproduced with permission from Sam Shames
Photo source: Reproduced with permission from Embr Labs
Logo source: Reproduced with permission from Embr Labs

Companies typically tailor the structure of a process plan to their unique needs, but all process plans share the same basic information as described in Table 11.2. A specific example of a process plan is given in Table 11.3.

The plans can be used to support several activities, including:

- **Cycle time calculations and line balancing**. The work will be assigned to different stations in ways that ensure a balanced line.
- **Resource requirements**. The number of stations, tools, and fixtures needed to meet demand and production rates can be calculated from this data.

- **M-BOM**. The process plan lists all of the materials that are needed to execute each operation. The M-BOM and process plan should contain the same information.
- **Routing and layout**. The process operations can be mapped to a facility floor plan to understand the transportation needed between stations and where inventory will be stored. The impact of layout can be evaluated by mapping on paper or through the use of simulation software.
- **Workflow simulation**. Process simulation software can use the data to run complex simulations of the facility and identify bottlenecks.
- **MRP and ERP systems**. The plan is entered into the MRP and ERP systems (systems used to manage the enterprise and manufacturing data), which are then used to release orders to the floor and order materials.
- **Auditing**. The plan can be used by the quality groups to check that the process is being executed as designed.

11.5 Standard Operating Procedures

The process flow and process plan are too general to enable an operator to know *exactly* what to do at each operation. The SOP (also called work instructions) should allow any operator – with the right skill-set and training – to be able to execute the operations laid out in the process plan. For the context of this book, we will focus the conversation on manufacturing and assembly SOPs. However, organizations

Table 11.2
Contents of a process plan

Label	Description
Operation number	The number that uniquely identifies an operation. Typically, the numbers are not sequential so it is easy to add operations without renumbering
Operation	Operation description
Area	Where the work is being done and what machine is used
Setup	Short description of what setup is required before the process
Process	Short description of the process
SOP	Formal SOP for each operation (or multiple operations)
Tooling and fixtures	Fixtures, jigs, and tooling needed to execute the process
Materials	Materials and parts used in the operation. The materials from the prior operation are implied
Setup time	Total setup time for the batch
Cycle time	Total cycle time for the batch (not including setup)
Labor	Average labor hours per part

Table 11.3
Sample
process plan

Production plan number			Key contact		Approved original		Approved changes		
FAB_0101			Jane Doe						
Part / assembly number			**Part name description**		**Date (original)**		**Date (rev)**		
FAB_0101			Urethane cast housing						
No	Operation	Area	Setup	Process	SOP	Tooling and fixtures	Materials	Setup (min)	Cycle time (min)	Labor (min)
---	---	---	---	---	---	---	---	---	---	---
10	Clean mold	Work-station 1		Inspect and clean the mold and set overspill pan	SOP_mold_prep_v1	Mold	Alcohol, wipes, small brush	0	5	5
20	Weigh and mix urethane	Work-station 2	Set up cups, stirring stick, cleaning supplies	Pour part 1 until 41.2g Add part 2 until 94.6g Mix completely for 5+ minutes	SOP_urethane_v1	Gram scale	Mixing stick, mixing cups, urethane	5	6	6
25	Degas urethane	Vacuum degasser	Set the timing on the vacuum system and line the volume with a plastic bag	Degas the urethane	SOP_degassing_v3	Vacuum degasser	Liner	1	3	3
30	Pour mix into the mold	Work-station 2		Pour the material into the mold slowly	SOP_FAB0101	Mold			3	3
40	Tap to remove bubbles	Work-station 2		Tap the mold for 5 min until bubbles stop coming to the top. May need to add additional material to top up	SOP_FAB0101	Mold	Metal rod		10	10
50	Cure	Work-station 3		Set aside for 60 minutes	SOP_FAB0101	Mold	None	1	60	1
60	Remove part	Work-station 1		Remove part from the mold	SOP_FAB0101	Mold	None	0	5	5
70	Cure	Work-station 4		Place part on worktable 3 with a label indicating the start time of cure	SOP_FAB0101	None	None	5	24 hours	5

Inset 11.1: Standard Operating Procedures

SOPs are not just for the manufacturing process. SOPs are written for almost all steps in product realization and include the following:

- **Pilot quality test procedures**: Testing for durability, life, and functional tests (Chapter 7).
- **Incoming quality control (IQC)**: Inspection requirements for incoming parts from suppliers (Section 13.3.1).
- **Setup sheets**: Setup process needed before executing an assembly/fabrication operation. In the case of fabrication, this can include loading tooling, cleaning equipment, and starting up the equipment. For flexible assembly lines, it could consist of setting up consumables (e.g., finger cots and glue), and the use of material handling equipment.
- **Calibration/preventative maintenance**: Instructions on calibrating and maintaining tools and fixtures to ensure high quality.
- **Aesthetic inspection**: The light source and measurement techniques for cosmetic inspection. For example, the SOP for a surface finish might show pictures of unacceptable and acceptable defects.
- **In-line quality check**: What is acceptable and not acceptable and how non-conforming products should be managed. For example, the SOP may specify how to run an on-board test with instructions to re-run the test if a failure is found.
- **Shipment audits**: The testing processes and acceptability criteria for finished goods including the sampling rates and lists of tests (each of which will have a unique SOP).

can also document any of their organizational processes in SOPs (Inset 11.1). Detailed SOPs ensure that:

- Any trained operator can execute the process and it will yield the same results.
- Learnings drive changes to the SOPs as the organization realizes where mistakes can happen.
- Quality checks and standards are consistently applied.
- The operators spend their time doing the work rather than spending time figuring out what to do.
- Any regulatory, environmental, and safety precautions are documented and followed.
- The operator has all of the necessary materials and equipment to do their job before they start.

As with process plans, companies typically tailor SOPs to fit their unique needs. However, most share the same basic structure:

- A title page with the SOP name, issue date, and approvals. A SOP should be a controlled document.
- A description of the resulting part, assembly, or product.

- A list of the required incoming materials, parts and assemblies.
- What equipment and environment should be used for the process (for example, it may require a cleanroom).
- A detailed description of how to execute the operations, including pictures and diagrams.
- A detailed description of what is acceptable quality.
- Instructions for how to handle any exceptions if the processes or quality checks fail.

Generating the SOPs serves two purposes. First, it ensures that the manufacturing engineering team thoroughly understands the processes before handing them over to the factory. For example, in the case of the urethane casting process, a detailed SOP will prevent organizations from forgetting to include stirring sticks in the M-BOM. Second, a SOP provides a baseline for driving improvements. If there is a quality failure, the team can determine whether the process was at fault (an insufficient SOP) or the operator was at fault (failing to follow the SOP). Either way, the process can be changed, operator retrained, and/or the SOP made more explicit.

SOPs are only useful if they are followed, updated, used for training, and audited regularly. Best practices in SOPs include:

- **Designing a product that barely needs SOPs**. The more you design your product to minimize potential errors (Section 10.3), the less complicated the SOP will need to be.
- **Making it easy to read and use**. Use more pictures and fewer words (think Ikea assembly directions). The more visual the SOP is, the more likely it is to be followed. One page of pictures is better than 10 pages of text.
- **Controlling the document**. The manufacturing engineering group needs to maintain the latest version of each SOP and communicate any changes to the operators. Whenever improvements are made to minimize failures, those changes should be reflected in an updated SOP.
- **Using SOPs for training and auditing**. The SOPs should be used as the basis for training, and the operators need to be periodically audited to ensure that they are following the procedures.
- **Validating the SOPS**. Start using SOPs as early as possible in the piloting phases. Ideally, a SOP should be drafted by the DVT stage and thoroughly tested by the PVT stage.

So far, we've stressed the importance of having highly detailed SOPs to minimize errors – it is always better to err on the side of too much detail rather than not enough. However, there are cases in which SOPs can suffer from being *too* detailed. If text is too long and complicated, operators may stop referring to SOPs after a while. If operators begin to rely on memory of what needs to be done rather than what is documented, they are likely to make errors.

If you're looking for help to hit the Goldilocks-level "just right" degree of detail for your SOPs, there are several technological solutions:

- **Electronic SOPs.** Gone are the days of large three ring binders full of SOPs that operators have to page through but most often sat collecting dust. SOPs are often incorporated into the ERP and factory floor management systems to provide the operators with the right information at the right time. When the parts arrive, the screens at the station automatically pull up the SOPs and transition to automated testing screens at the right points in the process.
- **Built-in testing.** If the product itself can prompt or highlight each quality test and prompt acceptance or rejection, it removes the need for looking at a paper SOP or remembering a complicated sequence. The product can have an on-board program that walks the operator through the assembly and testing process.
- **Automated optical inspection (AOI).** Combining cameras with automated image analysis can help take out the human variability in testing and interpretation of the results.
- **Augmented reality (AR).** This technology holds great promise for using visual prompts that are tied to the actual hardware being built. Using AR equipment, the operator sees both the product and the SOPs on the same screen with pointers that highlight the relevant part.

11.6 Material Handling

The materials used to create your product need to be tracked throughout their journey: they need to be moved between stations, stored for a time, or shipped back and forth to suppliers. The parts need to be routed to the right location at the right time and transported without damage.

The movement of material between workstations can be facilitated and controlled through travelers, Kanban systems, automated material handling, manual process or buffers. An often-overlooked aspect of moving materials between workstations is the way the materials are handled. When designing the process flow, teams should plan how all materials will be routed and handled, including how to mitigate the chance of damage including:

- **ESD-sensitive materials** have to be moved using special equipment that reduces the chance of static discharge damaging sensitive electronics. Material handling equipment can include ESD-shielding custom totes, or generic packaging, such as disposable static-shielding bags and grounded racks for moving large quantities of material around a factory.

- **Totes** to move products safely between stations. A product that can be easily damaged or is sensitive to dirt and debris may need custom totes designed to hold the product as it moves through a facility.
- **Films and wraps** can be used to protect delicate surfaces to reduce damage during handling.
- **Racks and holders** can be used to hold product during curing, cooling, and drying cycles.

Parts and subsystems also need to be protected from human contamination and damage. Typically, the SOP will specify the need for a variety of protective equipment. The protective equipment both protects the worker from harm (chemicals, lacerations, eye damage) but also protects the product from oils, debris or contamination. Protection typically includes:

- **Finger cots**. These are small covers for individual fingers and are used to prevent oils from being transferred to the product. Many factory workers find them more comfortable than full gloves.
- **Gloves**. Full gloves may be needed where either bio or other contamination is an issue or where toxic materials are being used.
- **Clothing**. Different coats/clothing may be required to reduce the risk of contamination, dirt, and ESD.
- **Hair covering**. Caps and hair nets reduce the risk of hairs and skin flakes getting into the product.
- **Shoe coverings**. Booties donned in a dressing area protect the product from dirt and dust brought in from the outside.
- **ESD protection**. This can include personal grounding strips to floor mats.

Summary and Key Takeaways

- ❏ In addition to designing the product, teams need to design the process to build the product.
- ❏ Design decisions can have a significant impact on the quality, cost, and efficiency of a production system.
- ❏ The process definition is maintained in several documents including process flow, process plans, and SOPs.
- ❏ Material handling can impact both cost and quality, so think carefully about how you will move your product throughout the process.

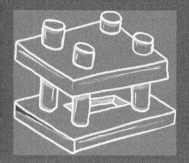

Chapter 12
Tooling

MANUFACTURING PLANNING

9. MAUFACTURING SYSTEMS

10. DESIGN for MANUFACTURING & DESIGN for X

11. PROCESS DESIGN

 12. TOOLING

13. PRODUCTION QUALITY

As production transitions from low volume to high volume, tooling, fixtures, and jigs are used to reduce cycle time, reduce the need for highly trained workers, and improve yields. While tooling is expensive to build, the cost to design and build a tool is balanced by the ability to produce tens of thousands of parts more quickly and more inexpensively. However, there is very little room for error in the design and production of tooling. The lead times can be extremely long and are often on the critical path in product realization. In addition, the cost of mistakes is high and tool modifications can take a long time. This chapter reviews the types of tools and fixtures used in a typical production line, how the design fits within the product realization process, and how to create a tooling strategy for a new product.

Product Realization: Going from one to a million, First Edition. Anna C. Thornton.
© 2021 John Wiley & Sons, Inc. Published 2021 by John Wiley & Sons, Inc.

If you are going to mass-produce a product, you will need custom tooling. It isn't cost-effective to 3D print or hand-build every part in a product. A plastic housing may cost $10 to print on an inexpensive 3D printer, but the same housing will cost $0.30 when injection-molded. In addition, the injection-molding process can produce several thousand units in the same time as it takes a 3D printer to print one. Finally, injection-molding a part with a highly polished tool can create parts with high gloss finishes and interesting surface textures, whereas most commercial 3D printers cannot. Even though the tool for the molded part may cost $15,000, you only have to make 1,500 parts for the tool to break even; for every part made after that, you save $9.70 in material costs.

The words "tooling" and "tool" apply to a broad set of activities and equipment used to manufacture parts and assemblies. In general parlance, a tool is an item or implement used for a specific purpose; the term includes everything from a custom injection mold to a hammer. When setting up a production system, the term "tooling" has a more precise meaning. We will use the following definition for a tool/tooling going forward:

> *"Tool" or "tooling" as a noun*: Specialized pieces of hardware – such as dies, fixtures, molds, patterns, and stencils – that are designed and built to be used for a specific product.

> *"To tool" or "tooling" as a verb*: The process of specifying, designing, and building the tools.

Tooling typically has the following characteristics:

- Built for **higher volume** production (100+ items).
- **High fixed costs** relative to the variable cost of each part.
- **Long lead time** to design and build.
- Requires **manufacturing expertise** to design, build, and maintain.
- Requires **regular inspection and preventative maintenance** to produce high-quality parts and assemblies.
- **Costs** a lot to change or modify (especially for hard tooling).
- **Designed specifically** for a part or product.

Different processes use different types of tooling (Table 12.1). Tools can be used to define part geometry, hold parts while they are assembled, or transfer patterns to a part. In addition to the word "tool," the industry uses a lot of jargon that you will need to become familiar with when you are speaking with manufacturers (Inset 12.1).

Here are some of the reasons that your team should think through your tooling strategy as early in the process as possible:

- Tooling costs can represent a **large percentage of the non-recurring budget** for a new product. Any reduction in tooling costs can directly impact the overall cash flow requirements for the product. For example, if several injection-molded parts can be combined into a single part, the total tooling costs can be dramatically reduced.

- Building and tuning tooling is on the **critical path** and impacts the overall time from the start of product realization to the start of mass production. Delays in cutting tooling or delays caused by needing to recut tooling to fix design issues can significantly increase the time to get ready for mass production.
- Tooling contributes to the **final quality of parts and assemblies**. It is critical to verify that the tool reliably delivers the functional and surface quality requirements. For example, if a deep-draw tool is not designed correctly, sheet metal can tear as it is formed or the surface can be damaged, leading to scrap.
- Tooling sets **the capacity of your production line**. If you build tools that cannot keep up with demand, you will need to spend extra to create additional capacity. For example, if you build a tool with only one cavity and your demand increases significantly, you may not be able to produce parts fast enough. However, if you overbuild tools – e.g., multiple cavities – and don't need the capacity, you will overspend on tooling and eat up capital.

12.1 Types and Their Uses

Table 12.1 introduced some of the terms for tools used in production. Each type of tooling is unique to a set of processes and has its own design guidelines. This chapter will describe a number of processes you may or may not be familiar with. To explain each in detail would take up an entire book. It is highly recommended that if you are not familiar with the processes

Table 12.1
Tooling types and terminology

Type of tool	Terminology and definition
Tools that define a part's geometry	• **Molds**. Blocks of metal with machined cavities which form molten material into a complex shape (injection molding, die casting). • **Dies**. Used to define a surface against which material is plastically deformed to create a shape (forging, stamping, transfer pressing). • **Patterns**. A copy of the part you want to produce. Used to create a temporary mold into which molten material is poured (sand casting, investment casting, urethane casting). • **Mandrels**. Forms used to align or form rotationally symmetric parts (composite manufacturing, metal spinning).
Tools that hold parts	• **Fixtures**. Hold parts during assembly. • **Jigs**. Hold parts while secondary processes are performed. • **Totes**. Transport products safely from one station to another.
Tools that transfer a 2D pattern to part	• **Stencils/plates/photomasks**. Used to transfer a 2D image to another surface. They are used in PCBA to transfer solder to the right location on a blank PCB. Stencils can also be used for printing on products or for etching.

Inset 12.1: Key Tooling Concepts

 When designing tooling, you need to be able to plan how to get the tool into and out of the part (for example, if you want to cut a pocket, you have to ensure the shank of the end mill doesn't hit the part while it cuts). Also, you have to get the material in and out of the cavity when injection molding. Here are a few key concepts that you will need to be familiar with, as they can have cost, quality and lead time impacts for your product.

Parting line. Most molds for casting or injection molding are made from two parts. The parting line is the line where the two halves come together and separate. The parting line can create a "flash," dimensional errors, or slight surface defects.

Draw line. The direction the tool moves in and out of the part, usually perpendicular to the parting line.

Draft angle. The perpendicularity of a wall to the parting line. If you have a perfectly perpendicular wall, the friction between the part and the wall can be so great that it will not be possible to get the part out. Most molds will have a slight angle to the walls to allow the parts to be easily removed.

Gates, risers, runners, and sprues. When pouring or injecting molten material into a cavity, you need a place for the material to enter (sprue), a way to get the material from the sprue to the parts (runners), a connection to fill the part (gate), and a place for excess material to go (riser).

Undercut. A geometry that prevents a part from being removed from a tool. In injection molding parts with undercuts, side actions or lifters are needed to move the tool out of the way of the part so it can be ejected, driving up costs. Undercuts in milled parts require multi-axis milling machines or multiple setups.

Ejector pins. Parts often need help getting out of a mold. Ejector pins are used to push the part out of the cavity so the mold can be closed for the next cycle.

Steel-safe. It is easy to remove material from a tool to make change, but it is tough and expensive to add material. Welding materials into a tool decreases the life of the tool and creates surface defects. The first version of a tool should be made so that the geometry of the resultant part is undersized. This steel-safe approach allows adjustments to be made by removing material from the tool. Once the tooling is tuned, the surface texture will be added and the tool polished.

First article inspection. SAE's (Society of Automotive Engineers) Aerospace First Article Inspection Requirement standard number AS9102B [98] defines the first article inspection (FAI), also called first part inspection (FPI), as providing "objective evidence that all engineering design and specification requirements are properly understood, accounted for, verified, and documented." FAI is especially important for tooled parts because once it is verified that the tool is producing the parts correctly, tooling modifications are stopped and full production can begin.

described here, or ones your factory are recommending, that you read up on them. Rob Thompson's book *Manufacturing Processes for Design Professionals* is an excellent book to learn the basics of a wide range of processes [55].

12.1.1 Molds

A mold is typically a single or multipart cavity that is filled with a liquid material that hardens either because of a chemical reaction (thermoset plastics) or cooling molten material (injection molding and die casting) (Figures 12.1 and 12.2). The term can also apply to a tool that only has one side and is used in vacuum forming, a process in which a thin sheet of plastic is heated and pulled over the tool to create a thin shell. Molds are designed to ensure that the part can cool (or cure) and solidify uniformly. For example, cooling too rapidly can leave a part with residual stresses which can cause subsequent warping of the part.

Typically, molds have other features in addition to the cavity (Inset 12.1). Sprues, gates, and risers need to be designed into the mold to ensure an even flow of material into the cavity. The mold is also not simply the negative of the part. Molten material shrinks as it solidifies, and the mold has to take into consideration that shrinkage and the cooling of the parts. Finally, molds need to be designed to enable the parts to be removed easily (e.g., with sufficient draft angles and locations for ejector pins).

Because of the cost and long lead times of multicavity molds, teams should minimize the number of times the molds have to be modified. Mold designers will often use simulation software to simulate whether the mold will create a high-quality part. In addition, tools are typically cut in a *steel-safe* mode, so any changes to the mold only require the removal of material, not the addition (Inset 12.1).

Typical problems associated with molds that teams will want to avoid include:

- Insufficient filling of the mold.
- Warping of the part after it is removed from the mold.
- Flash along the line where the two halves of the mold meet.

12.1.2 Dies

A die is typically used to plastically deform material into the desired shape. Large forces are applied to the stock material to form the material around the die. Dies are used in a number of applications, including forging and

FIGURE 12.1
Injection molding tool. *Source:* Reproduced with permission from Modern Mold and Tool, Inc.

FIGURE 12.2
Urethane casting mold and part.

stamping. Figure 12.3 shows a die from a progressive stamping process used to make bike chains.

The term "die" is also used to describe tools that are used – along with a punch – to cut shapes out of a piece of flat stock. Mechanical energy is used to apply forces beyond its ultimate yield strength, tearing the part out of the stock. For example, dies are used to cut and crease folds for boxes or in a

turret press to punch out specific shapes. "Die" can also be used to describe
a tool used in extrusion – typically, a metal disk with a shaped hole. As the
material is forced through the die, the part takes on the shape of the hole
as it is extruded.

Typical problems with dies that teams will want to
avoid include:

- Tearing of the metal if the material is deformed too
  much or corners are too sharp.
- Burrs and bending at the edges of the part (burrs
  and sharp edges are safety issues).
- Wear and tear on the dies.

12.1.3 Patterns

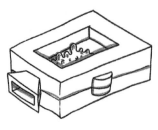

Patterns are used for several casting methods
including sand, investment, and urethane. Patterns are used to create a void into which molten
or liquid materials are poured. In the case of
sand casting, sand is packed around the pattern,
the halves of the mold separated, and the pattern
removed. In the case of urethane casting, silicone
is poured around each half of a pattern to create
the void. Pattern-based processes are used for low-volume parts that have
a lot of detail which would be difficult to machine. Design teams need to
ensure that the pattern is designed to allow for shrinkage, flow of material,

and even cooling. To avoid common problems with patterns, teams should consider the following:

- Designing parts so that critical features can be machined after casting. Casting often has a poor surface roughness, and because of uneven cooling, surfaces are never perfectly flat. If the part has critical mating features (i.e., fitting a bearing in a hole), then the feature has to be machined after casting. The pattern designer has to include excess material so there is sufficient material to remove. This allowance is called the finishing or machining allowance.
- Most material will shrink as it cools, so the pattern needs to be oversized.
- Parts will move as they cool, so features that stiffen parts (i.e., ribs) often need to be added to ensure that the part doesn't warp.
- Whenever possible, patterns should be designed such that the flash appears in areas of the design that are less visible and not critical.
- The surface finish of the pattern will transmit to the final part. If the surface finish is critical, then special attention has to be paid to the pattern's surface. Transmission of surface finish is especially critical when using additive manufacturing for fabricating the pattern because of the layer lines. Post-processing of the pattern may be needed to provide the desired surface finish.

12.1.4 Fixtures and Jigs

The terms "jigs" and "fixtures" are often used interchangeably. However, they have slightly different meanings. Both fixtures and jigs hold parts so they can be worked on, but fixtures are only used to hold parts to make assembly easier or to prevent damage, whereas jigs hold parts and also have guides to locate additional features. For example, a drilling jig holds the part while guiding the drill bit to the correct location. One of the jigs in Figure 12.4 was used to hold ten identical parts so all of them could be machined simultaneously and identically. The other was used to hold down a plastic part during milling. The machinist only had to set up the jig once instead of ten times, reducing the labor hours and cycle time. Traditionally jigs are machined out of metal, but additive manufacturing (3D printing) using rigid plastics is being used more frequently (Inset 12.2) because design changes can be accommodated inexpensively and relatively rapidly.

The fixtures and jigs are typically designed and built in-house by the manufacturing engineering staff. Because fixtures and jigs often need to be

Inset 12.2: Fixtures and 3D Printing

 While additive manufacturing (i.e., 3D printing) is not yet cost-effective at very high volumes, it does have a place in mass production, namely, making fixtures. Fixtures have historically been very expensive to create and hard to replicate, requiring expert designers and machinists. Additive manufacturing is now being widely used to produce a variety of tools including fixtures. These fixtures are relatively inexpensive to print (compared with machined parts), they can be easily replaced if damaged, and they have a much shorter lead time than traditionally made fixtures. The automotive industry, for example, is one of the largest consumers of additive technology, with a large percentage of their additive capacity used for tooling and fixtures.

FIGURE 12.4
Various machining fixtures.

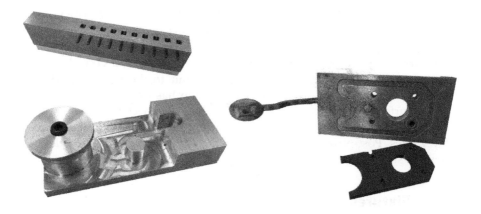

adjusted and reworked during product realization, it is too costly and time-consuming to ship the units back and forth between a factory and a supplier.

When designing fixtures and jigs, tooling designers need to think about:

- **How to locate the part in the fixture**. The fixture will typically have several locating features or surfaces to mate with the part and ensure that a part can only be loaded in the correct orientation.
- **How to clamp the part into the fixture**. The part needs to be held rigidly in the fixture to prevent movement during the work. A variety of cam lock, spring-loaded, hinge, or screw clamps are typically employed, depending on the shape of the part and the potential surface damage from the clamps.
- **How to locate features**. Jigs should have built-in methods to determine where secondary features will be created. For example, a jig may have a drill guide that locates a hole at the right location and angle.

After fixtures and jigs are designed and built, they need to undergo verification. Verification can be done either by measuring the locating features and machining guides, by ensuring that they are within allowable tolerances, or by making multiple parts on the fixture and measuring the variability in the resultant parts.

To ensure continued accuracy, fixtures and jigs must be regularly inspected, maintained, and re-validated to detect and correct any wear or changes in their dimensional integrity.

12.1.5 Other Types of Fixed Costs Associated with Manufacturing

Some processes don't use tooling but require significant upfront work to produce the part. For example, machined parts are may be mass produced when the design, loading, and part tolerances preclude using other near-net-shape fabrication methods such as metal injection molding, forging, or casting.

In the case of machined parts, the team will need to use CAM software (computer aided manufacturing) to create the computer numerical control (CNC) path used by the milling or turning equipment. The tool path is optimized to minimize cycle time while ensuring the part tolerances can be achieved. Good path planning can't be done by just pressing a button; it requires an experienced operator to plan which tools to use, the sequence of operations, and the tool paths. Any changes to the geometry can drive an expensive redesign of the CNC plan.

12.2 Tooling Strategy

For a given part, teams can use several different tooling strategies. Teams can choose soft or hard tooling, a single- or multiple-cavity mold, a flexible or dedicated tool, and a manual or automated process. The right answer for a given product depends on production volumes, cost targets, and quality targets for the product in question. Teams need to weigh considerations of time and budget to determine what kind of tooling to build and when.

12.2.1 Low Volume Production Methods vs. Soft Tooling vs. Hard Tooling

The first decision the manufacturing team needs to make is what kind of tooling to use and when to use it (Inset 12.3). The production methods used in engineering verification testing (EVT) are often different from those used in design verification testing (DVT), and production

Inset 12.3: Production Methods

 Teams will not usually start out with expensive hard tooling. The lead time on the tools is too long, and there are too many pending design changes. These are the different manufacturing methods that may be employed throughout the product realization process:

Prototype samples – additive. Early prototype samples will typically be built using additive manufacturing to mock up the part. The parts can be painted and finished to match the final aesthetics, but will likely not have the same mechanical properties (due to non-isotropic properties of the parts and differences in materials). The parts cannot be used for quality testing because they do not behave in the same way as the finished product.

Prototype samples – machined. A closer mechanical match to the final parts can be achieved by machining the parts out of the final material. These samples can be used to evaluate the mechanical properties, but they are not cost- or time-effective for final production. Design changes are relatively easy to make.

Low-volume production – urethane casting. Silicone molds are made from a pattern and parts are cast from urethane. This process can be used to create a short production run of 10–100 parts. The mechanical properties are close but not identical to those of the final product. In addition, these parts may require significant post-processing if they need to meet a high aesthetic standard. Urethane castings are beneficial for early production because they can be made quickly. However, they are not good for long production runs, because the silicone molds wear out, and they are not environmentally friendly since the materials are not recyclable.

Production-intent – soft tooling. A tool made of softer steel or aluminum can cost less to produce and is faster to cut than hard tooling, but it won't last as long. In addition, it is hard to get a high-quality mirror finish on the parts with a soft tool.

Production-intent – hard tooling. A tool made of hardened steel can last through the production of several million parts while achieving high aesthetic standards, making it the obvious choice for large production runs. However, it is the most expensive and time-consuming tool to build (taking 8–12 weeks to design and cut).

verification testing (PVT) because EVT doesn't use all production-intent parts. As the product volumes increase, designs become finalized, and more expensive and durable tooling will be built. For example, for plastic parts, teams might produce the parts first using additive manufacturing or machining to get parts quickly. During early stages, the design may be in flux, and it doesn't make sense to start making tooling – there is too high a risk that you will need to scrap the mold. For DVT, as volumes increase, urethane-cast parts might be used to more cost-effectively make parts in quantities of 10–100. Throughout early production until final volumes are understood, soft tooling (typically made of aluminum or soft steel) might be used. Once the sales volumes and production rates are de-risked, then the hard tooling (typically made out of hardened steel) is cut to minimize COGS.

Prototype methods such as urethane casting, machining, and additive manufacturing are typically used for low-volume production quantities. These methods are rarely used in final production (unless the product is made in quantities of less than 100) because they do not provide the same mechanical and material behaviors as hard tooling, and they typically create an inferior surface finish. With significant hand polishing and finishing, products made with these prototype methods can be made to look like final parts, but they will not always behave like them.

Hard tooling is expensive to manufacture, can take months to design and build, and is costly to adjust (called opening the tool). If teams need production-intent parts in a short time, or if the design is likely to be drastically changed, it might be appropriate to use prototyping techniques or build soft tooling as a short-term solution.

Soft tooling, predictably, is less expensive to produce than hard tooling, and it can be re-cut more easily. However, soft tooling tends to wear more quickly, and it cannot be hardened and polished to the same finish as a hard tool. In most cases, the soft tooling will need to be retired and a hard tool purchased at a later date as production rates increase. While it seems wasteful to design and create your tooling twice, there are cases when it makes sense:

- **There is too much design uncertainty to commit to hard tooling**. It is cheaper to scrap a soft tool than a hard tool, so if redesigns are likely, a soft tool is the best option.
- **Parts are needed as soon as possible**. If there is a large order that cannot be filled without parts, soft tooling can be used to fill immediate demand while you're waiting for a hard tool to be built.
- **The production volumes do not justify a hard tool**. Hard tooling is expensive, and amortizing the tooling over a low-volume run may not justify buying a hard tool.

The tooling approaches and their relative costs and benefits are summarized in Table 12.2. Figure 12.5 shows a hypothetical small plastic part entered into several US-based cost estimators to show the relative costs of each strategy at different volumes. It gives the total cost (amortized tooling plus part cost) at different volumes. At very low volumes, it is most cost-effective to print the part and, as the volumes increase, urethane parts cast in silicone molds become more economical. At high volumes, the price drops significantly when injection molding is used.

Table 12.2

Tooling options for plastic parts*

	Additive manufacturing	CNC	Urethane casting	Soft tooling	Hard tooling
Tool material	NA	NA	Silicone	Aluminum or soft steel	Hardened steel
Part material	Filament or resin	Stock	Thermoset urethane	Actual	Actual
Design change cost	Low	Low	Medium	Medium	High
Total volume of parts	1–5	1–5	5–100	50–10K	10K–1M
Hourly cost of equipment	Medium	High	Low	High	High
Cost of tooling	0	0	$100	$2–5K	>$5K
Tolerance tightness	Low	High	Low	Medium	High
Lead time	None	2–3 days	2–3 days	2–6 weeks	6–12 weeks
Cycle time per part	Long (5–24 h)	Long (5–24 h)	Long (24 h)	Seconds– minutes	Seconds– minutes
Similarity to production-intent parts	Poor	Mechanically similar	Similar but UV sensitive with poorer surface finish	Mechanically equivalent with poorer surface finish	Identical

* The technology used to build tooling continues to change and improve. The lead times for hard tooling, the speed of machining parts, and the costs of 3D printed parts continue to decrease; and what is expensive and poor quality at the time of writing this book may be very different within a few years.

12.2.2 Capacity

After the team decides what tooling to use, the second critical decision is the capacity of the tool – how many parts per hour can be built. For example, injection-molding tools can have one or more cavities depending on the size of the part and the production rate needed. Tools with fewer cavities are less expensive and can use lower-cost equipment (e.g., lower tonnage presses) but may be capacity-constrained if volumes ramp up significantly. Hard multicavity tools generally produce parts less expensively and have a longer life; but they have several drawbacks – they need to run in larger batches, and require higher upfront non-recurring costs.

In some cases, teams may choose to do a single-cavity soft tool to debug and ramp production and then build the multicavity tool once the design is fixed and a more concrete sales forecast is known.

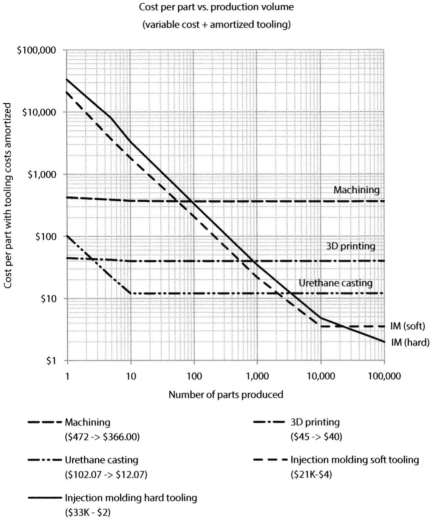

FIGURE 12.5
Example cost per part for a small plastic part using different strategies

12.2.3 Flexibility

Some tools can be built so that they can be easily modified and maintained. For example, inserts can be designed into stamping dies to allow for the geometry of key features to be changed or high-wear parts to be replaced. Inserts allow for a single mold to make a family of parts. This flexibility drives up the initial tooling costs; however, flexibility can reduce costs in the long term by minimizing the cost of design changes, sharing costs across multiple products, and reducing the time needed to repair tools.

Fixtures and jigs can also have flexibility built into them. Until a relatively short time ago, aircraft assembly used custom fixtures which depended on the fixture dimensions to hold and locate parts relative to each other. Figure 12.6 shows a custom fixture used to build a wing spar. The fixture is over 50 ft long and is designed to hold and facilitate the assembly of just one type of spar. More recently, aircraft assemblers are using flexible tooling. Flexible tools can be used to assemble a range of parts, and don't depend on the fixture dimensions to locate parts relative to each other. This has both sped up production and reduced costs. Boeing was able to use flexible tooling because they were able to create accurate features that allowed parts to be located relative to each other without the use of dedicated fixtures [99].

12.3 Tooling Life-cycle

When you are designing a new product, you know that the designs will go through several stages, from initial design through prototypes up to the final product. It should also be clear that the tooling required to create these prototypes and products will also have to go through several design and testing phases. Table 12.3 delineates the life-cycle of a tool from the initial design concept through retirement.

Step	Description
DFM of the design concept	DFM is applied to the parts to ensure that they can be reliably made (Chapter 10). Although designers can incorporate DFM rules by themselves, it is critical that an expert (CM or tooling supplier) also reviews the drawings. Ideally, the DFM reviews should be done as early as possible during product design to enable design changes to be easily incorporated.
DFM for tooling	After the parts are fully defined, the fabricator or tooling supplier will suggest to the design team relatively minor changes to improve manufacturability which won't impact the function of the part. These changes can include adding draft angles to allow for part removal from molds, changing the location of the ejector pins, or adding additional ribs to strengthen the part and reduce the risk of defects.
Tooling design	Manufacturing engineering or the tooling supplier will design the tool based on the modified part specifications. Tooling design can take several weeks, depending on the complexity of the part. For example, when designing the tooling you have to not only design the cavities to account for shrinkage, but you have to design the sprues, gates and cooling channels. The better the DFM rules that are applied to the part, the faster the tooling design will be. For example, if teams can avoid undercuts in injection molding or cores in sand casting, the tooling design will be greatly simplified.
Tooling production	After the tooling design is finalized, the tool can be built. The tooling production time depends on which technology is used. Cutting a hard-steel tool can take up to 12 weeks (Figure 12.7), whereas printing a fixture may take only a few hours. You should assume that the first parts off the tool may not perform as expected. Because it is expected that tools will be modified, tools are designed and built using a "steel-safe" approach (Inset 12.1).
First shots and FAI	The first shots are the first parts off the tooling and are subjected to FAI (first article inspection). The first shots are inspected and all dimensions checked against the dimensions and tolerances in the drawing. After the parts pass FAI, small production runs are made to support the pilot builds.
Pilot phases	The parts produced from the tooling post-FAI are built into pilot production runs. Problems with assembly, failures in durability testing, aesthetic issues, and user testing will highlight problems with the parts. The design team will then make any necessary changes to drawings and tools. Tooling may go through 2–10 iterations until the product meets performance specifications.
Polish and finish	After the geometry is finalized, the tooling will be polished, any surface textures applied, and a final pilot batch run (usually the PVT samples).
Qualification	The part fabricator along with the quality group must ensure that the tooling, as designed and built, reliably delivers the parts and features required by the final specifications. The process of verifying that the tooling is correct and identifying the process parameter windows is called *qualification*. Mold qualification can also include taking statistical samples to understand whether the part-to-part variability meets the design requirements. These samples can be taken from a single batch or across multiple batches.

Table 12.3 Steps to create tooling from design concept through production

(Continued)

Table 12.3
(Continued)

Step	Description
Production runs	The tools will be used to build large batches to support mass production. Typically, molds do not run continuously 24/7 (except for very high-volume producers such as Lego). Rather, a batch is ordered and put in the production queue. The tool is brought from storage, and set up in the machine; and after an initial warm-up cycle, the batch is produced. Typically, each production run will go through some form of inspection to ensure that the parts meet specifications. After the run is over, the tool is cleaned and put back in storage.
Preventative maintenance	Tooling will become worn and damaged through use and handling. All tooling should be on a regular maintenance schedule. A good preventative maintenance schedule includes: • Scheduled inspections and repairs to ensure good performance. • Requalification of the tooling to ensure consistent quality. • A clear set of procedures to execute preventative maintenance. • Well-maintained records of completion and results. • Clearly defined responsibilities for maintenance.
Replacement	At some point, tools will become too worn to repair. Manufacturing engineering and the factory should have a replacement plan for each tool to allow sufficient time to get the new tooling produced.

12.4 Tooling Plan

Very early in the product design phase, the operations and design teams need to decide what tooling is needed and when. Hopefully by this point in the book, the value of early planning has been well established, but here, specifically, are the reasons to think carefully about the tooling early in product realization:

• **Tooling can represent a large part of your NRE budget**. Each tool can cost tens of thousands of dollars. When budgeting a project, you don't want to realize that you need an expensive tool just when cash flow gets tight.

FIGURE 12.7
Cutting tooling.
Source: Reproduced with permission from Modern Mold and Tool, Inc.

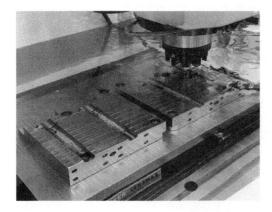

Checklist 12.1: Tooling Plan Checklist

How much is the tool going to cost relative to the part costs and volumes? As previously stated, hard tooling has the lowest per-piece cost at very high volumes, but at low volumes, CNC machining may be a better choice. However, these trends are changing even during the writing of this book. Soft tooling is becoming cheaper and easier to source, so it may soon be the case that soft tooling is advisable even for relatively low volumes.

❑ What material and manufacturing processes are used for each part (Chapter 10)?
❑ What are the likely volumes (batch sizes)?
❑ What is the likely volume growth?
❑ What mechanical tolerances and surface quality are required?
❑ What are the possible tooling strategies for each part?
❑ What are the relative costs of your options for tooling (Chapter 8)?

How fast do you need the part? Hard tooling might be the most cost-effective, but as we've said, it has the longest lead time.

❑ When are parts needed for each pilot and for the start of mass production?
❑ What cash is available to purchase tooling and when?
❑ Is there an option to use a different method in the short term until the long-lead tooling is available?

How many more design changes and what kinds of design changes do you expect? If the design changes are only going to be cosmetic, then starting the hard tooling first may be the right choice. If the teams are iterating on the design but need parts that behave like final parts, CNC machining might be the right starting point. The ultimate volumes may be very high, but if there is a risk that the tool will need to be scrapped, it might be worth building soft tooling until the design is stabilized.

❑ How likely are design changes after the start of mass production?
❑ How expensive will those changes be?

❑ What is the relative cost of doing soft tooling and then hard tooling vs. starting with hard tooling and having to make modifications to the tools?
❑ Are these parts being designed to be used in several product lines?

What fixtures and jigs are going to help with assembly? The best designed product won't need any fixtures or jigs; however, design for assembly (DFA) can't always remove all difficult assembly steps. If teams can think through the assembly fixtures and jigs ahead of time, parts can be designed to either avoid the need for fixtures or with features that make the design and use of fixtures and jigs more manageable. Based on analyzing the assembly process, teams should propose where fixtures and jigs may be needed. The assembly factory can provide input into these issues.

❑ Are fixtures needed to hold parts while they are assembled?
❑ Is the fixture used to define the geometry of the final assembly?
❑ What features need to be located using a jig?
❑ Is the assembly process going to be manual or automated?
❑ Is there a risk of damage during assembly?
❑ Are the parts awkward or difficult to hold?
❑ How error-proof is the assembly process?

Where and how are parts and assemblies likely to get damaged in transit? Totes can be expensive and can have a long lead time. Teams need to think through how the material will be transported.

❑ Where and how are parts and assemblies being moved between stations, storage, distribution, and suppliers?
❑ At what steps in the assembly are parts and sub-assemblies subject to damage from handling, vibration, and contamination?
❑ What damage is likely to increase rework and degrade downstream product quality?
❑ What material handling equipment already exists?

- **The lead time on tooling is very long**. You can't just decide at the last minute that you need another tool because that could delay your project by up to three months.
- **Tooling can set the overall capacity of your production line**. You don't want to overspend on tools and build capacity you don't need, but you also don't want to create a bottleneck.
- **Tooling is a major contributor to the COGs of the part**. You will always need to balance the cost of the tool against the cost of the part. Because production volume has such a large effect on the total part cost (variable plus amortization of tooling), tooling choices will be highly dependent on the sales forecast.

Creating a tooling plan is a cross-functional effort; for example, marketing and sales teams need to provide the sales forecasts, operations needs to predict capacity requirements, and the design team needs to evaluate design changes that reduce the tooling budget.

There is never one single right or obvious solution; however, discussing all potential outcomes and their relative risks can help teams to make an informed decision. For each process (fabrication, assembly, or test), the team needs to determine what tooling the factory needs and what strategy is going to be taken for that process. They need to ask themselves a number of questions on the topics as listed in Checklist 12.1.

Summary and Key Takeaways

- ❏ Tooling is a significant driver of cost, quality, and schedule of product realization.
- ❏ There are several tooling strategies ranging from low volume/soft tooling to high volume/ hard tooling.
- ❏ Choosing the right tooling at the right time involves weighing many competing constraints.
- ❏ DFM will reduce your tooling costs.
- ❏ It is critical to plan your tooling strategy as early as possible.

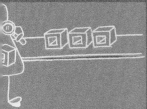

Chapter 13

Production Quality

MANUFACTURING PLANNING

9. MAUFACTURING SYSTEMS

10. DESIGN for MANUFACTURING & DESIGN for X

11. PROCESS DESIGN

12. TOOLING

13. PRODUCTION QUALITY

Once the process for making the product is defined, the team needs to design how that process will be monitored for quality. Processes will degrade over time, suppliers will make mistakes, operators won't follow procedures, and tooling will wear. Production quality is maintained through quality testing, process control, and ongoing continual improvement. This chapter describes the typical quality control points and how they are implemented.

Product Realization: Going from one to a million, First Edition. Anna C. Thornton.
© 2021 John Wiley & Sons, Inc. Published 2021 by John Wiley & Sons, Inc.

Maintaining and improving production quality is critical to the satisfaction and safety of your customers, the reputation of your product, and the financial viability of your business. In 2000, the NTSB (National Transportation Safety Board) started an investigation into tread separation failures in Firestone tires. In the US, these failures ultimately caused 271 fatalities and over 800 injuries. After significant investigation, the NTSB determined that poor quality and process control in the Decatur plant was a major contributor to poor bonding of layers in the tires [100]. Firestone ultimately had to recall over 6.5 million tires, costing Ford – the major customer of the tires – over $3 billion. More rigorous process quality control would probably have saved these companies billions of dollars, and more importantly, lives.

As you're envisioning bringing your dream product into being, you might be tempted to think that getting high-volume production running will be the endpoint of your design and planning process. But take a lesson from Firestone: your team has to design and plan your quality control process just as carefully as you design and plan for your product and production line. After your product and the production system have been verified and validated during the pilot process (Chapter 7), it is still necessary to continue

---

Inset 13.1: Quality Management Systems

 A QMS is the overall governance structure (in the form of policies and procedures) that defines how quality will be managed throughout the organization. There are several industry standards for QMSs, including ISO 9001 [34].

A well-conceived QMS will have the following characteristics:

- Documentation and control of the quality plan, procedures, and testing.
- Measuring quality at critical points in production (either simple AQL sampling, defect rates, or SPC).
- Identifying the causes of defects and driving quality control improvements.
- Application of a continual improvement process to address quality issues and identify areas where improvements in material handling, process standardization, and design for assembly can be implemented (Section 19.2).
- Ongoing training of operators and auditing of the production floor to ensure that a quality process is being maintained (Section 19.4).

If you are supplying a product to another business, your customer may require your company to have a QMS system in place. A QMS can be as simple as a document outlining your organizational structure and processes used to manage quality, or as complex as a full ISO 9001 deployment. If you are using a contract manufacturer (CM) to build a majority of the product, you can piggyback on their system. The documentation, control, and application of a QMS system is a regulatory requirement for many industries, especially medical devices.

to control quality throughout production. Just as engineering teams designs the product and manufacturing methods, and manufacturing engineering designs the production system, quality engineering needs to design the system to ensure quality. Managing quality is more than just implementing some testing in the production line. It requires defining an overarching quality management system (QMS) (Inset 13.1), documenting the quality control plan, and driving continual improvement.

This chapter will focus on the design of the production quality test plan, starting with methods used to measure quality (Section 13.1), then detailing how to use those quality measures. Sections 13.2 and 13.3 will outline where quality can be controlled throughout the production line. Section 13.4 will describe how the production quality system is documented.

13.1 Measuring Quality

When you impose quality controls, how you measure quality is as important as what you are measuring. Teams have a wide range of options when selecting how to measure what you think is important. For example, a dimension can be measured with a $0.50 ruler or a coordinate measurement machine (CMM) that costs tens of thousands of dollars. Functional testing can be done by checking to see if the product turns on or can rely on complex self-test algorithms that analyze the performance. There is an entire science of how to measure called *metrology*. How you choose to measure your quality will depend on a number of factors:

- What is the chance that the production system will cause a failure?
- What is the impact of that failure (minor, major, or critical)?
- What is the cost of controlling the quality (both fixed and variable)?

As with many of the topics in this book, we will give a quick introduction to the terms that you will need for this chapter and some of the more frequently used techniques.

13.1.1 Metrology Terms

Metrology is the science of weights and measurements. Below are several key concepts that teams should understand when measuring quality.

- **Validity** indicates how well the equipment measures the value you are concerned about. Often you can't directly measure what you are interested in. For example, many medical devices that transport fluids are checked by pressurizing the system with air, since a liquid can't be used because it would contaminate the product. Air (a compressible fluid) will behave differently than a liquid (incompressible fluid). The test may find small holes that the viscous fluid wouldn't permeate (Type I error or false positive) or the test may not find weaknesses that develop as the system maintains pressure over time (Type II error or false negative).

- **Accuracy** indicates how close the measurement is to the actual value. If a 5.00 kg weight is measured 10 times, and the average measurement is 4.9 kg, the instrument used is less accurate than an instrument that measures an average of 4.99 kg.
- **Precision** indicates how close the measurements are to each other. For example, if you have the following measurements: 4.5, 5.5, 4.75, 5.25, and 5 vs. 5.1, 5.05, 5, 4.9, and 4.95. The two have the same accuracy (their average value is the same) but the second set is more precise than the first. Precision is essential when you only want to measure something once and get an answer that indicates if the part is good or not.
- **Gauge R&R** (gauge repeatability and reproducibility) is a formal process to determine the precision of a testing system arising both from the equipment and the people using it. Gauge R&R takes a series of measurements by several people to understand the inherent variability in the measurement due to the equipment and that introduced by different people.
- **Calibration** is a formal process that uses industry standards such as those defined in the US by NIST (National Institute of Standards and Technology) to ensure the accuracy of a piece of equipment [101]. Equipment is typically calibrated on a regular schedule and using a controlled SOP.

13.1.2 Dimensional Measurement

How you measure dimensions of parts is a large part of the field of metrology. Parts need to be checked to see whether their dimensions fall within the acceptable range of variation (i.e., their tolerances).

For example, when two aircraft body sections are joined to create a fuselage, the geometry of sections is critical. If the sections are not aligned, assembly will result in steps or gaps which increase drag and increase fuel consumption. Aircraft manufacturers spend a lot of time and money measuring the complex geometry of fuselage sections to ensure that their shapes are correct before they are assembled. On a small scale, Lego also does a great job of controlling the feature sizes of their Lego bricks – the Lego bricks the author had in the 1970s fit with the bricks her daughters used 40 years later. However, a 1–2% size variation in Lego blocks may not create any noticeable difference in their ability to snap together,

FIGURE 13.1
Pictures of gauges including a bore gauge, a pin gauge, thread gauge, and radius gauge

whereas such a 1% discrepancy could have a significant impact on an airplane flight performance.

The quality control teams have a wide variety of tools and methods at their disposal to check the geometry of parts. These tools range from simple gauges to complex CMM equipment. Which tool is appropriate for which part will depend on the accuracy required, the cost, and cycle time to achieve the measurements. Several types of tools can be used to assess the dimensions of parts, including:

- **Gauges**. These include custom gauges or standard gauges such as feeler gauges. Several standard gauges are shown in Figure 13.1. Gauges are used to verify that a dimension falls within the allowable tolerance (sometimes called go/no-go gauges). They are typically made of highly durable and low-wear materials. Many gauges are now being built using additive processes to reduce the cost and build time (Inset 12.2). The cycle time for using a gauge is typically short.
- **Rulers or tape measures**. These low-accuracy measuring tools are used only when the test doesn't require significant precision (e.g., typically for large dimensions such as boxes).
- **Vernier calipers or dial calipers**. These can be used to generate more precise and accurate dimensions but are highly dependent on user training and consistent use. Also, they require time to take the measurement and document the results.
- **Coordinate measurement machine (CMM)**. CMMs accurately measure the locations of features in 3D space. They can be run manually or be computer-controlled. CMM often requires parts to be taken off the line to be measured because taking measurements with a CMM typically has a long cycle time. Handheld non-contact CMMs which enable measurement in the production line are becoming more prevalent. CMM includes an

extensive range of technology, including using contact (touch probes) and non-contact (photogrammetry and optical) measurement methods. Taking CMM measurement is typically extremely time-consuming and the equipment is very expensive. However, it can provide accurate measurements of complex geometry.

Picking the right metrology tool is based on several factors:

- How accurate and precise do the measurements need to be?
- Is it a pass/fail test, or do you need to take a variable measurement?
- Is the measurement going be taken on the production line (100%), or will the sample be sent to a measurement lab?
- How complicated is the part? Do you need to measure complex surfaces or simple dimensions?

13.1.3 Golden Samples

Ideally, the pass/fail test should be based on quantitative or measurable results. For example, "does a pin gauge fit in a hole?" or "is there sufficient charge on a battery?" However, in some cases the acceptance may be harder to quantify. For cases when it is hard to quantify the acceptable quality, "golden samples" are used to provide a baseline for evaluation. Golden samples are units or outcomes that are used as an example of what "good" is, and are typically used to check image quality, aesthetics, color matching, and allowable surface defects. Golden samples, approved by the product team, define acceptability and are typically signed or marked by the product leader to indicate their approval as the baseline for quality checks.

13.2 Tracking Quality

Once you have determined *how* you are going to measure quality, then you need to determine how frequently and in what way that information will be used to drive quality.

13.2.1 Pass/Fail Acceptance

The simplest type of quality assessment is a 100% pass/fail test. Each unit or subassembly is evaluated/measured/assessed and either accepted or rejected. The pass/fail may be done on a test fixture (does the Wi-Fi connect?), a gauge (does the pin fit in a hole?), or a visual inspection (are there scratches?). The factory may report on the daily or weekly rejection rate if requested. If a part fails, it is sent back for rework or it is scrapped.

13.2.2 Acceptance Sampling

Acceptance sampling or AQL (acceptable quality limit) sampling is used when it is too expensive to 100% check all parts, if there is little risk of a quality defect, or if products are destroyed or damaged during testing.

AQL sampling began during World War II and was used to ensure the quality of bullet manufacturing. The AQL standards were formalized and documented in 1963 in the MIL-STD-105D (MIL-STD are military standards used by the US defense organizations) and then updated in 1989 as MIL-STD 105 E [102]. The ANSI (American National Standards Institute) standard – the civilian version of the MIL standard – ANSI/ASQ Z 1.4 [103], was published in 1995. The MIL standard and the ANSI standard are virtually identical, with the benefit of the MIL standard being that it is a publicly available document, whereas the ANSI standard has to be purchased.

The foundation of the acceptance sampling process is the AQL, or the percentage of product you are willing to have as defective. In addition, you pick the inspection level (I, II, or III) which indicates, in ascending order, the criticality of the failure. Based on the batch size, the standards provide the minimum number of samples you need to test from each batch. The tables also indicate how many failures can be tolerated in the sample before the batch of parts or products is considered a failure.

Based on the number of rejected parts in the sample and the acceptance levels, a batch might be rejected. Depending on the agreement with the factory and the severity of the failure, the entire batch may need to be re-inspected at 100% or the batch quarantined and corrective action implemented.

13.2.3 Statistical Process Control

Teams are often tempted to "inspect in" quality, meaning that they rely on inspection to sort out the bad product from the good. When you inspect in quality, every aspect of the part and process is checked, and the systems that do not conform are scrapped or reworked. However, assuming that you can inspect-out bad product is dangerous. According to Juran [69] – one of the founders of the quality movement – visual inspection is only 87% effective at finding quality problems. Also, sorting by inspection is highly dependent on the acceptability limit. If the acceptability criteria are incorrectly set, inspection will either under- or over-reject parts.

You do not want to find yourself in a situation in which you have so many defective parts that you need to inspect 100% of the parts produced. The only sustainable and cost-effective way to improve quality is to find trends in those defects, then improve the underlying processes creating them. Ensuring that the processes are stable and capable reduces the likelihood of unexpected quality issues. The first two steps in process control are the definition of the SOPs (Section 11.5) and the maintenance and calibration of tooling and equipment.

The third step is applying statistical process control (SPC) to processes to track the value of a certain product or process characteristic over time (Figure 13.2) [104].

SPC is used to understand not only whether the quality is acceptable but also whether there are outliers (products that fall out of the statistically likely range) or a trend (a consistent increase or decrease that may indicate a fundamental change in the process). For example, manufacturing quality teams might see a trend in a dimension. The dimension might still fall within the tolerance limit, but the graph clearly indicates that the quality is degrading. This allows the team to address the underlying cause of the drift before you start rejecting parts.

FIGURE 13.2

Example attribute SPC chart counting defects per batch

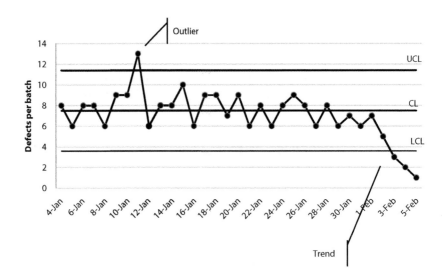

At the simplest, SPC charts can be divided into two types: variable vs. attribute. In variable SPC, the process measures a variable (e.g., the length of a part or the strength of a signal) either on statistical samples or on each part/product. In attribute-based SPC, the teams track variables that have one of several discrete possible values; for example, the pass/fail rate of a camera test or whether a screw is missing in an assembly.

SPC charts can be broadly applied, but be careful about being over-zealous: the over-application of SPC charts can take focus away from what is critical. In addition, it is not enough to commission a SPC chart; you have to clearly delineate who is responsible for analyzing and acting on the information from the chart. An effective SPC system has the following characteristics:

- **Prioritizing what is critical**. Putting SPC on every feature dilutes the impact of each chart. The teams should focus on the variables or attributes whose risk (cost and probability) of failure is high.
- **Interpreting the data to identify trends**. Too many SPC charts are generated and then never looked at again. The data itself will not solve a problem; it is the action taken based on the data that improves the product.
- **Triggering corrective action**. Outliers and trends should trigger root cause analyses and putting in a plan to fix it. Typically quality engineering will do the initial root cause analysis and will engage the right team to fix it. If it is an equipment failure, then manufacturing engineering will take lead; if it is a supplier, then supplier management will drive the corrective action.
- **Having processes to drive variability down**. Teams should seek to identify the common causes of variation and reduce them to drive quality improvement.
- **Communication**. The charts should be communicated to the right teams (manufacturing engineering, quality engineering, continuing engineering, and supplier management) to keep everyone aware of the quality trends.

13.2.4 Part vs. Process Control

SPC can be applied to either the underlying processes (e.g., temperatures, flow rates, etc.) or to the outcomes (e.g., the dimensional accuracy). It's important to remember not simply to measure outcomes; while measuring the product will help you to find issues that will impact product performance, you may not know *why* the failure happened. Ideally, not only the part or product features should be measured, but also those process parameters that drive quality. A few examples of process control include:

- Tracking and controlling temperatures and pressures in injection molding.
- Inspecting, maintaining and replacing tools and capital equipment regularly.

- Providing a template with a cut-out used to ensure the correct location of labels during assembly. The template is placed on the surface of the product, and the label applied inside the cut-out.
- Using temperature and process monitoring equipment for soldering.
- Controlling the temperature and age of solder paste in a surface mount technology (SMT) line.

13.3 Production Quality Test Plan

The prior two sections describe how to take measurements and how to use those measurements to drive quality. This section outlines *where* to put the quality control points in a facility to maximize quality while minimizing the cost and cycle-time effects of quality testing.

If you only check the quality of a product after it is completely assembled, you get very little insight into the cause of a given failure; you likely have to take apart the product to hunt for the error, and either rework or scrap the whole product. In addition, the 100 units still in production when you find a faulty product at the end of the line likely all have the same problem. Tracing the source of problems is often time-consuming, and while you are finding a problem, you continue to build defective product.

Finding defects before the finished product is built and packaged reduces overall scrap and rework and improves material flow. It also enables quality teams to more quickly identify areas for improvement. Figure 13.3 shows the points in a process where quality is evaluated in a typical electromechanical assembly facility. The quality control steps are typically done within the production line.

- All purchased materials are put through incoming quality control (IQC) (Section 13.3.1).
- Fabricated parts are subjected to FAI (first article inspection). Then SPC is used to monitor quality trends (Section 13.3.2).
- Printed circuit board assemblies (PCBAs) are subjected to several tests, including in-circuit testing (ICT), automated optical inspection (AOI), and functional testing (Section 13.3.3).
- The production line consists of a series of assembly or workstations. Some of the subassemblies will be tested and sent to rework if issues are found (Section 13.3.4). The final product will also be tested.
- During production, the product will be checked to ensure it meets the aesthetic requirements (Section 13.3.5).
- A small subset of the products is sent for shipment audits (Section 13.3.6) and ongoing production testing (OPT) (Section 13.3.7).

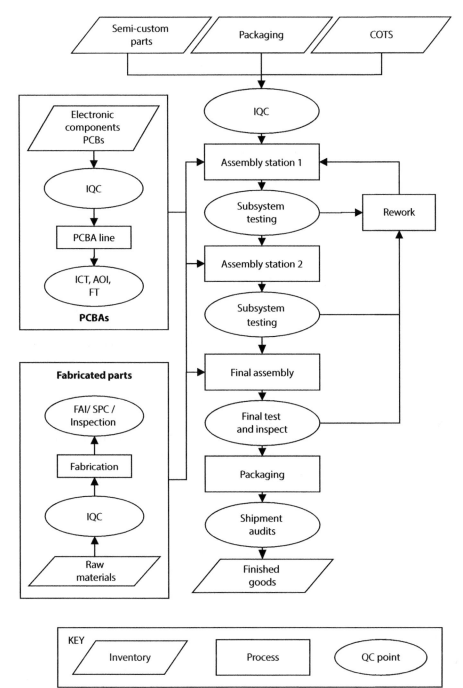

FIGURE 13.3
Process flow of a
consumer electronics
production system
with quality control
points

The next few sections will discuss each of these quality control points in more detail.

13.3.1 Incoming Quality Control

All of the materials, whether raw materials or complete subsystems, will come from outside suppliers. When material arrives from the supplier, it is typically subjected to IQC (incoming quality control). The quality of the final product is highly dependent on the quality of the incoming material. If parts don't match drawings, they won't assemble, and if components don't perform as expected, the product won't function. It is better to catch problems in incoming materials before you build them into your product.

In some cases, IQC is managed by the supplier, and in some cases, IQC is executed at the factory. Having your supplier perform the IQC before shipment enables the materials to be delivered directly to the factory floor. However, this approach is typically limited to production lines with very high volumes, where storing material for IQC would be overwhelming (e.g., an automotive plant) and where the supplier is a highly-trusted partner.

IQC can be as simple as checking whether the part number matches the purchase order, or it may involve comprehensive dimensional, functional, and aesthetic checks. In some cases, a few samples will be pulled from each batch and destructively tested (meaning they can't be sold). In most cases, the acceptance or rejection of the shipment is based on the AQL sampling rates and criticality of the defects (Section 13.2.2).

The type of quality testing at IQC will depend on the risk of *not* catching a problem and the cost of the quality control:

- **The impact of a defect on assembly and quality**. For example, a cover with important surface finish quality standards will require more intense IQC than an interior screw.
- **Cost of inventory**. IQC takes time and hence drives up inventory and the effective lead time to get the parts.
- **The reliability of the suppliers**. Some suppliers are more consistent than others (Section 14.5).
- **The cost of IQC**. A full QC check may require both time and expensive equipment.
- **The cost of destructive testing**. Some testing may destroy the parts.

13.3.2 **Fabricated Parts**

Fabricated parts are uniquely made for a product, and the geometry of those parts is critical to the final performance of the product. If fabricated parts are outsourced, they are typically checked as part of the IQC process (Section 13.1.1). Whether you or an external supplier make the parts, those parts will be checked using one or more of the following methods:

- **First article inspection (FAI)** is a formal and comprehensive inspection of the first (or first several) part from a batch.
- **100% inspection**. Features on every piece are checked to see whether they fall within acceptable tolerances. This is not typically done for high-volume/large-batch production.
- **Statistical process control (SPC)**. Parts are sampled and certain features are measured; the values are plotted on a time series. These charts are used to track whether there is a change in the fabrication process that may impact quality.
- **Process control**. Rather than measuring the parts, the process can be monitored to ensure that it will produce acceptable parts.

13.3.3 **Electronics Quality Control**

If the product includes electronics, the boards and electronics will typically be tested after populating them with components, reflow soldering, and manual assembly. Several methods may be used to test electronics:

- **Automated optical inspection (AOI)** uses digital cameras to determine whether all of the parts are on the boards and if the parts are oriented correctly. AOI is done after population but before reflow. Errors can easily be fixed if found before the components are permanently soldered to the PCB.
- **In-circuit testing (ICT)** checks whether the integrity and conducting paths of the boards are correct. It does not check whether the board functions correctly. It typically is a "bed-of-nails" test fixture which has a series of test pins that are held in a fixture so they mate up with the test pads on a PCBA. Automated equipment runs a current through the pins and makes sure the unit exhibits the right behavior. ICT is typically done before manual assembly.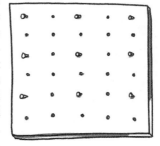
- **Functional tests (FT)** are more sophisticated tests than ICT or AOI. The boards are mounted to either a golden sample (i.e., ideal product) of the

rest of the device or a simulator (the computer pretends to be the rest of the device). The functional test fixture runs the board through the core functions and makes sure the boards are working as intended. Typically this is done just before the completed boards are assembled into the product.

A picture of a typical electronics testing facility is shown in Figure 13.4.

13.3.4 Subsystem and Final Functional Testing

During the assembly, critical functional performance can be tested either at a subsystem level (in-line testing) or after final assembly (end-of-line testing). For example, a subassembly might be tested to ensure that it is sealed correctly, or a motor tested to ensure it operates correctly before assembling the housing around it. In-line functional tests are done for several reasons:

- **It may not be possible to test for functions at the end of the line.** For example, once a motor assembly is built into a product, it may not be possible to check the torque because you don't have access to the motor anymore.

FIGURE 13.4
Electronics testing facility. *Source:* Reproduced with permission from Shenzhen Kaifa Technology

• **The cost of rework is too high after the final assembly.** Often after final assembly, it is not possible to rework a product. For example, if the housing is ultrasonically welded shut, it would need to be discarded if the product is defective.

Most products will also have some tests done once the whole product is assembled. End-of-line testing can be as simple as turning the product on to make sure it works, or as complex as time-consuming custom tests of individual functions. Some products may require a *burn-in* process. Burn-in means repeatedly running the product through its functions to find any early-life failures. Burn-in is very expensive because it takes time, equipment, and space. It is typically used for safety-critical issues, where there is a small but critical chance of early-life failures, or the cost of premature failure is very high. For example, if a critical safety circuit used to cut out a motor has a potential of early life failures, you might stress test that component to weed out any products that are likely to fail.

Whether in-line or end-of-line testing, functional testing will probably require custom equipment, fixtures or built-in testing to facilitate, standardize, and reduce the cycle time of testing. The testing equipment can range from simple go/no-go fixtures that determine whether a feature falls within tolerances, to sophisticated automated testing equipment that collects copious amounts of data to evaluate the performance of the product. For example, a smartphone is subject to over 50 automatic tests to check every aspect of the product performance.

Your team can reduce its dependence on testing equipment through the use of built-in testing. For example, a battery-powered device that connects to Wi-Fi might run a check on the battery charging circuitry and then check if it successfully connects to the Wi-Fi. The operator would just wait for the device to pass the test and only intervene if it shows a failure of one or more of the tests. On the other hand, the device might just prompt the operator to execute a series of tests. For example, it would ask the operator to plug in the charging cable and see if a light goes on. The advantage of built-in testing is the ability to use the same testing protocols in the case of a product return or warranty. Also, changes to the internal software are a lot less expensive than changing hardware-based test fixtures.

13.3.5 Aesthetic Inspection

Products and packaging should be checked to ensure that the final product adheres to the aesthetic requirements outlined in the specification document and aesthetic inspection documents. Aesthetic inspection is typically done visually – remembering that visual inspections are only 87% effective – and in limited cases using automated equipment. Typically, the production aesthetic tests are a subset of the pilot aesthetic tests described in Section 7.3.2. However, the tests are streamlined and automated where possible to reduce cycle time and the dependence on individual judgment.

13.3.6 Shipment Audits

Shipment audits are a third-party check of product quality before the final invoice is paid for the shipment. Typically, the master service agreement (described in Section 14.4.4) will include a clause that enables you to engage with a third party to perform the audits on your behalf. Shipment audits pull several units from the shipment – usually based on AQL sampling guidelines – and conduct a full aesthetic and functional check on the product. If the defect rate exceeds the agreed-to AQL levels, then the factory/CM is typically responsible for re-testing and fixing 100% of all products before payment is approved.

Shipment audits are highly recommended because:

- **The onus is on the supplier to correct the problem**. If the damage is found after the product is FOB (free on board), suppliers can argue that damage was created in transit or at the customer site. If the auditor can find the problem before it leaves the factory, the manufacturer is responsible for rectifying the defects.
- **Reduced cost of shipping back non-conforming units**. It is expensive to ship defective units back to the country of manufacture, and in some cases, it may be challenging due to customs and regulatory issues.
- **Fewer defects get to the customer**. By testing just at the end of the assembly line, you are catching quality testing issues in the factory before more defective product is built. If you wait until the product arrives in your country to test it, you continue to build bad product while you are shipping the product to your distribution center and inspecting it.

There are any number of third parties who can be hired to do shipment audits, but not all shipment auditors are the same. Some will just do a cursory check, while others will be able to help with root cause analysis and implementing changes in the factory. Typically, the second are more expensive.

The shipment audit document will typically involve:

- The sampling plan based on AQL levels agreed to by the factory (see your master service agreement)
- The aesthetic requirements being evaluated and their level of criticality (minor, major, critical)
- The list of functional tests done on the product and their level of criticality (minor, major, critical)

Depending on the number of defects found and the criticality of those defects, different responses will be triggered. For example, any critical safety failures will automatically trigger a 100% inspection of all batches, a quarantine of any product not delivered, and potentially a recall of product already sold. Minor failures may just trigger rework of the defective product, and a corrective action request sent to the factory.

13.3.7 Ongoing Production Testing (OPT)

Some quality issues are of a high enough safety or performance risk to warrant ongoing life-cycle and durability testing while the product is in production. For example, if a critical component could fail and cause a safety risk, even if the component passed reliability testing during piloting, it may be necessary to sample several parts in each batch and destructively test them. For example, the blades on the food processor described in Inset 7.8 should probably be tested for fatigue failures periodically. Small changes in metallurgy and manufacturing methods could have increased the risk.

OPT is expensive and time-consuming because the tests usually involve life-cycle testing, specialized equipment, and destructively testing samples that have to be discarded. The quality engineering teams should balance the probability and cost of a defect occurring against the impact if a problem were found by the customer.

13.4 Control Plans

In the same way as we document the operations and their associated information in the process plans, the quality control points, methods, and equipment need to be documented in the *process control plan*. These documents (typically one per part or subassembly) are controlled documents (meaning they can't be changed without approval). Table 13.1 shows an

Table 13.1
Example of
a process
control plan

Control plan number		Key contact	Approved original			Approved changes		
CONTROL_FAB_0101		Jane Doe						
Part/assembly number		Part name description	Date (original)			Date (rev)		
FAB_0101		Urethane cast housing						
			Characteristic		Measurement method			Reaction
Part/ process number	Operation/ process name	Area/ equipment	#	Product/process characteristic	Spec/tolerance	Measurement method	Sample (size/ frequency)	Control method	
60	Remove part	Workstation 1	1	Number of bubbles	No more than three visible bubbles. No visible bubbles on critical mating surfaces.	Visual	100%	Defect records	Determine if parts can be used and scrap if not. Increase time degassing. Evaluate pouring methods and tapping.
60	Remove part	Workstation 1	2	Complete fill of the gate	Gate height >0.25"	Visual	100%	Defect records	Check part for damage and scrap if necessary. Re-evaluate filling method.

example of part of a process control document for the urethane casting example we have been using.

Most industries use a similar format for their process control plans. The plans contain information about the product, which process and workstation they are used on, and then describe what is being controlled and how. Almost all templates include the following information:

- **Headers** delineate the part or assembly in the control plan, revision control, team member names, and approvals.
- **The part and process numbers** correspond to the numbering in the process documentation (Section 11.4) and enable cross-referencing between the two documents.
- **The characteristics** being measured are listed for each process. These may be functional or dimensional characteristics, or the characteristics of a process (e.g., temperature or speed). Each process may have more than one characteristic and each characteristic gets a unique number.
- **Specification/tolerance** describes what is considered acceptable. These numbers should be consistent with the drawings or specification documents.
- **Measurement methods** define how the characteristic will be measured. For example, a dimension can be measured very accurately using a CMM machine or roughly using inexpensive calipers.
- **Sample size** tells the factory how frequently the parts are measured. Sample sizes range from 100% of all product to just the first of a batch.
- **Control method** defines how the operator/machine will ensure the control of the feature or process; for example, an SPC chart, defect chart, or online records.
- **Reaction plan** defines what the operator should do if the specification is not met. It includes short-term actions to address the defect. The reaction plan can include rework and any paperwork or quarantine that must be completed for non-conforming parts. In addition, it should provide longer-term solutions to stabilize the process.

The process control plan will evolve throughout the production life of a product. Sampling rates can be reduced for quality characteristics that are expensive to measure and show no failures. Any quality issues that emerge later in the process may result in additional quality control.

Summary and Key Takeaways

❑ Quality can be controlled in production using a wide range of tools, from 100% visual inspection to destructive testing of small samples to comprehensive functional testing.

❑ Quality control points are used throughout the production system to ensure that problems are caught early in the process.

❑ QMS plans need to be documented and maintained in a controlled document.

❑ The process control plan documents all of the quality control points for the parts and the assembly process. These are living documents that are updated over the product's life.

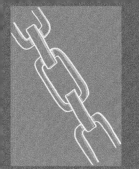

Chapter 14
Supply Chain

PRODUCTION PLANNING

14. SUPPLY CHAIN

15. PRODUCTION PLANNING

16. DISTRIBUTION

It is not possible to design, make, assemble, and test all of the components of a product in-house. All companies will need to select suppliers to provide certain components and services. Determining what to outsource and which suppliers to select involves trading off between cost, time, intellectual property, capability, cash flow, and quality. Once a supply chain is in place, it has to be managed. This chapter reviews the fundamental make-vs-buy decision-making process, how to select suppliers, and how to define and manage the supplier relationships.

Product Realization: Going from one to a million, First Edition. Anna C. Thornton.
© 2021 John Wiley & Sons, Inc. Published 2021 by John Wiley & Sons, Inc.

Manufacturing anything today requires dependence on an extensive network of suppliers. Your suppliers might be next door or might be half the world away. It is just not possible to completely design, make, transport, and support a product within the walls of your own organization. Say you are designing a simple children's toy with sensors and lights. Even if you make the housing and assemble it in-house, you are unlikely to make the LEDs needed to light the product. What you decide to keep in-house and what you decide to outsource will depend on a number of factors, including your capability, monetary resources, need for speed, and what you are trying to make. Materials and work that can be outsourced during product realization fall into several broad categories:

- **Contract manufacturing**. Unless you already have your own factory, you will likely opt to outsource the fabrication, assembly, and packaging of your product to a factory that specializes in building products for other companies. These services are called *contract manufacturers* or CMs.
- **Material, part, and subsystem purchases**. You will need to buy raw materials that go into your product. You might fabricate your own parts from raw materials you purchased, or you might opt to purchase entire subassemblies that leverage the expertise and capacity of an outside supplier.
- **Product-specific tooling fixtures and testing**. Most products will require the design and production of custom tooling, fixtures, and testing equipment. Tooling design and fabrication require special skills and knowledge.
- **Capital equipment**. It is unlikely that a product development team will build the equipment needed to make their product as well as making the product itself. In the rare case that manufacturing equipment is built in-house, that process can be treated as its own product development process.
- **Distribution**. Teams might choose to outsource some or all of the processes of getting a product from the factory to the customer (Chapter 16). These suppliers are often called third-party logistics (3PL) suppliers.
- **Services**. Launching a product involves more than just creating a computer-aided design (CAD) and manufacturing it; it requires packaging design, marketing, manufacturing support, logistics, customer support, quality planning, legal, and consulting. While it might be tempting to save money by doing everything in-house, hiring outside help can speed up product realization and ensure better outcomes, ultimately saving money.

Managing your supply chain is more complicated, however, than simply picking a supplier and sending them a check. Here are some of the decisions you will need to make:

- What are you going to make and what are you going to buy? (Section 14.1)
- What kind of relationship are you going to have with your supplier? (Section 14.2)
- Are you going to assemble your product yourself or are you going to hire a contract manufacturer? (Section 14.3)
- How are you going to select suppliers? (Section 14.4)
- How will you contract with your supplier? (Section 14.5)
- How will you manage your supply base? (Section 14.6)
- Will you have one or more suppliers for the same part? (Section 14.7)

14.1 Make vs. Buy

The first decision when designing a supply chain is "make vs. buy." Does the company hire outside organizations to provide a service or a part, or does the company do the work internally? This question applies to components, subassemblies, processes, and services. There are hundreds of books on supply chain strategy; we won't aim to cover all of that theory here. But here are the major factors you will need to consider:

- **Relative cost**. Sometimes it is cheaper to hire an outside company. They have access to cheaper materials, can share capacity across many clients, and are generally more efficient.
- **Internal capabilities**. Sometimes teams have capabilities that you want to leverage, and other times you don't have critical capabilities. For example, very few start-up companies will have an in-house legal team to educate them on intellectual property (IP) law; they will need a contract with an outside law firm. On the other hand, if you have a battery expert in your organization, you will want to use that capability.
- **Resource availability**. If your team is small, you will not have the bandwidth to manage all of the processes. For example, smaller companies typically outsource distribution because the core team does not have time, space, or resources to manage all of the shipping.
- **Fractional resources**. If you only need certain resources for a very short time, it might not be worth hiring someone full-time and then have them sit idle. It is easier to hire an outside consultant than to add to headcount. For example, you might hire external consultants to generate the packaging concepts at the start of the process or to do a detailed design for manufacturing (DFM) analysis later on.
- **Capital**. A process might require too much capital to build capability internally. Even if purchasing the equipment would likely have positive return on investment down the road, you might simply not have (or not be able to borrow) enough cash to do so. For example, most hardware companies will outsource injection molding because it is too expensive to buy, install, and run the presses.

- **Speed**. Hiring outside teams can speed up the process of completing a product design. For example, it might be faster to hire an outside consulting firm to build the firmware. They might have a large team that can finish in weeks, whereas you might need months if you do the job internally.
- **Space and zoning**. Some manufacturing processes require a great deal of space for equipment and assembly. Obtaining real estate can be costly, and even if you have the real estate, ensuring that it is zoned correctly can be expensive and time-consuming.
- **Strategic value**. You might not want to outsource a technology to a supplier if the technology contains critical IP or if you want to build the expertise in-house for future projects.

14.2 Types of Supplier Relationships

 You won't have the same relationship with every supplier in your supply base. If you use a CM, someone in your company will know your CM well, spend time in the factory, and work with them to solve problems. On the other hand, you might have no idea where your Bluetooth modules are coming from and only work through a distributor's website. Each type of relationship has different costs and benefits (Table 14.1) and includes, in descending order of integration:

- **Strategic partnerships**. These typically involve two companies co-developing the product. These partnerships are critical to the success of both companies, and both companies take significant business risks.
- **Business-essential/key suppliers**. These suppliers spend significant non-recurring engineering costs (NRE) developing custom subsystems or capabilities that are critical to your product. While these relationships are difficult to untangle, you can find alternative suppliers if necessary.
- **Semi-custom/transactional**. The next tier down builds semi-custom and fabricated parts (e.g., motors, batteries, or injection-molded parts). This relationship involves some NRE on the part of the supplier. The parts need to be verified, but it is relatively easy (but time-consuming) to find an alternative vendor and qualify them.
- **Arm's lengths/expendable**. These are suppliers whose company-specific capabilities are not critical and can easily be replaced within days.

Table 14.1
Characteristics of different supplier segments

	When engaged	Purchasing method	Difficulty of change	Co-development	% of total COGS	% of part count	Example
Strategic	At the start of product development	Long-term contractual relationships	Almost impossible to change. Supplier is critical to the product and very difficult to untangle	Working closely together through the whole process	High	1%	Aircraft companies outsourcing a section of the aircraft
Business-essential	Detailed design	Formal request for proposal (RFP)	Expensive to redesign core components and requalify the part	Engaged in later phases of development. Working through product realization	Medium	5%	Bicycle manufacturer co-designing a new shifter configuration to fit a specific frame (core technology not changed)
Semi-custom/ fabrication	Product realization	Request for quote (RFQ)	Need to requalify new parts and may need to pay for new tooling	Some engagement to identify DFM opportunities	Low	15–30%	Consumer electronics company selecting a custom battery vendor
Arm's length	As needed	Order placement	Very low. Just the cost of changing the BOM and ordering process	None	Very low	60–70%	Buying screws from fastener distributor

Vignette 8: Working with a Contract Manufacturer

Adam Craft, Chief Production Officer, Hydrow

I'm Adam Craft, Chief Production Officer at Hydrow™, Inc. where my team focuses on hardware, production, quality and supply chain while producing Hydrow, the Live Outdoor Reality™ Rower. My degree is in mechanical engineering and I've been working in product development and manufacturing for over 30 years. I've been fortunate to work for a number of companies both large and small over the span of my career, including at Hasbro, iRobot in its early growth, Jibo, and Hydrow.

All of these companies worked with contract manufacturers in Asia, primarily China and Taiwan. Working with a contract manufacturer requires a long-term strategic partnership but the nature of that relationship can vary greatly depending on how big you are.

As a larger established company, working with a CM means that you will likely:

- **Work with mature CMs**. Often the CMs you're working with are relatively large, with mature manufacturing processes and quality controls in place. Larger CMs are vertically integrated, so much of their supply chain is controlled in-house (i.e., they will have their own fabrication and tooling capacity)
- **Engage with more than one CM**. A large company wants to ensure both the best costs and make sure you have enough capacity. Large companies will typically work with more than one CM so you can direct new products to different CMs depending on volume requirements and factory skill sets. This can foster competition between CMs and help you with pricing and other demands.
- **Have local staff (from your company) in-country**. This team manages the day-to-day relationship with the factory and can quickly troubleshoot issues. You will likely have dedicated staff who work at the CM (in the factory itself).
- **Be able to be relatively demanding**. As a large favored customer, you typically get preferential access to the CM staff, first dibs on capacity, and competitive pricing.

As a smaller company or start-up, your relationship with a CM is different. Most importantly, when you are a large company, the CM knows you can generate volumes and profit, but when you are a small company they are taking a risk too. All CMs want to catch the next FitBit or GoPro that comes along – but there's associated risk that the new company will fail. As a result, you're often selling yourself at the same time as you're evaluating CMs during your RFQ process. Working with a CM as a small company means that you will likely:

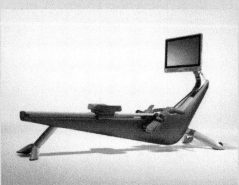

- **Work with smaller CMs**. You may need to manage more of your supply chain and/or the CM may also subcontract out more portions of your project since they may not have all expertise in-house.

- **Pay the CM to do more of the design work**. You're often leveraging the CMs development and production staff because they can do it faster and cheaper and you probably don't have the internal expertise. This saves cost and time, but be careful about who owns any IP in this scenario.
- **Engage with only one CM**. You want to be an important customer and be able to get the attention you need. As a small customer at a larger CM, it's easy to get lost in the shuffle when the next push from one of their larger customers becomes critical.
- **Travel a lot or have someone on the ground**. You still need to have your own representation on the ground at the CM to ease communication and to be able to jump in when something doesn't look right – even if all intentions are good, no one knows your product like you do and an unexpected change in process or cost could have unintended consequences if your team is not aware.

Ultimately, a small company has to partner with their CM. Both sides need to see the value and understand the risks when you want to push the process and timeline, as is highly likely when you're making your first (or only) product. At times it will feel like you are paying for everything; however, CMs are investing their resources heavily in your program and suffer as much as you do if you delay your program and miss dates. They only start to make money when the POs start coming in. They make their real money at high volume – not your first shipments.

All relationships with CMs, whether you're in a large or small company, will benefit when you have:

- **A partnership where both sides benefit instead of a customer/supplier relationship**. Granted, you are indeed the customer and they are indeed the supplier, BUT things will not always go smoothly and the relationship will be tested. Best to have that happen from a position of strength!
- **An executive to call when you need to escalate**. At some point resources will disappear, subsuppliers will let you down, and quality issues will crop up. You need a contact who will listen and can step in to make a difference.
- **Representation in-country/in-factory**. There's no substitute for someone who is personally vested in your company who knows all about what's important and what's not. Even if you fully trust your CM, having someone there will smooth communication, resolve issues in the same time zone, and be a good conduit for any other issues back to your main office.

Text source: Reproduced with permission from Adam Craft
Photo source: Reproduced with permission from Hydrow
Logo source: Reproduced with permission from Hydrow

FIGURE 14.1
Contract
manufacturing
facility in China.
Source: Reproduced
with permission from
Shenzhen Kaifa
Technology

14.3 Owning Manufacturing or Using a CM

The most significant make-vs.-buy decision is who is going to manufacture your finished goods – whether you are going to own the factory or work with a contract manufacturer. Many companies, even large ones, use contract manufacturers, either in-country or overseas. When hiring a CM, you have a choice ranging from small mom-and-pop facilities to large well-known ones such as Foxconn or Kaifa (Figure 14.1). Each option has benefits and downsides (Inset 14.1).

The primary reason for engaging a CM is to avoid the overhead of owning a manufacturing and assembly facility yourself. Renting or owning space, getting the necessary permits, purchasing the equipment and infrastructure,

---

Inset 14.1: Supplier Tiers

Suppliers will often be referred to as Tier I, Tier II, and Tier III. The term "supplier tier" is used in two ways, depending on the context of the conversation.

Where in the supply chain. This use of the term "tier" refers to how far the supplier is from the design/manufacturing team. Typically, an organization will only interface with a Tier I supplier who will, in turn, have a supply chain of Tier II suppliers, who will have a supply chain of Tier III suppliers.

Quality and size of the CM. When talking about tiers in relation to CMs, the term typically indicates the size and capability of the supplier. Tier I refers to the highly capable, large, and typically more expensive CMs (e.g., Foxconn and Jabil). Tier II CMs are smaller, but have similar technological capabilities to a Tier I. Tier II will often focus on specific product types, and are generally less expensive than Tier I. Tier III CMs are understood to be very low quality, use older equipment and less sophisticated material management and quality processes. Tier III might be a perfectly reasonable choice if the sophistication and price sensitivity warrant a lower-cost supplier. When dealing with a Tier III, you may run into language issues.

and then hiring the skills to manage everything in-house is often too expensive given the likely revenue of a product. Also, remember the old adage that "time is money": CMs can start producing almost immediately, whereas building your own factory is likely to be very time-consuming.

Secondly, if you want to take advantage of lower costs in developing countries (energy, material, and labor costs), hiring a CM might be the only feasible solution. In China, for example, it is complicated and expensive for non-Chinese firms to build their own dedicated facilities. Setting up facilities in China requires organizations to set up structures such as a Wholly Foreign-Owned Enterprise (WFOE, commonly pronounced "woofie"), which allows a foreign firm to produce in China and export elsewhere (typically through special economic zones). Setting up a WFOE is very expensive and time-consuming, and almost certainly not worth the time and investment for smaller firms.

The third benefit to hiring a CM (either local or overseas) is that the CM typically has access to a network of suppliers who can quickly and easily provide tooling, parts, and other capabilities. Think of the analogy of hiring a contractor to do your house. One of the benefits of a top-notch contractor is that they have a team of vetted electricians, plumbers, and finish carpenters. The contractor manages the complex coordination of the trades and handles the payments. You could hire each of these trades yourself, but the overhead and time (and the potential for miscommunication) are not worth the potential savings. The same applies to CMs: they will have long-standing and trusted relationships with their injection-molding factories, component suppliers, and PCBA assemblers. They can coordinate the supply chain at a lower cost and more efficiently than you could.

Another advantage to hiring a CM is that they have capabilities that they share among many customers (e.g., purchasing, quality control, and HR). Working with a CM allows the customer of the CM to purchase fractions of a person's time rather than having to hire a full-time person who will not be fully utilized. CMs also have purchasing power for components. You might only need 10,000 capacitors, but because the CM buys in bulk for a range of customers, they can purchase a million capacitors at a discounted rate, then pass those savings on to you.

Finally, CMs have experience with designing and producing similar products, so they do not have to climb the learning curve for each project. There are also downsides to hiring a CM, which include:

- It can be challenging to do cost-downs on products once the price is quoted. Because the profit for CMs is often a percentage of the total COGS, their incentives to reduce cost are not aligned with yours.
- When a problem arises in quality, it can be tricky to determine whether the root cause lies with you or with the CM. Because the CM incurs a financial penalty if they are at fault, they might not be incentivized to find or disclose problems within their facility.

- By hiring a CM, you are not building your own internal manufacturing capability. If you imagine that your current product might be part of a large and ongoing product line, it might make sense to build your own factory.
- It can be time-consuming to execute design changes with the CM.
- You are locked into their supplier base. Unless you allocate parts as consigned or assigned at the start of the negotiations, it can be challenging to force a change to a new supplier.
- There is always a risk of intellectual property loss either accidentally or maliciously. At the time of the publication of this book, there was quite a bit of conflict between the US and China over the theft of intellectual property by Chinese partners.
- For overseas CMs, there is always the risk that geopolitical forces might change tariffs or import/export laws, so costs might change suddenly and dramatically.
- CMs (especially lower-cost ones) might not be aligned with your team's ethical values in their treatment of workers or the CM's impact on the environment.

14.3.1 OEM Vs. ODM

If your organization has decided to use a contract manufacturer, your CM can take responsibility for none, some, or all of the designs of packaging, mechanical and electrical components, firmware, and software. The relationship with a CM will fall in the continuum between being the original equipment manufacturer (OEM) to a pure original design manufacturer (ODM).

- **OEM (original equipment manufacturer).** You do all of the design, and the factory builds exactly what you have specified. The CM sources some parts, manufactures some components, and assembles and tests the product. In an OEM relationship, the CM will typically do DFM reviews with you but will not make major changes to the design.
- **ODM (original design manufacturing).** On the other hand, you might want your contract manufacturer to design some or all of your product. In most ODM cases, the CM is provided with the industrial design, the functional specifications, and any critical technology. The CM will take ownership of the mechanical design, electrical design, and packaging. They then build the product and label it with your name. Many small appliances are designed in this way (i.e., coffee makers). The "guts" of the product are identical between brands but the industrial design and user interface are defined by the company.

There is a large gray area between pure OEM and pure ODM. For example, you might do all of the mechanical design, but hire the CM to design your PCBAs. There is no right answer to the OEM vs. ODM decision. Depending on the team's needs and capabilities, finding the right balance of between OEM and ODM is based on several strategic decisions:

- **How much of the technology do you need to own?** If the product differentiator is the industrial design, but the guts are shared across many product lines, then taking an OEM approach can be the most efficient.
- **How fast do you want it?** Hiring a CM can significantly speed up the process of design and production. They may have experience designing similar products and already know where to get the components and how to test the product.
- **What skills do you have vs. what they have?** For example, the CM might have expertise in camera design, and while a camera is essential to your product, it is not the key differentiator. It will be faster and probably yield better results to hire the CM to design the optical subsystems, while your team designs the differentiating features.
- **How tied do you want to be to the CM?** If your CM designs major components of the product and thus owns the IP of those components, it will be hard for you to change CMs, should you want to. If you want to design the next-generation product yourself, you might not have access to the bill of materials (BOM), drawings, and process plans for your first-generation product.
- **What do you want to learn vs. what do you want to outsource?** It might be critical for next-generation products for you to learn about key technologies and develop in-house expertise by managing the design process.
- **Cost control and tradeoffs**. Your organization will not be able to make real-time tradeoffs between design decisions and costs if you leave the design to the CM. Because of this, the CM might make decisions that are good for their manufacturing costs but not necessarily good for your customers.
- **Do you want the ability to control quality issues?** If you outsource the design, it might be challenging to find the root cause of a failure.
- **Are investments part of cost offset?** Some CMs are making venture investments in products by providing design services at a significantly discounted rate. The CM then will own a portion of your company. This can make moving factories extremely difficult but can alleviate cash flow problems.

In general, ODM will be relatively cheaper and faster than OEM to get a design into production, but you lose some control and ownership of the product.

14.3.2 How to Manage a CM

Selecting and managing a contract manufacturer can be a full-time job for several people. If you do not have prior relationships with factories or expertise in managing a CM, you have a range of options from going it alone to complete outsourcing of the relationship with your CM.

- **Do it yourself**. It has become much easier to contact and contract with a CM without external help. However, if you do not have the internal expertise, there can be a very steep learning curve.
  - *Pros*: You learn and manage the relationship yourself, possibly at lower cost.
  - *Cons*: It takes a lot of time, and it is easy to make big mistakes that have long-term consequences.
- **Hire a broker, middle-man, or trading company in-country**. These companies (or more typically individuals) are hired by you to take your design to their CM network and get the product built. While this can appear to be a simple solution, there are many downsides. While there are a large number of ethical and capable brokers, the industry is full of horror stories of brokers taking the money and running. If your broker runs off with your deposit (contract or no), the factory will make you pay again. You may not be able to find them and it is difficult to bring them to court in a foreign country. Also, you do not own the relationship with the factories – the broker does. If there is a quality issue, you may have difficulty getting it resolved.
  - *Pros*: Local expertise and a single point of contact.
  - *Cons*: No control over the CM relationship, no ability to change factories, and the risk of unethical behavior
- **Full-service US-based OEM supplier**. The OEM provides end-to-end services starting with completing the design through handling the distribution logistics. Sometimes they will not charge you full rates for NRE costs in trade for equity or an increased COGS.
  - *Pros*: They can help solve cash flow issues and provide a single site for full service.
  - *Cons*: You lose control of your design, do not own the factory, or control your product cost.
- **Operations consulting firm**. These firms perform the duties of an operations team. You get the equivalent of a VP of operations, a factory support team, and a supplier management team. These companies tend to be expensive, but you can gradually transition to owning more and more of the responsibility. Working with an operations consulting firm is a great way to learn about what needs to be done, then your team can take on those responsibilities for the next-generation product.
  - *Pros*: The consulting firm manages the relationship with the CM, helps you build internal capability, and provides expertise.
  - *Cons*: Increased costs and not learning by doing.
- **Internal CM start-up support**. Some CMs are now providing end-to-end services to companies in exchange for equity.
  - *Pros*: You are provided with one-stop shopping.
  - *Cons*: This option ties you to the CM, the CM owns part of your company, it is difficult to control costs, and you have very little leverage or negotiating power.

14.4 Supplier Selection

Once the decision to outsource a product, part, or service is made, the teams need to create a short-list of suppliers to ask for quotes. The process of selecting a supplier typically occurs in six steps, outlined in the next sections. The same basic processes apply to selecting a CM as a vendor; the CM process is longer, involves more paperwork and lawyers, and is based on more factors than just being the lowest cost.

14.4.1 Define What You Want to Purchase

The better you can define what you want to buy, the more accurate the price quote will be, and the more likely it is that you will get what you want. Asking the supplier for a quote can be as simple as asking for a price for a single part number or as complex as a 100-page request for quote (RFQ). Identifying what you want to purchase is more than just saying, "I need a battery." It involves answering questions such as:

- "How much customization is needed?"
- "What is the relative importance of cost versus quality?"
- "What capacity in Ampere-hours is needed?"
- "What form factors will physically fit your product and be producible and cost-effective?"

14.4.2 Decide Where You Want to Make the Part

Many companies in developed countries struggle with the question of manufacturing locally vs. using a low-cost overseas supplier. People may want to buy locally, but it can be hard to leave significant COGS reductions on the table. For some, the extra cost of producing locally is worth the reduction of tariff risks, ease of communication, and ethical issues. On the other hand, when facing stiff competition, some opt for the lower COGs available in other countries. This is not to say that all local manufacturers will be more expensive or all overseas suppliers will have poorer quality control. Every supplier is different. Table 14.2 gives a generalized comparison of local and global suppliers, but you should do your own research to understand the pros and cons of your potential suppliers.

Table 14.2

Comparison of local
vs. global suppliers

	Local	Global	Comment
Cost competitiveness	⇓⇓	⇑⇑	Local suppliers tend to have a higher COGs.
Quality control	⇑	⇓	In most cases, the quality from local companies is more reliable than from overseas, but there are many excellent suppliers in Asia and other lower-cost locations.
Short shipping time	⇑	⇓	Getting product from overseas can be time-consuming.
Supply chain stability	⇑	⇓	Strikes and other disturbances can cause delivery issues if ports are shut down or weather causes halts to shipments for some overseas suppliers.
Intellectual property management	⇑⇑	⇓⇓	There is more of a risk of intellectual property loss overseas, and sharing IP might be a contractual requirement.
Coordination and design iterations	??	??	These depend on the facility. While it is easier to time communications with vendors in the same time zone, CMs in low-cost locations are often responsive and flexible outside of working hours and might be willing to put more resources on a project to ensure its timely completion.
Resources put on the project	??	??	
Environmental	⇑	⇓	Some low-cost locations have practices that significantly damage the environment. Then again, so does the USA at times. You'll want to carefully research environmental practices before choosing a CM.
Labor ethics	⇑	⇓	Many factories in developing countries have had worker mistreatment, poor working conditions, and pay issues that might reflect poorly on your product.
Tariff uncertainty	⇑	⇓	Geopolitical uncertainty will change the tariffs and trade agreements over time. What might be cheaper today might not be tomorrow.

⇑ - better ⇓ - worse

14.4.3 Select from Whom to Get Quotes

Finding a list of suppliers to approach can be a daunting task. Searching online for the term "bearing" or looking on Alibaba can return hundreds of websites but very little guidance on where to go. Asking around, doing teardowns of other similar products, or hiring a supplier consultant can ensure that you have the right selection of companies to evaluate. Here are several ways to identify good suppliers:

- Some consulting companies can coordinate the supplier selection process for you. They have pre-vetted lists of suppliers they know well. Many of these firms show up at hardware meet-up meeting and conferences. They also tend to have the best blogs.

- Do a teardown of a product that contains similar technologies. The labels inside might give you ideas for suppliers you can use.
- For a lower-cost commodity product, Alibaba can be a good starting point.
- Most states have non-profit organizations like MassMEP or NYSERDA who have the mission of promoting local manufacturing. They can often recommend local vendors [105, 106].
- Investors or advisors might have relationships with existing vendors. When selecting your team of advisors, look for people who have these connections.
- Several companies have arisen who act as manufacturing brokers for fabricated parts. They will take your drawings and find the appropriate vendor for you from their networks. These are different from consultants because they just take your drawing and shop around to get the best price. They act more like a clearing house than a partner.
- Ask around, go to hardware meet-ups, and talk with other product teams.

14.4.4 Compare Quotes

When selecting between quotes, don't just look at costs. Factors such as supplier quality, material availability, and purchasing terms are often just as important – if not more so. Also, there are non-tangibles such as the historical quality of a supplier, willingness of the supplier to work with and teach the team, recommendations and referrals, and ease of communication.

It is always a good idea to get quotes from a range of suppliers, from small firms to larger big-name ones. Often, organizations are torn between hiring big-name firms with reputation and experience versus hiring a smaller firm whose prices are lower and are more likely to focus on you as a customer (Table 14.3). In general, you should aim to get at least three quotes, have a clear scope of work and timelines, and get references for every manufacturer. You should compare the quotes using an A2A (actual-to-actual) analysis described in section 14.5.4.

Small firm	Large firm
• Less potential for bait-and-switch (you interview the senior person and get the junior one).	• Known designers and capability. Long track record of big-name products.
• Your project is more critical to their portfolio and will get their focus.	• Capacity and bandwidth. They have extra staff they can throw on to a project for a short period when it is necessary.
• Smaller firms typically have a lower overhead.	• You get the cachet/reputational boost of having a well-known firm engaged.
• Small firms are hungrier for your business. They are looking for a big win and will go the extra mile.	• Will sometimes do work for equity.

Table 14.3
Pros of hiring small versus large firms

14.4.5 Tour the Facility, Meet Staff in Person, and Get Samples

Depending on the criticality of the part or service, the team will want to get a sample part for evaluation and visit the facility. Section 14.8 gives a detailed list of factors to consider when deciding whether a factory is appropriate or not. In the case of service providers such as marketing firms or 3PL (third-party logistics), getting recommendations and meeting the supplier in person is an excellent way of learning whether your team can work with the supplier.

14.4.6 Choose Between Suppliers

Once you have collected all of the data through the above steps, you will need to choose which one to go with. There is never such a thing as a perfect supplier. There will always be tradeoffs between cost, quality, schedule, ease of working together, trust, experience, and contract terms. It is still essential to get multiple quotes or proposals to enable the team to learn, for itself, which of the various factors is more important.

Ultimately, the team will need to make tradeoffs. The shortcomings and inherent risks of each supplier relationship should be identified early and managed explicitly throughout the product realization process.

14.5 Documents

Suppliers are your partners. In the best case, you will create an agreement in which the success of your product will drive their sales as well as yours. However, it is always necessary to clearly define your relationship and document it in legally binding agreements. For simple purchases, such as commodities from known suppliers, the documents are very simple: just a quote, purchase order, and invoice.

For strategic suppliers and contract manufacturers, you will need to go through a much more detailed process to formalize the relationship. Comprehensive contracts with precise costs, promises, and consequences are critical to keeping that good relationship. You do not want to wait until there is a disagreement to find out that you haven't written your contracts well. The most critical contract, your manufacturing services agreement (MSA), often takes multiple iterations. This section describes all of the documents that get generated, starting with initial contact through payment to returning a defective product. Typically, the supplier management or procurement team takes the lead in this process. Figure 14.2 shows the flow of

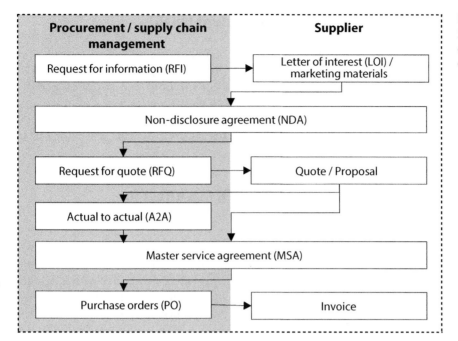

FIGURE 14.2
Supplier
engagement
document flow

documents used to formalize and manage the supplier relationship. As with all legally binding documents, it is imperative that you retain legal counsel and have a professional review your documents. Ignore the advice of counsel at your own peril!

14.5.1 Requests for Information (RFI)

The project manager will initiate interaction with the supplier though a request for information (RFI) about products and services supplied by a company. The RFI starts the conversation between you and a supplier. The RFI will include a basic description of what you are interested in purchasing without including any sensitive information. It is used to determine whether there is sufficient overlap between your needs and their capabilities to proceed to the next step.

For example, for components, you might ask about their ability to customize a motor or similar products to yours they have made. If you are sending a RFI to a CM, the RFI also asks the CM whether they are interested in bidding for the business. The RFI has to market your product to them. If you don't have a clear market or enough money to get through the product realization process, they aren't going to waste their time bidding on the project. As such, the RFI needs to include enough information about your product for the CM to decide whether your product is worth the time to respond to a RFQ (request for quote).

14.5.2 Non-disclosure Agreement (NDA)

An NDA is typically the first document that both organizations sign – whether they eventually agree to work together or not. The NDA is used to protect sensitive information for both the supplier and the development team. Also, it can be used to protect patent rights because it keeps the disclosure from being considered public. The next documents (RFQ or RFP) often include sensitive information that needs to be protected.

14.5.3 Request for Quote (RFQ) / Request for Proposal (RFP)

The next step in the process is writing a formal RFQ or RFP. An RFQ or RFP is a comprehensive document that outlines what the supplier is being requested to quote on. Both documents provide the supplier with a detailed outline of what should be quoted, along with information about the product, potential volumes, and schedule and quality targets. Here are the differences between them:

- **RFP.** Request for proposal is used when the design team does not have a clear plan for what is being quoted. It provides the supplier with a relatively high-level set of requirements and asks the supplier to submit a detailed proposal of how the needs will be met and what the cost and terms are. RFPs are used more for services and for ODM designs than for OEM.
- **RFQ.** Request for quote is used when the design team has a clearer understanding of what they are asking for. An RFQ will include a very detailed set of specifications, bill of materials, and product designs. These are generally used for OEM, fabricated, and semi-custom parts.

The more detail you provide in the RFQ or RFP, the better the supplier will be able to accurately respond to it. For example, if you just list "battery," they will not know if they need to quote on a simple AA cell or a customized rechargeable battery with non-standard connectors. When faced with uncertainty, the supplier will often pad the resulting quote with higher COGS.

When getting quotes, you should insist on an open-book quote in which you get a breakdown of each cost element in the BOM, all overhead charges, and a detailed breakdown of all NRE charges. Checklist 14.1 provides a starting list of elements you might include in your RFQ.

Checklist 14.1: RFQ Checklist

Information You Provide to the Supplier

Information about the business
Information about the company – e.g., revenues, expected sales, marketing information, and press releases
- ❏ Description of the product, its functionality, and photos and videos of any working prototypes
- ❏ Description of any existing partnerships with third parties that the CM will need to be aware of
- ❏ Description of the design team and any board members who lend credibility to the project
- ❏ Forecast of sales with data to back up claims
- ❏ Your funding situation: how much money you have raised and by whom

Product definition
- ❏ Bill of materials
- ❏ Manufacturing methods for fabricated parts
- ❏ Assembly and part drawings
- ❏ Color, material, and finish (CMF) document
- ❏ Specification documents
- ❏ List of SKUs and product variety
- ❏ Packaging concepts
- ❏ List of assigned and consigned parts

The scope of the NRE to be executed by the supplier
- ❏ DFM activities
- ❏ Design responsibilities: PCBA, mechanicals, firmware, packaging
- ❏ Number of pilot runs and number of samples in each
- ❏ Sourcing responsibility
- ❏ Certification responsibilities
- ❏ Tooling

Quality testing requirements
- ❏ Quality targets (AQL levels)
- ❏ Quality test plan (pilot and production)

- ❏ Certification requirements
- ❏ Shipment audits

Other
- ❏ Exclusion clauses to prevent the supplier from building the same product for a competitor
- ❏ Distribution requirements
- ❏ Access to the factory by third parties

Information You Ask the Supplier to Provide

Costs
- ❏ BOM costs by part
- ❏ Overhead costs (labor, profit, scrap, etc.)
- ❏ COGS
- ❏ Landed costs (if they are handling distribution)

Payment amounts and timings
- ❏ NRE costs
- ❏ Tooling costs
- ❏ Piloting costs and timing
- ❏ Sample costs and times

Financial terms
- ❏ Payment terms for tooling and NRE costs
- ❏ PO timing and payment terms
- ❏ Fees or charges

Requests for other information
- ❏ Cost-down sharing policy
- ❏ Volume pricing
- ❏ IP protection methods
- ❏ Return material authorization process (RMA)
- ❏ Distribution capability
- ❏ Examples/samples of related products from their portfolio

14.5.4 Quote and A2A Process

Based on your RFQ, the supplier/CM will provide a detailed estimate or quote. Each supplier may provide you with the information in very different forms (even if you give them a template to fill out). Ideally, the supplier will provide open-book costing that lays out the cost for each part or subsystem, the markups, and the terms and conditions. An open-book estimate allows you to compare quotes as well as understand the major factors driving the cost in each quote. Depending on how the quote/estimate is worded, it might or might not be binding (i.e., whether or not the CM is committing to providing the product at that price). Make sure you read the fine print in the quote before agreeing to grant the supplier the business.

Once all of your suppliers have returned their responses, you need to compare them by doing an A2A (actual-to-actual) analysis. This often requires significant work to line the quotes up to understand the differences. For example, some suppliers lump all overhead costs into one large percentage, while others break overhead down into categories. Some suppliers will include the cost of tooling in the part costs, while others will break it out as a separate charge. Building an A2A is more of an art than a science. The goal is to identify the tradeoffs between the suppliers. For example, a well-known Tier I supplier might have higher overhead rates and costs, but you would have more confidence in their quality. You will need to use this information in conjunction with your financial models and cash flow models (Chapter 8) to decide what the best supplier is for you.

14.5.5 Master Services Agreement (MSA)

Based on the A2A, the total costs, terms, and your tours of the facilities (see Section 14.8), you will make a decision and award one of the CMs the business. After you select a supplier, you and the supplier need to formalize the terms of the contract. This relationship is defined by the master services agreement (MSA), which outlines the terms and conditions for how you will work with a contract manufacturer. An MSA is like a pre-nuptial agreement; in the best cases, you will not need one, but if you do need one you will be glad you have one. Many companies start working with CMs with the understanding that "we will work out the details later." This is fine if all goes well or there is no confusion, but can cause major headaches if things don't go well. The later you leave signing the MSA, the less leverage you have to walk away from the contract. Also, it opens you up to late changes in terms by the CM.

Checklist 14.2 outlines some of the terms that need to be negotiated as part of a MSA. This book should not be used as legal advice. The book introduces many of the concepts so you can understand what everyone

Checklist 14.2: Typical Terms in an MSA

Intellectual property and information ownership
- ❏ Who owns the design, tooling, BOM, IP, etc.
- ❏ Management of IP
- ❏ Access to manufacturing information such as test results, production rates, SOPs, etc.
- ❏ The scope of contract/statement of work

Responsibility for design of parts/subsystems
- ❏ If they are designing your product, are you obliged to stay with them for mass production or can you send their design out to bid?
- ❏ Responsibility for designing and supplying tools, fixtures, and test assemblies
- ❏ Samples and their costs included in the contracts
- ❏ Certifications
- ❏ Minimum production volumes

Quality management
- ❏ AQL levels
- ❏ How the RMA process (returned material authorization) works
- ❏ Corrective action and continual improvement activities and responsibilities
- ❏ Responsibility for the quality of incoming parts
- ❏ Shipment audits
- ❏ Third-party access to the factory to do the shipment audits on your behalf or review production
- ❏ Certification responsibility

Costs and payments
- ❏ Access to actual factory costs (usually refused)
- ❏ Cost of parts, notification on cost increases, and markups
- ❏ Material authorization process for long lead-time parts
- ❏ Purchase order timing and payment agreements
- ❏ Cost-down targets and cost/benefit sharing

- ❏ Stipulation of which currency the transactions are done in and how currency fluctuations will be managed
- ❏ Periodic review of pricing

Procurement and approval of supplier changes
- ❏ Supplier identification and contracting process
- ❏ Material procurement processes
- ❏ Approvals required if parts are substituted

Distribution
- ❏ When you take ownership of the product (when it becomes FOB)
- ❏ Delivery responsibility
- ❏ Customs and paperwork responsibility

Contract terms
- ❏ Term of contract
- ❏ Renewal terms
- ❏ Method of dispute resolution
- ❏ Venue of Law
- ❏ Indemnification clauses
- ❏ Continued existence of contract should either party be acquired, sold, etc.
- ❏ Conditions under which contract can be unilaterally terminated – e.g., bankruptcy, non-performance, breach of IP or confidentiality requirements
- ❏ Non-competition – will the factory agree not to build product category X for Y years?

Unexpected costs and how they are approved
- ❏ Cost of running design or process changes during mass production
- ❏ Payment for obsolete inventory
- ❏ Cost of rescheduling runs after delays

is talking about and why it is important. Your legal counsel will recommend additional or different clauses based on your exact product and their experience. It is very important to listen to your legal team when building the MSA.

The terms in your MSA can have a significant impact on cash flow, deliveries, cost, and quality. Most of this information should be quoted in the RFQ process to avoid surprises later during the MSA negotiation process. We will repeat this again. **Do not rush the construction of this document, and make sure you understand all of the implications of the various clauses. Listen to your lawyer!**

14.5.6 Purchase Order

Once the MSA is completed and you are ready to start buying product from your supplier or CM, you will need to give them a PO. A PO is a commercial document that defines an offer from a buyer to purchase from a seller. The purchase order, when accepted by the seller, is a legally binding contract between the two parties. A purchase order typically includes the following:

- Quantities
- Price
- Items
- Discounts
- Payment terms (e.g., "all upfront" or "net 30 days")
- Date of shipment
- Any associated terms and conditions (e.g., if the delivery is late, what the cost will be, or if the payment is delayed, what the penalties will be)
- Purchase order number
- Billing addresses
- Shipping addresses

For the seller, having a binding purchase order is essential because they need to buy materials, allocate resources, and potentially turn away other business to ensure fulfillment of your order.

14.5.7 Material Authorization (MA)

You will probably need to purchase certain materials even before you sign the purchase order to account for long lead times. For example, a part might have a 180-day lead time, but the PO might only be signed 60 days before you would like to begin production. To ensure that there is sufficient material to begin production, you will need to order the long lead material 120 days in advance of the PO. The material authorization (MA) process will have different terms than the PO; it might require a larger down-payment on the materials, depending on your MSA agreement.

14.5.8 Invoice

After the PO is created, the supplier will send you an invoice. An invoice details what you owe your supplier and when payment is due. It is typically sent in response to either a contractual agreement (payment terms on tooling), to a purchase order, or to a material authorization.

14.5.9 Return Material Authorization (RMA)

If there are quality issues, the purchaser can send materials back for refund or replacement. Usually, you can't ship the material back without authorization – the supplier has to approve your request to return materials. Before an RMA is finalized, there are typically extensive discussions about who is responsible for each type of quality failure.

14.6 Managing Your Supply Base

Even with only a few suppliers, managing the supply chain is more than just selecting the suppliers and sending them a check.

As the size of the supply base grows and becomes more complex, organizations need to formalize the teams and processes used to manage the supply chain. As companies grow, they need a way to determine who are "good" suppliers and those that should be replaced. For example, they might give the suppliers they've worked with one of three certification levels: bronze, silver, or gold. Poorly-rated suppliers are identified and replaced. Gold suppliers are given preference in the selection process and often are given longer-term contracts. Suppliers are coached by your supplier management group on how to become better suppliers with the incentive of getting more business from you.

Many supplier management systems rely on globally recognized certifications as a way of ensuring that suppliers have the right capabilities. According to the ISO certification organization, "Certification can be a useful tool to add credibility by demonstrating that your product or service meets the expectations of your customers. For some industries, certification is a legal or contractual requirement" [107]. However, being certified does not mean that the factory will produce high-quality products. You must still evaluate the capability of the facilities yourself (Section 14.8).

The most frequently referenced certification is ISO 9001 [34], which is used to certify that the facility has a documented quality management process and that the process is followed. Some certifications are general, whereas others are tailored to specific industries. For example, AS9100 [39] defines the quality management systems for the aerospace industry. Other industry-specific certifications include:

- Medical devices ISO 13485 [36]
- Software engineering: ISO/IEC 90003 [108]
- Environmental Standards ISO 14000 [109]
- Health and safety ISO 45001 [110]
- Information security management ISO/IEC 27001 [111]
- Lab certification ISO/IEC 17025 [112]

14.7 Single vs. Dual Sourcing

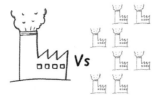 When starting to produce a new product, you will typically have one supplier for each part. However, as volumes grow and you discover risks with your suppliers, you might want to dual-source.

In single sourcing, only one supplier will provide a part or service. If the supplier needs to be changed, 100% of the orders are transferred to a different supplier. In dual or multiple sources, two or more suppliers provide identical services for the product at the same time.

Whether single or dual sourcing should be employed depends on several factors:

- **Capacity**. In some cases, a company might be required to dual-source because of capacity constraints in the supply base.
- **Risk mitigation**. Dual sourcing is used to ensure the constant availability of materials. With the present geopolitical uncertainty, disruptions in labor, and political instability, dual sourcing (e.g., one plant in Mexico and another in Thailand) can reduce the risk of parts shortages.
- **Cost competition**. Supplier management organizations will often maintain a dual source to apply cost and quality pressure to the supply base. Depending on the relative performance of each supplier, a larger percentage of orders will be transferred to the better of the suppliers.
- **Build supply chain capability**. If the product line is expanding or additional capacity is needed, new suppliers are often allocated to low-risk product, then given more orders as they gain your confidence.

Dual sourcing is not without problems. Having a single supplier has many benefits:

- **Inventory management and forecasting**. When there are two suppliers, the potential variability in inventory ordering increases dramatically. It is necessary to maintain sufficient inventory at each site. Having a single supplier reduces the complexity and chances of errors.
- **Supplier management overhead**. Adding a supplier to the supply chain increases the overhead of the whole organization, including an increased load on engineering (to track quality issues), procurement, and supplier management teams.
- **Quality control**. No two facilities will produce the same part in exactly the same way. They will have different machinery, operators, and incoming materials. If there is a quality problem, tracing the source of the issues can be more difficult with two suppliers than with one.
- **SKU (stock-keeping unit) traceability**. A single source of supply reduces the complexity of batch traceability and tracking.
- **Supplier trust**. The best suppliers are your partners. Creating dual sourcing can reduce trust and cooperation. Suppliers are more likely to implement cost savings and improvements if they do not think their work will be immediately handed to their competition.

14.8 Touring a Factory

The first time you go out to tour a factory, it can be overwhelming. You might have to put on special equipment or you will be told not to wear makeup or jewelry. Your hosts will walk you through the line, stopping at key stations to show you their most impressive technology but rushing past the things they don't want you to dive into. There will be workers and inventory and machines in seemingly random places (and most likely coffee will be terrible).

Your supplier management team will need to make a quick assessment of whether this is a "good" factory and whether your company is going to put the fate of your product in their hands. It is a scary and daunting task, even for those who have visited many facilities. There is just too much to evaluate without pre-planning what you are going to look at and how you are going to compare your suppliers.

Checklist 14.3 provides a basic primer on what to look for and why each item is important. While the list is not comprehensive for a supplier audit, it will describe many of the important things to look for and the likely risks. No factory is perfect: there will always be tradeoffs between cost, quality, schedule, attention, and so on. There will be issues, but by knowing them and the risks they pose, teams can actively manage around the risks and shortcomings.

Checklist 14.3: Visiting a Factory

☐ **Cleanliness**. Cleanliness is an easy first measure of a CM's overall capability. Debris, dust, and metal chips in the assembly and work areas are key indicators of the attention the factory pays to quality overall. Metal filings pose risks for key bearing or sealed surfaces, and dust and dirt can short out sensitive electronics. Oil and debris can get on product and inside the packaging.

☐ **Protective equipment**. The CM's attention to safety is another measure of their quality. Operators should be wearing the appropriate personal protective equipment (PPE). Primarily, they should be protected against hazards and injuries (e.g., using safety glasses). Also, importantly, shoe covers, finger cots, hair nets, and coats can significantly reduce the debris the operators can transmit to the product. You need to order the right equipment. For example, a factory was having debris issues. Someone decided to order new lab coats. However, A quick run of a fingernail across the embroidered logo resulted in tiny fibers coming off which contaminated the product.

☐ **Environmental controls**. Variations in temperatures and humidity as well as dirt and debris (and insects) from the outside can impact moisture-sensitive devices. Temperature-sensitive materials can be dramatically affected by even small variations in temperature. Humidity trapped under conformal coatings can damage the performance of key electronic components. Your CM should have high-quality HVAC systems and monitor the environmental conditions.

☐ **Operator training**. Based on their training, only certain operators should be allowed to perform critical tasks. For example, in electromechanical systems, only some operators are qualified to perform the most delicate soldering jobs. The back of each operator's ID tag should have their training qualifications listed.

☐ **ESD control**. Electrostatic discharges can damage unprotected boards and other electronics. The right ESD equipment should be in place, including wrist straps, grounded equipment and storage materials, and ESD mats on the floor.

☐ **Soldering quality**. Not all soldering is the same – good quality soldering will reduce failures and warranty returns. Soldering is critical, especially for antennas or other communication devices that are subject to significant vibration. When you tour the facility, you should not see any solder drops; the area should be clean and the operators trained. There should be effective quality control to catch any soldering failures (through AOI, visual inspection, or other tools) about quality.

☐ **SOPs**. Each station should have a clear set of standard operating procedures. These procedures should be up-to-date, easy to read, and actually followed. If the SOPs are tucked in a corner or covered in dust, they are probably not being referred to. Good SOPs have clear pictures and are relatively simple to read. SOPs are important when the product has a complex assembly or manufacturing process.

☐ **Work area organization**. Each station should be clean and organized with well-labeled materials. It should be clear where the incoming and outgoing material is supposed to be. If more than one station is doing the same process, then the setup and flow of materials should be identical.

☐ **Inventory management**. The state of the incoming warehouse and material storage areas should be an indicator of how well materials are managed. Are the bins organized and clearly labeled? Does the factory use a FIFO (first-in/first-out) policy for inventory? Is it easy to see where inventory is running low?

☐ **Inventory segregation**. Does each product have a dedicated inventory storage area? Does the facility have a secure storage place for high-value parts? Proper inventory segregation can ensure that your inventory isn't used for another customer. Segregation is particularly critical when the parts have high value or have IP you don't want leaked.

- ❏ **MRP (material resource planning) system**. The factory will have an MRP system. It might be as simple as a spreadsheet. The speed and accuracy of the system are critical to ensuring that all of your inventory is available for your build. A good test is to ask the factory to pull a material readiness report while you watch. If they have to get back to you in a day or so, the MRP system is not responsive. The more complex the supply chain, the more critical an MRP system.

- ❏ **Material handling**. Materials should be appropriately protected when transported. Ideally, custom totes should be used for more delicate products. The totes should be stackable and clean. Each tote should be marked with the right travelers to ensure material traceability.

- ❏ **Quarantine**. Products that have not passed quality assessments should be appropriately segregated to ensure that they are not mixed back into production. It is very easy for poorly labeled rejected product to be mistakenly assembled into final goods. While the CM typically assumes the cost of the re-manufacture, the delay can hurt sales and revenue.

- ❏ **IQC**. Incoming quality control is where all purchased parts and subassemblies are inspected and accepted for production. The IQC area should be well organized, the equipment well-maintained and up-to-date, and the SOPs clear and up-to-date. Ideally, the quality records should be electronically maintained.

- ❏ **Ethics**. Manufacturing facilities, especially those in developing countries, sometimes have a reputation for treating their workers, the environment, and the community badly. Social Accountability International has a standard, SA8000 [113], that evaluates the ethical practices of a company. In addition, there are a number of firms that will perform an audit of a factory that has not been certified to this standard.

- ❏ **Equipment quality and age**. The age and quality of machinery can have an impact on the quality of your product. For surface mount technology (SMT) processes, newer Japanese and European equipment is typically more accurate and has better manufacturing controls than older Chinese versions. If you are doing simple controller boards or low-precision injection molding, quality might not be an issue, unless the product requires tight process control. SMT quality is critical for components requiring a high degree of process control, repeatability, and accuracy.

- ❏ **Preventative maintenance and process control**. Even the best equipment will produce poor quality if the equipment is not maintained, the incoming materials is not controlled (e.g., thermal control of solder paste), and the process SOP is not followed. The manufacturing engineers should be questioned on their understanding of the processes and how to control it.

- ❏ **Best practice assessment**. Does the factory follow best-practices in their processes? For example, when removing sprue from an injection molding part, the operators should use a heated knife if the part will be subject to cyclical stresses.

- ❏ **Quality reports**. The factory should have quality records easily available. Ideally, the trends and quality metrics should be posted centrally on the factory floor, updated, and reviewed daily with the factory workers. In addition, those records should be available electronically.

- ❏ **Corrective action**. The facility should maintain an active corrective action process to track failures, perform effective root-cause analyses, implement the improvements, and monitor ongoing quality. These corrective action plans should be documented and regularly updated.

- ❏ **Costs**. Cost-down initiatives should be implemented continually by means of second sourcing, volume discounts, or re-negotiation with suppliers. It is critical that any change is approved by the product owner before being implemented.

Summary and Key Takeaways

- ❑ Careful and intentional design of your supply chain is critical to the success of your product.
- ❑ Teams need to decide what will be outsourced and what will be kept in-house based on the relative cost, quality, lead time, and strategic requirements of the product.
- ❑ Hiring a CM is a complex process that takes time and thought.
- ❑ Selecting a supplier requires understanding your own needs, identifying a candidate list, and then comprehensively comparing their offerings.
- ❑ Suppliers need to be managed actively.
- ❑ There are many documents that need to be executed to ensure that both parties are in agreement on the relationship. These documents can have severe legal and financial impacts and need to be written in conjunction with legal advice.
- ❑ Prepare before you visit a factory.

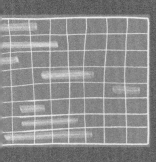

Chapter 15
Production Planning

PRODUCTION PLANNING

14. SUPPLY CHAIN

15. PRODUCTION PLANNING

Once the product, the supply chain, and the production system are designed, it is time to start building your product to fulfill demand. However, because of uncertainties in demand combined with long lead times, planning what is going to be produced when, and when materials need to be ordered is complex. This chapter describes how lead times, forecasting, and production planning are interrelated.

16. DISTRIBUTION

Product Realization: Going from one to a million, First Edition. Anna C. Thornton.
© 2021 John Wiley & Sons, Inc. Published 2021 by John Wiley & Sons, Inc.

Apple, in 2014, was not able to meet the demand for the iPhone 6 [114] and many people expecting to buy them for Christmas couldn't get the product. There are many hypotheses on why they couldn't meet the needs – one was that they had supply chain issues, which slowed the production process. Selling out because you underestimated demand is not uncommon. Every year there is a hot new toy – Hatchables or Tickle-me-Elmo – that is not produced in enough volumes, leading to panicked parents and chaotic pre-holiday shopping. You can imagine that the producers would have liked to snap their fingers and had more of the hot toys available to ship the day.

On the flip side, companies can sometimes overestimate demand and end up with too much inventory. For example, in 2017 Snapchat launched their Spectacles, a pair of $130 sunglasses that could take photos. They significantly overestimated customer demand and ended up with hundreds of thousands of unsold units in a warehouse and a $40 million US loss [115]. In 2018, Apple's prediction of iPhone X sales were way over actual sales – a sharp contrast to the situation four years earlier – and the company had to slash production [116].

If customer demand could be satisfied instantaneously (with no production lead time or inventory), there would be no need to plan for production – an order could be taken and the product produced on demand. However, because of lead times and cash limitations, companies need to balance the costs of over-building and having excess material against the risk of under-building and missing sales. Companies need to plan what they are going to buy and build ahead of actual orders.

Significant literature is dedicated to forecasting market size and demand, planning for lead times, and balancing inventory against demand uncertainty. This chapter introduces the basic concepts of production planning and how production planning is managed during product realization.

15.1 Production Planning Concepts

Production planning is the process of translating "I think the market size is $M" to "This is how many units of each SKU will be produced in first-shift on September 1" down to "This is how many Wi-Fi modules I need to order six months ahead of production." Production planning is a complex multistage process involving many functional groups and is highly dependent on the supply chain, lead times, and the relative importance of COGS and cash flow.

To understand how production planning occurs, you need to understand four key concepts of:

- The production planning stages (Section 15.1.1).
- How sales forecasts become committed orders (Section 15.1.2).
- How production forecasts become a production schedule (Section 15.1.3).
- What lead time impacts the planning timeline (Section 15.1.4).

Before discussing how these concepts interrelate, we need to review some terminology we are going to use in the next few sections:

- **MOQ.** Minimum order quantity is the order size you need to get a price break.
- **Safety stock.** The amount of material you want to maintain for changes in forecasting and yield.
- **Inventory carrying costs.** The money you need to spend to hold excess inventory.
- **Write-off.** If you can't use inventory, you need to scrap it, and write off the inventory.

15.1.1 Production Planning Phases

Creating a production plan is much more complicated than simply figuring out what needs to be built, then ordering the materials and producing them. The planning process typically occurs in multiple stages. Many companies use the same framework of cascading planning processes to move from the early strategic planning to the day-to-day planning of what is build when and shipped to whom. These planning processes (Table 15.1) don't just occur once but continue to be updated as better information is gathered about the actual demand. The time-frame and frequency of these plans will be highly dependent on the industry. When designing a new military aircraft, the planning horizon may be decades, whereas the newest Christmas kid's toy may be weeks.

- **Strategic manufacturing plan** is typically executed years ahead of starting the actual product development. It is used to ensure that the right infrastructure is in place to support the long-term growth plans. For example, if a car company decides to expand their portfolio to include electric cars, they need to plan for the new facility several years in advance of building the new product. This process is typically done at the executive level.
- **Sales and operations plan (S&OP)** aligns the long-term sales and marketing forecasts with the manufacturing capacity plan. It is a cross-functional effort that is typically revisited every month. S&OP is used to plan for shorter lead-time capacity such as tooling and the expansion of the existing production lines. This is typically done as a cross-functional effort between marketing and operations and is done six months to a year ahead of demand.

Table 15.1
Production
planning process

	Lead time	Outcome	Participants	Update frequency
Strategic manufacturing plan	24 months	Identify and budget long-term capital investments and factory build-outs	Executive committee	6–12 months
Sales and operations plan (S&OP)	6–12 months	Identify tooling requirements and incremental capacity increases	Senior operations and marketing team	Monthly or bi-monthly
Master production schedule	3–6 months	Long-lead purchase, communication to suppliers	Operations	Bi-weekly
Material requirements planning (MRP)	1–3 months	PO plan, material orders	Procurement and factory operations	Weekly
Purchasing and production activity control (PAC)	1 week	Factory orders	Factory	Daily

- **Master production schedule (MPS)** creates the medium-term plan of what will be built, including specific information about each SKU. It is typically done three to six months ahead of demand. This plan drives the purchase of very long-lead materials and is communicated to the suppliers to ensure that they have enough capacity.
- **Material requirements planning (MRP)** estimates what has to be ordered by when. Material ordering is typically managed through dedicated software. This is done a month or two before starting actually producing the product.
- **Purchasing and production activity control (PAC)** manages the flow of materials into the facility and what orders are released into the production facility to start production. Done a week or so before the production starts.

15.1.2 Sales Forecasting and Orders

When a product is first being conceived, one question will permeate all conversations: *How many are we going to sell?* Quantifying the likely sales volumes and how fast sales will ramp up is key to deciding whether or not a product is commercially viable, how much capacity to build, and how to design the supply chain.

Throughout the life of a product, the marketing and sales teams will produce a rolling forecast of how many units (and what mix of SKUs) will likely be sold through which sales channels (Inset 15.1). These estimates can be done with planning horizons of several months to several quarters to several years (or in the case of a completely new aircraft, a decade). The *long term sales forecasts* will be estimated when the product concept is initially proposed and is used to determine whether the product has a feasible business model and what production capacity will be needed to support it.

Inset 15.1: Sales Channels

Sales channels are the different paths through which you can sell your product. For example, bicycle components can be sold:

- Directly to the bike manufacturers, to be sold on to the customer as a complete bike
- Online through distributors who then sell to customers
- Online through the manufacturer's website
- Through a bike shop

Each channel will have different lead times, payment contracts, and revenue to the manufacturer. In addition, they will each have different impacts on how much inventory has to be held to ensure that the various customers can be served in a timely fashion.

As the customer demand is better understood, the *medium-term forecast* is created, which is used to plan for tooling and long-lead materials.

As the date of a shipping to your customer approaches, the sales and marketing teams will begin to get more accurate information about how many units each channel will likely purchase including tentative orders (*preliminary sales*). These estimates are not real until they become a *committed order*. Committed orders arrive in the form of purchase orders (POs) from other businesses or in the form of payment from an individual customer. Often, committed orders will come with a pre-payment as insurance that the order will not be canceled.

Not infrequently, customers ask to change their orders after a purchase order is placed. Product mix, volumes, and target delivery dates can be adjusted several times. While, legally, you are only obliged to deliver what is on the PO, most companies will accommodate late changes as best they can to keep customers happy. The final outcome is *shipped orders*. Depending on the structure of the supply chain these may be sent to a distribution center to be stored in finished goods or directly to the customer.

15.1.3 Production Forecasting and Planning

Production forecasting is different from sales forecasting. The production forecast is a prediction of what the likely production plan is going to be. Before the product is produced, the product team will create a *long-term production forecast* that is used to scale production facilities and plan for capital equipment needs.

Once the product is in production, the *production plan* communicates what the factory will likely be asked to build of each SKU. The production plan is generated on a rolling basis, typically three to six months ahead.

One of the most important reasons to create a production plan is to ensure that the right materials are available in time for production. The product can only be built and shipped if every part is in stock and ready for assembly. To complicate matters, the lead time for some of your parts will be much longer than the time your customers give you or the lead time of the PO you place with your factory. For example, some IC components can have 120-day lead times but you only place your order with the factory 60 days ahead. This means you have to estimate your demand and buy the materials four months in advance of placing the actual orders (*production forecast for long-lead items*).

The exact production schedule isn't known until the actual POs are placed and the factory can begin purchasing materials and planning the day-to-day build plan. The PO commits the CM to produce and commits you to purchase the product ordered from the factory. A PO usually includes upfront payment for part or all of the order (usually enough to cover the material purchasing).

15.1.4 Build-to-Order vs. Build-to-Stock

How sales forecasts are translated into a production plan is impacted by how customer orders for different channels are fulfilled. Depending on the lead time, the forecast, and the sales channel, some products are build-to-order, and some are filled from inventory using a build-to-stock model.

Build-to-Order

Build to Order

This term is used for products that are shipped directly from the production line to the customer. In build-to-order, production planning is aligned directly with customer orders. Build-to-order is used where the customer is very tolerant of long lead times (e.g., purchasing an aircraft), the COGS are very high, the product is customized for a customer, or the production lead time is very short. In the late 1990s, Dell changed the way customers thought about computer customization by switching from a build-to-stock to a build-to-order model. Before Dell's innovation, most laptops were one-model-fits-all, and could only be customized after the customer purchased them (or custom configurations, where they existed, took a long time to deliver). Dell was able to streamline their operations to enable customers to get a custom-built computer (within certain common parameters) within a week [117].

In build-to-order manufacturing, a forecast is still required to estimate:

• **Manufacturing capacity and capital investments**. If a product line is capacity-constrained, the factory needs to plan for long-lead capacity and capital investments.

- **Inventory of long-lead materials**. Some materials will have a lead time significantly longer than the customer delivery promise. In this case, the factory will need to buy long-lead items ahead of demand to ensure that inadequate inventory doesn't impede the delivery schedule.

Build-to-Stock

In build-to-stock manufacturing, the team uses production planning to estimate how much finished goods inventory is needed to supply potential orders. Units are built and placed in finished-goods inventory rather than going directly to a customer. When a customer order is created, the product is pulled from inventory and shipped to the customer. Build-to-stock is used in several situations, including:

- **The lead time to produce is longer than the customer is willing to wait**. For example, if you want a custom color and interior package for a car, you may need to wait several weeks, but the dealer lot will have inventory on hand (in popular combinations and colors) if you want to drive away in a car that day.
- **Demand is variable due to seasonal trends**. In this case, the production system builds extra inventory every month and holds it in anticipation of a peak-demand season. For example, some medical products are used more during specific seasons – flu vaccines in the fall, and ace-bandages when the weekend warriors come out from hibernation in the spring.
- **Shipping takes a long time** (by ocean freight) and shipping lead time is longer than the customer is willing to wait.

Combined Models

In many cases, organizations will do a blended model of build-to-order and build-to-stock. When there is an expected peak in demand (e.g., holiday sales), a company may need to build up inventory to ensure that customer demand can be met. However, confirmed orders are built to order and delivered directly to the customer.

15.1.5 Lead Times

The timing of the planning processes is highly dependent on the order lead times. To plan, it is necessary to understand the *total* order lead time – how long it will take between receiving a customer order and getting the product into the hands of the customer. For a build-to-stock system, the customer order is the purchase order and the delivery time is the time to get

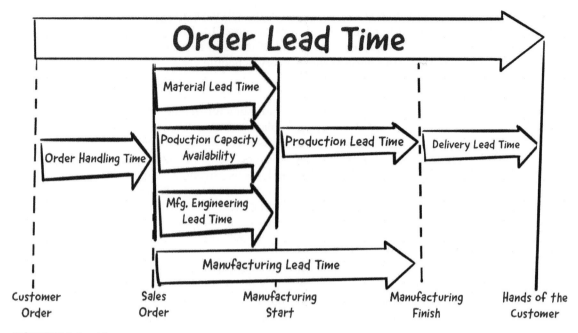

FIGURE 15.1 Lead time

the product to the distribution site. The order lead time (Figure 15.1) is driven by several factors:

- **Order handling time** is the time between when the customer order is received and when the order is entered into the sales system (as a sales order). The order handling time can include quotes, invoicing time, and entering the order into the MRP/ERP systems.
- **Manufacturing lead time** is the time between when the sales order is entered and when the order is completed and ready to ship. The manufacturing lead time includes both the time to prepare to start manufacturing and production time.
  - *Time from sales order to manufacturing start.* In order to start manufacturing, all of the materials have to be ready and there has to be enough capacity to start the process. Typically, the material lead time for high-cost components is the "long pole in the tent," but the time from sales order completion to manufacturing start also can be limited by other factors including:
    - *Production capacity availability.* There may be a lead time before the factory has capacity available. If the machines are fully occupied with building product for other customers, the next-ordered product may need to wait until capacity is available.
    - *Manufacturing engineering lead time.* There may be a need to design and build custom fixtures or tools to support the production. For example, if the product requires a custom extrusion, the factory can't start production until the die is designed and built and the batch of custom extrusion produced.

  – *Production lead time* is the time from the start of manufacturing to when the finished goods come off of the line. This lead time is dependent on the number of stations, the buffers between stations, and the cycle time for each station. Other factors can include testing time and any rework required.
- **Delivery lead time** is the time to package, transfer the product to the main carrier, and ship the product.

15.2 Forecast to Order Timeline

The prior sections outlined a dizzying set of strategic planning, sales forecasting, and manufacturing planning activities. This section ties all of these together. Figure 15.2 shows the timeline of sales forecasting and production planning steps and how these are refined over time as better information about actual demand becomes available. Forecasting is not a single one-and-done process. Initial estimates will be highly uncertain (as indicated in Figure 15.2 by the wide range of potential volumes relative to the expected volumes) and will be updated several times until final orders are shipped.

Long before orders are placed, the capital equipment and infrastructure for the production line has to be designed and built. The executive team plans the need for production capacity through the strategic business plan (very far in advance) and the S&OP plan (in the nearer term). These plans are done in conjunction with the sales and marketing groups, who provide the long-term sales forecast, and the operations groups, who determine the long-term production capacity needs based on this forecast.

Once the product is designed and the launch date is known, the sales and marketing teams can have conversations with potential customers (specifically distributors and retail) to determine a medium-term forecast. From this data, they'll have more information about how the market is responding (or is likely to respond). Based on these refined numbers, the production team will revise its production plan, and the operations and purchasing team will use the material-requirements planning process to plan how much material will need to be ordered by when. Long-lead items are ordered by procurement (or your contract manufacturer) using the material authorization process. Executing this phase well ensures that there is sufficient material (but not too much) to support the range of potential demand.

As preliminary sales are confirmed, the production team will create the production plan to meet the demand, and the procurement team will sign the POs with the factory using the purchasing and production activity control. Often the POs will have to be placed ahead of confirmed and committed orders, in anticipation of future demand. The POs are typically set in stone

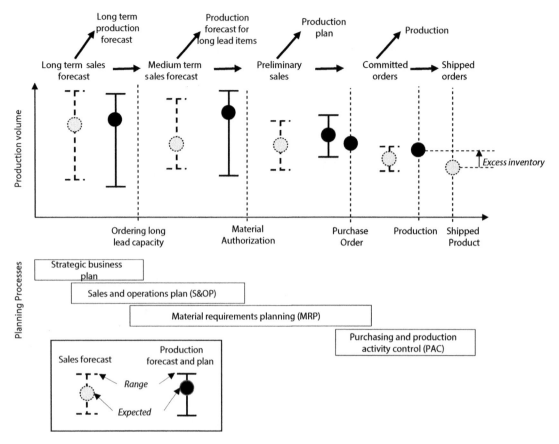

FIGURE 15.2 Sales and production forecasting timeline

but the factory can usually adjust (within reason) the final production plan as it is executed.

Once the product is built, the inventory is either shipped to customers or is placed in finished goods. During this whole process, the production planning team will evaluate and revise the upstream plans for the next production cycle, adjusting based on the amount of excess finished goods inventory and the accuracy of prior sales forecasts and new demand.

15.3 Complicating Factors

Production planning is complicated by several other factors besides demand uncertainty and lead times. First, the factory wants to have level production with as few dips and spikes as possible. Second, as we've mentioned, some materials have lead times that are significantly longer than the typical eight-week PO. This requires teams to plan and pay for very long-lead items upfront using a process called material authorization. Third, teams also need to balance the cost of components against the cash requirements for buying in bulk. Fourth, unexpected spikes in demand or constrictions in supply can lead to part shortages.

15.3.1 Production Leveling

To complicate matters further, the production planning needs to account for demand variability, planning ahead for spikes and dips in orders. You may need to ship two units on Monday and then 100 units on Tuesday. In a perfect world with no lead times, you would only need to build every day what you need that day. However, if your line capacity is 50 units per day, you will need to build extra on Monday to make sure you have enough for Tuesday's delivery. In general, the production system operates best on a steady-state basis. It is inefficient for a facility to produce 10 units one day, then 100 the next, then 10 again, rather than 40 per day every day. Variability in demand can be caused by a number of factors:

- **Production ramp**. During the first few weeks or months of production, the factory may not be able to produce at full production rates because, while production verification testing (PVT) worked out many of the production issues, still more will be found as production rates continue to increase. If you need a large volume of product to support the initial launch, you may need to start building far in advance of the first sale to ensure you have enough product.
- **Day-to-day demand variability**. Unless your customers are willing to receive product late in the event of a peak in orders, you will need to average the production rate to ensure that there is enough already built to meet large customer orders.
- **Seasonal and promotional peaks**. The factory will need to build ahead of known spikes in orders. Depending on your product and your target customer, you may have purchasing spikes around Christmas, national holidays, Mother's Day, or the back-to-school season. The worst thing a marketing department can do is surprise the factory with a marketing promotion, because that may create a spike in demand that dwarfs the supply.
- **Factory schedule**. Be aware of national holidays and local customs where your factory is located. For example, if you are manufacturing in China, you need to anticipate that production shuts down for at least two weeks during the Chinese New Year (CNY) in January/February. Typically, companies will ramp up production before CNY and then have a slow restart after the return of the workers.

- **Time and method of shipping**. Depending on how you will be shipping your product, the timing of shipping can impact the production schedule. If the product cost depends on shipping a full container (rather than less-than-truckload), the planning process has to take into consideration the time to produce all of the products to fill the container and the time for that container to travel by boat to its final destination.

15.3.2 Material Authorization

When working with a CM, you will typically place a PO 8–10 weeks ahead of when the product is to be shipped to you. At the time the PO is issued, you will pay a portion up front. This lead time and deposit gives the CM enough time and cash to order materials in time for your PO. However, there may be materials whose lead time is significantly longer than 8–10 weeks. In that case, you may need to approve an order for and pay 100% for parts at the time of the long-lead order. You will need to estimate the likely demand and order enough to cover future production.

There are inherent risks associated with purchasing very long-lead items:

- **Under-ordering materials can result in shortages**. These shortages can lead to additional expediting costs, buying materials on the spot market (Inset 15.2), or missing sales opportunities.
- **Over-ordering materials impacts cash flow**. Buying large quantities of long-lead materials eats up cash which can't be reclaimed until the product actually sells. For some components, this may mean holding inventory for up to a year. If you dramatically overestimate your demand, the stock may need to be scrapped and the loss written off.
- **The design team changes the bill of materials (BOM)**. Especially early in product launch, design changes can make inventory unusable. If you can't use the inventory, you may need to write off the inventory and take a loss.

15.3.3 Balancing Part Cost Against Inventory Carrying Costs

The price of components generally drops with higher volumes. Your procurement team will need to weigh the risk of buying too many units at a lower per-unit price versus too few at a higher price. Fundamentally there will always be a tradeoff between low inventory and low price, as shown in Table 15.2.

Figure 15.3 is an example of how different ordering frequencies can impact the overall cost of a product at the expense of inventory carrying

Inset 15.2: Spot Markets

Inventory shortfalls can grind production to a halt. When faced with shortage of electronics and some commodities, you may be able to buy them on the "spot market" – often at a significant markup. The spot market is the collection of channels through which you can buy commodities for immediate delivery. They include:

A distributor or electronic parts search engine. Distributors may have small volumes of materials in stock for immediate sale. It may be necessary to purchase from several vendors to get enough supply for the production run. Coordinating the purchasing and shipping of the materials from the US to an overseas factory can be challenging.

Electronics markets. Often it is possible to buy the components in a physical electronics market such as the Shenzhen electronics market in China or Yongsan Electronics Market in Korea. You can hire a company that will send someone to the market to search for parts. However, it is not possible to guarantee that the parts purchased there are not counterfeit, and when buying parts for highly regulated products (e.g., medical or safety-related), it is not legal to substitute parts that cannot be traced.

Alibaba. This website is a clearinghouse for electronics components. What is listed on the site may or may not be available, though, and (as with the Shenzhen electronics market) the components may lack traceability.

Paying to get to the head of the line. Some suppliers will charge you a premium for expediting your parts and getting you to the head of the line in their production queue. While this option isn't really considered the spot market, it has similar lead time/cost tradeoffs.

costs (and cash flow). The cost curves in the top graph for a simple electronic component were pulled from an online distributor. The bottom graph shows the cost savings as the order frequency decreases from weekly to every two, four, and eight weeks. As the order frequency decreases, two things happen. First, the cost per part goes down (in this case from $1.27 to $0.92). The marginal savings with each doubling of order size decreases and the cost bottoms out at around $0.90. The second thing that happens is the inventory carrying costs increase linearly. This is the amount of cash that is tied up in inventory because you have a lot sitting around.

	Low MOQ/high price	High MOQ/low price
Cash/inventory	✓	
Risk of write-off	✓	
Cost per part		✓
Risk of material shortages		✓

✓ - better.

Table 15.2
Tradeoff between MOQ and price

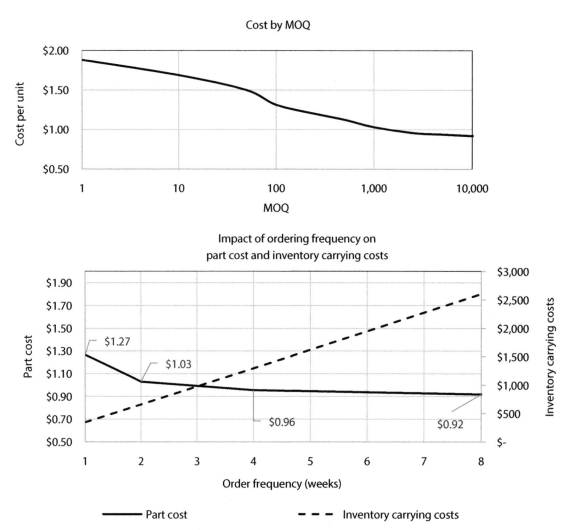

FIGURE 15.3 Order frequency impact on parts cost and inventory costs

If a company does not have infinite cash available, there is a tradeoff point. At some order frequency, the savings in COGs don't justify the inventory carrying costs and complexity and the inherent risks associated with carrying a lot of inventory. Unfortunately, there is no universal inflection point that can be calculated to optimize the tradeoff point. Instead, each organization needs to look at the tradeoff between COGS, available cash, and risk of writing off inventory, and choose what works best for them.

15.3.4 **Unexpected Market Conditions**

There are times when certain materials are in worldwide shortage due to economic or geopolitical circumstances. These can unexpectedly increase lead times significantly. For example, through 2017 and 2018, there was a worldwide shortage of dynamic random-access memory (DRAM) modules, because the increased production of cell phones and the increased demand for cloud computing and data storage had exceeded manufacturers' capacity [118]. This caused a twofold price increase for DRAM, and long (and costly) wait times.

It's not just rapidly growing technologies that suffer these shortages, though; shortages can also occur in what would be considered relatively simple commodity components. In late 2018, there was a significant increase in lead times for resistors and capacitors commonly used in consumer products. In some cases, the lead time for these parts increased to almost a year [119]. The rise in demand and limits on critical raw material likely drove these constrictions on supply. Teams should keep abreast of material shortages and market trends to ensure that they are not caught short. You may be able to get product on the spot market (Inset 15.2), but you can't guarantee it. The worst outcome is that you need to redesign your boards to replace the parts you can't get.

15.4 Shorter Lead Times are Better

A good rule of thumb is, if you have a choice between two equal options, pick the one with a shorter lead time. Anything you can do to reduce the lead time on parts or what it takes to get your product produced will reduce risk, cost and cash flow. Shorter lead times mean that you:

- **Reduce the need for finished goods inventory**. The shorter the lead time, the closer you can get to a build-to-order model.
- **Reduce the number of planning cycles**. If you don't have to place any material authorization orders, you don't have to do the calculations to plan for them.
- **Reduce your cash impact**. The faster the time between when you have to shell out for parts and POs, and when you get your payment, the less cash you have to tie up in inventory.
- **Reduce the risk of supply chain disruptions**. Long lead items are long lead because either supply is constricted or the manufacturing process (and the lead time to get the raw material) is long. The shorter the lead time, the less chance of disruption.

Summary and Key Takeaways

❑ Lead time for production, materials, and demand uncertainty make production planning complicated.

❑ Forecasting and production planning is an ongoing process of refinement and adjustment.

❑ Purchasing teams need to order materials ahead of production and balance having enough inventory on hand with having the cash required to hold the inventory,

❑ Teams will need to continually trade off COGS, cash flow, and the risk of part obsolescence against customer demand and lead times.

❑ Everything is easier with shorter lead times. Given the choice between two suppliers with equivalent cost and quality, always pick the one with shorter lead times. It will make planning much more manageable.

Chapter 16
Distribution

PRODUCTION PLANNING

14. SUPPLY CHAIN

15. PRODUCTION PLANNING

Getting the product to the customer can be as simple as drop-shipping the product directly from a local factory or as complex as going through several distribution centers across many international borders. You'll need to think carefully about the impact the distribution system will have on resources, cost, space, and lead times. You will probably outsource some or all of your distribution system.

16. DISTRIBUTION

Product Realization: Going from one to a million, First Edition. Anna C. Thornton.

© 2021 John Wiley & Sons, Inc. Published 2021 by John Wiley & Sons, Inc.

Once your product is built, packaged, put in a master carton and is sitting on a pallet at the factory shipping dock, the product may still have a long road to ultimately get to the consumer. It has to get to an airport or (sea)port to your main carrier, out through the country of manufacturer's customs, onto transport, across the air or ocean, through customs at the destination country, off the transport, onto trucks to the distribution centers, to the store, and ultimately to the customer. Depending on how much you are willing to pay, this process can take as little as a few days or as much as several months, and it can involve several companies and hand-offs.

It is not enough to design and build the product – teams need to plan for how the product will make this journey. Unless you have negotiated delivery to a customer in your MSA (master services agreement), most factories hand over ownership of the product, a process called free on board (FOB), after it is packaged into a master carton and audited for quality. The process from that moment until the moment the product gets into the hands of the customer is termed "distribution." The distribution system has to be designed long ahead of the first mass-produced batch. Also, a program for reverse logistics (when a product has to be shipped back because the customer is not happy) also has to be designed.

Several terms are used to describe the process of getting a product to and from a customer – operations, logistics, distribution, and shipping. While they are related, they have particular meanings (Figure 16.1). Operations encompasses a broad range of tasks to ensure that products are produced, distributed, and shipped efficiently, including supply chain design, process planning and factory layout. Logistics are the activities related to the planning and information flow about how materials are moved. Distribution is the physical process of moving material from manufacturing location to the customer, including order fulfillment (the process of linking inventory

FIGURE 16.1
Relationships between terms for operations and distribution

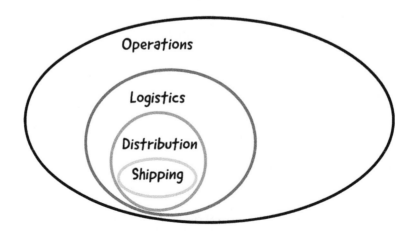

Inset 16.1: Who Can Coordinate Transportation of your Product from the Factory?

The contract manufacturer (CM) can take responsibility for handling part or all of the distribution process (for a fee). They may have the capability in-house and often it is easier for them to manage it than involving a third party. The responsibility of the CM for distribution will be negotiated and formalized in your MSA. If you want to manage your distribution, you can contract with a number of vendors including:

- **Third-party logistics (3PL, or sometimes TPL)**. These are companies hired to manage and execute some or all of the distribution and fulfillment services.
- **Freight forwarder**. This type of company organizes shipment requiring multiple carriers. Also called a non-vessel operating common carrier (NVOCC), forwarder, or forwarding agent.
- **Main carriers**. Providers of air, sea, and land transport.

to orders and directing shipments to the right customer) and getting the product across borders. Shipping is the process of physically moving product from one point to another.

This chapter will focus on how to design a distribution system. There are many options available for distribution systems, so the right choice for your product will depend on many factors described in this chapter. You can choose to manage the whole process yourself or outsource part or all of it (Inset 16.1).

16.1 Distribution Process

The distribution process is a multi-step process that will vary depending on where the product is made, where the customers are, what distribution methods are used, and so on. Figure 16.2 shows a simplified version of a standard distribution system.

The elements of the distribution system and the terminology will be discussed in the sections below.

16.1.1 Loading onto Transport

The product starts at the factory where it is packaged and labeled for shipment. Typically, the factory is responsible for packaging and labeling the shipment in accordance with the PO. The material is then loaded onto truck or rail transport to go from the factory to the main carrier (either air or sea).

FIGURE 16.2

Distribution system

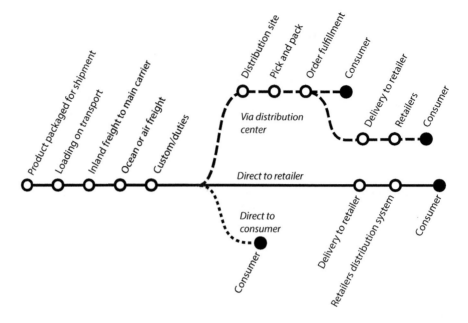

If the ownership of the goods is handed over to the buyer (you), then one of two sets of rules applies:

- **FOB – free on board**. The seller (the factory) has to load the goods onto the inland freight (e.g., the truck to the airport).
- **EXW – ex-works**. The buyer (your company) is responsible for loading the goods on the inland freight.

While this may not seem to be a critical distinction, it is, in fact, an important one. You don't want people standing on the dock with your boxes, saying, "It's not my responsibility to lift them on the truck." This is analogous to having a large package shipped to your home: delivery companies are usually unequivocal with owners about whether the delivery is "curb-side" or whether they will bring it into the house. This distinction typically matters to the seller because they only want to deliver the service they are paid for and are insured for. It only matters to the buyer when there is a gap in coverage (outside in the rain on the curb with a new refrigerator and no contractors on-site to bring it into the house).

The choice between the two depends on who is managing the next step in the distribution system. You just want to make sure you have coverage for all of the transfers.

16.1.2 Inland Freight to Main Carrier

The product is then transported to the main carrier, typically at an airport or seaport. For both modes of transport, space has to be booked well ahead of time to ensure that the materials can be shipped immediately, or you may have to pay for storage. Depending on where the shipment is starting and where it is going, you may be required to fill out export paperwork. If manufacturing is done in China, production will typically occur within a special economic zone (SEZ). These zones were set up by the Chinese government to ease the paperwork and cost of moving goods in and out of their country. Most products manufactured in China incorporate goods imported from other countries, and they are then are exported. SEZs reduce both the paperwork and the cost of importing and exporting goods, among other financial freedoms and benefits. If the goods are being sold in China outside the SEZ, they need to be officially imported into China.

16.1.3 Ocean or Air Freight

If a product is shipped overseas, it is shipped in one of two ways: by air or by sea. By air, the delivery time is typically 1–4 days, but delivery by sea can take 14–30 days, depending on the route and how much you are willing to pay. For air shipment, the product is packaged in pallets, master cartons, or individual inner packs. By sea, the product is packaged in a dedicated container, or for smaller lots, on pallets in a "less than truckload" (LTL) container that you may share with other companies.

Here are the most common options for shipment sizes:

- **Containers** are standardized structures which are used to transport product via sea, rail, and road. Standard ISO shipping containers are 8ft (2.43m) wide, 8.5ft (2.59m) high, and come in two lengths – 20ft (6.06m) and 40ft (12.2m). If the product is shipped by container, the container can be packed at the factory, loaded onto  a truck, and transferred directly to the ship. Once through customs, the whole container can be loaded on a truck and shipped by land to its final location without needing to be unpacked. Shipping via container significantly reduces distribution costs but requires you to ship in large quantities by sea.
- **Less than truckload (LTL)** shipment is used when you don't have enough material to fill a whole container. LTL is a partial load in a truck or container. Orders from multiple customers are combined by the carrier (or freight forwarder). Typically, the lead time on LTL is longer and the costs per unit are higher than 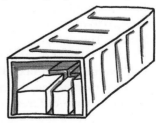 for a full container because shipments have to be consolidated, packed, and then unpacked after transport.

Ocean freight is significantly lower cost than air freight, but its lead time is substantially longer, and the product is subjected to more thermal, loading, and vibratory stresses during transport than it would be during air freight. Also, unexpected delays such as dock strikes can hold up materials for months, as has happened several times over the last few years.

Some products can't be shipped by air freight because either air transport isn't cost-effective (the cost of shipping by air exceeds the product margin) or because the shipment contains materials that can't be sent via plane due to safety hazards (e.g., explosive or toxic chemicals). The rules for shipping batteries and hazardous materials are continually becoming more restrictive [95].

16.1.4 Customs

Once the shipment crosses an international boundary, the receiving country will impose a variety of fees and will check to see whether the product is legal to import. During the import process, the product will be checked for correct certifications, labeling, codes, and other requirements. The paperwork required for getting the product through customs is complicated, and small errors can lead to significant delays. Most organizations importing goods to the United States use customs brokers. US Customs and Border Protection (US-CBP) describes the role of customs brokers as ". . . assist[ing] importers and exporters in meeting Federal requirements governing imports and exports. Brokers submit necessary information and appropriate payments to US-CBP on behalf of their clients and charge them a fee for this service" [120]. You can hire brokerage companies for other countries as well. The fees involved in shipping across international borders include:

- **Unloading, port, and storage fees**. Incoming ports will charge for material handling and storage. These fees include the cost to take the container off the ship and store it until it is ready for delivery.
- **Customs**. The process of clearing a product to come into a country. Customs and Border Protection will ensure the legality of the product and verify that the right tariffs have been paid. A custom broker will charge you to facilitate getting your product through customs.
- **Tariff**. Most countries impose tariffs on goods imported from other countries. In the US, the tariff rate is set by the Harmonized Tariff Codes, in which imported/exported products are classified under the Harmonized Tariff Schedule (HTS) [121]. The HTS defines which tariff rate should be charged to each product. Each country will have its own version of the HTS.
- **Duties**. The actual amount spent to import or export products based on the tariff rates and value of your product.
- **VAT (value added tax)**. Depending on where the product is landing, the country may collect taxes on your product in addition to duties. Some countries have different names for this tax, for example, in Canada it is called a GST or goods and services tax.

16.1.5 Delivery to the Consumer

Customers can get their products through a number of pathways including:

- **Direct-to-consumer**. In this model, as the name implies, the product is shipped directly from the factory to the consumer. In this case, the product is shipped from the factory in a small shipping box. Most of the time, direct-to-consumer uses air freight. This has the advantage of a short shipment time but is the most expensive. The author has bought many new products through start-up company websites which were shipped to her directly from the factory.
- **Direct to distributor or retailer**. Containers, pallets, or master cartons are shipped to distributors or retailers. In the last case, the retailer will take responsibility for the last mile delivery to the retail location. Most consumer electronics products that you buy through big-box retailers are shipped this way.
- **Via distribution center**. In this case, products are delivered to a distribution site owned either by your company or by a 3PL provider. The warehouse is typically located near the company or centrally located to facilitate distribution to customers (either consumers or other distributors). Order fulfillment is done through the warehouse. Companies that sell directly to the consumer or through multiple sales channels often use this delivery model.

16.1.6 Costs for Distribution

When contracting with distribution providers, you will delineate what services they are providing and who is paying for what part of the distribution costs. Each step described has costs associated with it (e.g., payment of duties and loading costs), and the responsibility for that payment will depend on the contracts with the factory, 3PL suppliers, or freight forwarders. To simplify and standardize contract language, the International Chamber of Commerce (ICC) developed *Incoterms* (International Commercial Terms) [122]. When you review contracts with your distributors, make sure that you understand the terms in the contracts and that you have covered all of the distribution steps. You don't want to find out when your product is sitting on the dock that you are responsible for paying a very large bill for the duties. Worse, you don't want to find out after a container gets damaged that you were responsible for buying shipping insurance.

The broad categories of expenses incurred during the distribution process can include:

- **Loading costs**. The cost of hiring companies to manage the transfer of materials from buildings to transportation and between modes of transportation: for example, factory to transport, air to land transport, truck to the loading dock.
- **Transportation costs**. The cost of moving materials using land, air, or sea transport.
- **Insurance**. The cost to cover potential loss or damage to goods during transportation.
- **Fees, duties, and customs**. If transferring products over international borders, the buyer is typically responsible for paying all duties and customs. The exception is when you use a DDP (delivered duty paid) transport contract.

16.2 Outsourcing Distribution

Managing each step in the distribution process is time-consuming, complicated, and typically out of the scope of most small-to-medium firms. Unless distribution is part of their core strategic capabilities, most firms will outsource it to distribution companies. These companies are broadly called third-party logistics (3PL) suppliers. They will provide some or all of the services needed to support distribution, including:

- **Getting the product from the factory to the plane, ship, or truck**. Depending on where manufacturing is done and where the product is going, it may need to go over borders to get to the main carrier (e.g., China to Hong Kong).
- **Coordinating sea shipments** and passing through customs in the final country. This service typically also includes the customs broker process.
- **Transport to the 3PL's distribution location**, or to your customer's location from the port. This transport is typically by land.
- **Order fulfillment and pick-and-pack**. In some cases, the packaging may be done and additional accessories added (from a different supplier) in the country of sale. Also, firmware and/or software may need to be tested or re-flashed at the pick-and-pack facility.
- **Warehousing**. 3PLs will have existing distribution facilities that they can share across many customers.
- **Tracking and traceability**. 3PLs can provide the information services around tracking serial numbers (SNs) and orders, so your company does not have to build that infrastructure.
- **Reverse logistics**. 3PLs may offer to handle the return of defective products and the order fulfillment for replacement products.
- **Interfacing with the factory for orders and fulfillment**. Some 3PLs will manage the entire ordering and order fulfillment process from end-to-end.

The benefits of hiring a 3PL provider include:

- **Surge capacity**. Often the shipments from your factory to your distribution warehouse won't arrive at a steady rate; instead you will receive big shipments once a month and will need to process large amounts of inventory quickly. If you hire a 3PL, you don't need to scramble to find part-time workers each time a shipment comes in or recruit your entire company to help.
- **Consolidation**. 3PLs will ship your products along with others to get a full truckload or container.
- **Negotiation**. Because they are buying transportation capacity in bulk, 3PLs can often ship at lower costs.
- **Software and processes**. These companies have the software and systems to manage orders and track products, so you won't need to expend the time or resources to build your own.
- **Claims**. Filing for freight claims and insurance claims can be very time-consuming, but 3PLs are experts at handling these.
- **Single-point payment**. Rather than hiring several transportation suppliers and managing the hand-offs between each one, your company only needs to write one check.
- **Security**. For high-value items, 3PLs have the security systems in place to protect the product.

As you might imagine, there are downsides to 3PLs as well:

- **Learning**. As companies grow, they typically bring managing the distribution back in-house to streamline processes and reduce costs. If you originally work with a 3PL provider, your company won't be gaining the capabilities and infrastructure to build your own distribution system when you decide to bring it in-house.
- **Oversight and visibility**. You cannot control your operations as tightly as you might wish if you outsource them to another company. Also, you might find it more difficult to ensure your desired level of quality control.
- **Cost**. Because the 3PL companies need to make a profit, they may charge you more than it would cost to run the identical operation yourself. If your company is able to create a well-run in-house logistics department (generally only possible for large companies), you will probably save money over using a 3PL.

16.3 Distribution System Design

When designing a distribution system, the operations team needs to ask and answer several questions for each of the steps in the process (Checklist 16.1).

Teams need to make tradeoffs when choosing their distribution methods. These tradeoffs include:

- **The cost of shipping: bulk vs. individual**. Individually shipping 100 units in separate packages will be more expensive than putting 100 units on each of 20 pallets and shipping by ocean freight. However, the product gets to the end customer more quickly if shipped by air in small lots. You will hold less inventory if you ship in small volumes.
- **Getting it fast or keeping costs down**. Drop-shipping from the factory cuts out several days in transit. Short delivery times may be necessary depending on the shelf-life of the product, customer delivery promises, and the production schedule.
- **Shipping hazardous or perishable materials**. Some materials cannot go by ocean freight if they need to be climate-controlled or are perishable, whereas others cannot be shipped by air because of limitations on hazardous material such as batteries and magnets.
- **Inventory costs vs. the speed of order fulfillment**. If orders are fulfilled from finished goods inventory, then the inventory will need to be held somewhere. It may be cheaper to keep stock at the factory and drop-ship it out from there rather than maintaining inventory at an intermediate location.
- **Demand variability and fulfillment**. Factories like to have a steady production rate each day – it decreases the complexity and the cost of setup and breakdown of the production line. However, if demand is highly variable or uncertain, it may be necessary to hold inventory to allow steady production rates. Cheaper shipment methods (e.g., ocean freight) can be used when the stock is not needed for a while. The distribution system will be very different for a built-to-order product (for which you don't hold inventory) than for finished goods held in inventory and shipped in bulk.
- **Shipping costs vs. space constraints**. Your company may not have the space to handle large shipments that come in from the factory every four weeks. While, on average, there might be 100 units shipped per day, the distribution warehouse may only have space for 1,000 units. However, on the first of the month, there may be a 3,000 unit shipment being dropped off at the loading dock. Start-ups who do not think this through end up with boxes stacked in the offices, hallways, and spare bathrooms because there was insufficient space allocated for surges in incoming inventory.
- **Shipping costs vs. labor costs**. As with space, the company may need a large number of people to process the incoming material once a month, but it is not economical to hire full-time staff to only work two days a month.

Checklist 16.1: Questions to Ask when Designing a Distribution System

Final packaging for shipping
- ❑ What is the final packaging – individual, master carton, pallet, or container?
- ❑ Are different SKUs packaged together to save on space, or segregated to facilitate easy inventory management?

Transfer of the product from facility to shipping location
- ❑ Does the CM or the 3PL take the product from the factory to the main carrier?
- ❑ Are you producing in a SEZ?
- ❑ Who is handling the paperwork for transferring the product in and out of the SEZ?

Airfreight/container shipping
- ❑ What transportation mode is being used?
- ❑ Are you using small batch (LTL) or full containers?
- ❑ Who is booking space on the transport?

Customs
- ❑ Who is getting the product through customs?
- ❑ What paperwork is needed?
- ❑ How will the products be certified to get through customs?
- ❑ How are products coded for the right tariffs?

Transportation from the dock to distribution
- ❑ How is the transfer coordinated?
- ❑ What shipment method is used?
- ❑ Who is responsible?

Distribution warehouse tasks
- ❑ If products need to be repacked, tested, or assembled at the distribution site, who is doing it where?
- ❑ Where is the inventory held?
- ❑ How are the master cartons unloaded and products repackaged for the customer (individual or distributor)?

Order fulfillment
- ❑ Who is coordinating inventory and customer orders?
- ❑ What method to is used to ship to the customer? Air or ground?

Other
- ❑ Who manages the reverse logistics if product is returned, and what process is used?
- ❑ How are you insuring your product?

- **Simplifying distribution vs. cost**. Some CMs will provide transport of products to the shipping dock and manage the customs paperwork. This simplifies this first step because you don't have to coordinate between the CM and another carrier. Factories are experienced in facilitating both processes. However, you will need to pay for them to do this and they often charge a premium for the service.
- **Costs vs. building distribution capability**. If distribution is part of the core capability and competitiveness of your company, then keeping distribution in-house is the best option.

Summary and Key Takeaways

- ❏ There are many options for how you can design your distribution system.
- ❏ 3PL service providers can be hired to provide some or all of the distribution and reverse logistics services.
- ❏ Your team will need to evaluate the tradeoffs of costs against speed and complexity when deciding the distribution method and to whom to outsource what.
- ❏ It is best not to leave distribution planning until the last minute. When hiring outside organizations, you want time to evaluate and compare quotes without being under pressure to get the product to the customer.

Chapter 17
Certification and Labeling

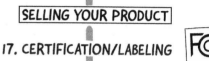

SELLING YOUR PRODUCT

17. CERTIFICATION/LABELING

 18. CUSTOMER SUPPORT

19 MASS PRODUCTION

Almost all products will require some certifications to allow the product to be sold. The certification process can be time-consuming, complicated, and expensive. In addition, the product has to be correctly labeled to be legally sellable. If there are many countries of sale, the certification process is all the more involved. This chapter reviews the common types of certifications required, how to learn what certifications you will need, and how to determine what information needs to be displayed and where.

In addition, this chapter discusses the labeling process and how to uniquely identify your product to reduce the risk of counterfeit and gray market sales.

Product Realization: Going from one to a million, First Edition. Anna C. Thornton.
© 2021 John Wiley & Sons, Inc. Published 2021 by John Wiley & Sons, Inc.

Take any moderately complex product and look at it closely. Look at the manuals, the labels on the gift box, and the labels on the product. There will be logos, terms, bar codes, serial and part numbers, warnings, text, and other data. All of this information is legally required to be on the product before it can be sold.

To legally sell a product in a given country, the seller and manufacturer need to obtain the right certifications and document those certifications on the product, packaging, or manuals. Not getting the proper certifications or incorrectly labeling the product can result in hold-ups at the border or, in the worst case, fines and recalls. The rules for both labeling and certification are subject to interpretation, can change over time, and are highly dependent on the country of sale, the product classification, and the end-user. The first section of this chapter will discuss testing and certifications typically needed, and the second will review how to document those certifications, along with other information on the product, packaging, and manuals.

This chapter does **NOT** tell you what certifications you need or don't need for your specific product. It is intended to give you a road map for what questions to ask and how to find answers. You will need to do extensive research (or employ your own experts) and get legal advice on what is required for your specific product in your countries of manufacture and sale.

17.1 Certifications

 All but the most basic products will require certifications in order to be legally sold in a given country. Certification requirements are continually changing in every country in the world. It is best to engage with an expert to determine the exact certification requirements. Getting educated before you talk to an expert will help you ask the right questions. This text gives some example certifications and guidelines; however, these are examples only, and teams need to do their own research and due diligence. The rules and guidelines for certifications have changed even in the two years that this book was being written and will continue to evolve and grow. It is always best to do the research and leave nothing to chance (Inset 17.1).

Your certifications will be based on several characteristics of your product and your intended user:

- **What are you selling?** Depending on the product's technology, its type of installation, and its functionality, different certifications will be required. For example, if a contractor installs the product during housing construction, it may be subject to building codes.
- **Is the product regulated?** Many product types require significant oversight, testing, and documentation: medical devices, products used in aerospace, and products in food preparation, to name just a few.

LOGO

INPUT: 20Vdc ⎓ 2.25A
Model Number: RB104-B
FCC ID: FR1A323456

Made in China
www.logo.com

‖‖‖‖‖‖‖‖‖‖‖‖‖‖‖‖‖‖‖‖‖‖‖
1030B459AT130495

FIGURE 17.1
Sample label

- **Who is your user?** Selling to young children or other vulnerable populations increases the stringency of regulatory requirements. For example, there are limitations on materials a child might put in their mouth or choke on.
- **Where are you selling it?** The regulatory requirements can be vastly different from country to country and even state to state in the US. For example, in the US, California regulations such as Prop. 65 [123] are typically treated as if they are US regulations because it is not economical to build products separately for the California market.
- **What channels are you using?** Some retail and business customers impose additional requirements above and beyond the legally required certifications. For example, some distributors and retailers require a UL (Underwriters Laboratory) [124], equivalent safety certification, or additional channel-specific testing. This is required to ensure the products they are selling are safe and tested.
- **How are you shipping it?** Air transport restricts the transport of some batteries, chemicals, and magnets.
- **Where are you building it?** Products sold in the US (and several other countries) are required to have a label stating the country of manufacture. If you want your product to be labeled "Made in America," you will have to adhere to stringent sourcing rules.

Inset 17.1: It is Always Better to Ask Permission When Dealing with a Certification

Action in the face of uncertainty and risk falls into two categories. For each decision, teams need to ask themselves, "Is it better to ask forgiveness, or is it better to ask permission?" In the case of anything regulatory or safety-related, it is *always* better to ask permission. Cutting corners, not getting the proper certification, or pleading ignorance will hold no water with customs officials, consumer safety organizations, or the law. Failing to adhere to legal requirements can lead to significant delays, fines, and, most importantly, putting your customer at risk.

- **How much are you selling?** Some regulations are only applicable when the volume of products sold or the weight of certain materials exceeds a certain threshold. As long as you are under the threshold you don't need the certification. However, be advised that these thresholds often change, and what passed a few years ago may not be legal today.
- **What materials are in your product?** Some materials are restricted for environmental and safety reasons (for example, lead in electronics or paint).

17.1.1 Types of Certifications

Localities, states, countries, and economic groups all can impose certification requirements (Inset 17.2). However, certifications mostly fall into several broad categories:

- **Intentional and unintentional EMF (electromagnetic field).** All electronics emit an EMF, which can interfere with normal communications channels and can have potential health hazards. These certifications are overseen in the US by the FCC (Federal Communications Commission) [125] and in Europe by Conformité Européenne (CE).
- **Safety.** Multiple regulations apply to the safety of the device. The most well known is UL (Underwriters Laboratory) [124] in the US. Other examples include ASTM's (American Society for Testing and Materials) children's toy certifications [30]. The CE mark used for European products also tests for multiple safety issues.
- **Material certifications.** These are required for export/import/transport. For example, batteries need to comply with USDOT (US Department of Transportation) regulations if they are going to be shipped (often referred to as USDOT-UN3480) [126]. The IATA (International Air Transport Association) also has regulations recognized by the airline industry for shipment of hazardous materials such as magnets (to ensure they don't interfere with flight controls) [127].
- **Environmental certifications.** These are intended to reduce the impact of products on the environment and on climate change. Environmental regulations cover a range of issues, including:
  - *Toxic/hazardous materials.* Many countries have restrictions on the sale of toxic or hazardous materials such as lead. For electronic devices, this is covered by RoHS (Restriction of Hazardous Substances) requirements [95].
  - *Chemicals.* Some chemicals may be regulated or require a special warning either on the product or in the manual, as in the California Proposition 65 requirements [123].
  - *Energy.* Vampire devices (those that draw a constant low current when in standby), as well as the efficiency of charging circuits, are increasingly being regulated [128].

– *Recycling*. Europe is imposing increasingly stringent regulations on recycling. For example, the WEEE (Waste Electrical and Electronic Equipment) Directive sets requirements on recycling electronics [129].

- **Proprietary communication certifications**. These give you the legal right to use a certain company's proprietary communication protocols (e.g., Bluetooth [130] and ZigBee [131]). To advertise these capabilities on your product, you must pass certain certifications before using that company's logo on your packaging or advertising. Individual components in products may come pre-certified when you buy them from suppliers; these pre-certifications will speed up the certification process. However, for certain products or technologies, the completed product may also require certification.
- **Industry-specific requirements**. These are imposed based on the type of product. For example, medical devices are subjected to Food and Drug Administration (FDA) regulations. Aerospace and automotive products have industry-specific certification requirements developed by industry consortia and the government (e.g., Federal Aviation Administration [FAA] and National Transportation Safety Board [NTSB]).
- **Export certifications**. Some products need to be vetted to ensure that you are not exporting technology critical to a country's defense. For example, in the US the ITAR (International Traffic in Arms Regulation) [32] may limit the export of product with encryption or drone technology.
- **Voluntary certifications**. These include Fair Trade and socially responsible certifications [96]. These certifications help with the branding of your product and communicate your commitment to specific values.

Inset 17.2: State vs. Federal vs. International Regulations

In the US, some rules are imposed by the federal government and apply across all states and territories (e.g., FCC and FDA regulations), while others are required by individual states. It is logistically costly to produce and control the sale of some versions of a product for just a single state, so as far as your business is concerned, state regulations become federal regulations. For example, California Proposition 65 labeling [123] is only legally required in California; however, most products affected by Prop. 65 and sold in other states comply with the regulation even if the products are not being sold in California.

Other regulations imposed internationally have also become worldwide de facto regulations. For example, RoHS [95] was imposed by the European Union (EU) for all electronic products produced after 2006. RoHS specifications require that all materials going into a product be RoHS compliant and be produced in a RoHS-compliant facility. Maintaining a separate non-RoHS product version for the product not sold in the EU would require companies to have a completely different bill of materials (BOM) and manufacturing facility. The cost savings from using non-compliant parts are typically outweighed by the cost of maintaining two products and factories. As a result, RoHS has become a de facto international standard for most electronics.

17.1.2 How Do You Learn What You Need?

If you type "what certifications does my product need?" into your internet browser, you will get an overwhelming number of results. There are several places to start to learn what certifications and labeling will be required for your product:

- **Look at similar products**. Find a product with the same market (e.g., Europe), user profile (e.g., adults), and features (e.g., Li-ion batteries). Review the certifications that appear on the device, in the software, in any manuals, and on the packaging.
- **Search the CPSC (Consumer Product Safety Commission) website** [31] for similar products which were recalled due to failure to conform to federal and local regulations.
- **Call a certification lab**. They will review your product and give you a quote for what they think is necessary. Compare several quotes and identify what appears in every quote (things you almost certainly need), question any differences (items subject to interpretation), and ask a lot of other questions.
- **Talk to your factory/contract manufacturer (CM)**. Your CM has probably produced products with similar certification requirements, and you can learn a lot by talking with them.
- **Hire a certification consultant**. This is probably only necessary in cases of safety-related devices, products that need to adhere to FDA regulations, or unique products whose classification is ambiguous.

Using one or more of these methods will give you a list of what certifications your product needs. Once the list is understood, then teams need to determine:

- **Who issues each certification?** Some certifications require testing by a qualified lab (e.g., FCC), some by a corporate body (e.g., Bluetooth [130]), and some by the factory (e.g., RoHS [95]).
- **What is the lead time for the certification?** Some certification labs can run the tests and tell you whether you passed within a few days, but it may take several weeks to process the paperwork. You don't want to delay shipment because you didn't start the certification process early enough.
- **What is the timing and maturity of the product samples required for testing?** Some certifications can be issued using samples from early pilot runs, and others certifications require products built on the production line used for mass production.
- **What additional documentation is needed to issue the certificate?** For example, some certifications require a copy of the quick-start guide.
- **How many samples are needed?** Each different certification may require several samples to be tested; plan ahead to make sure you set aside enough samples.
- **Does the certification require a representative in a foreign country?** For example, some regulations require an authorized representative to manage compliance. There are companies that will fulfill that role for a small fee.

17.1.3 What is the Process for Obtaining a Certification?

Processes for getting a certification include the following steps:

1. **Get quotes for needed certifications**. Compare multiple quotes to understand what is likely required. Also, check the labs' backlogs and lead times for testing.
2. **Understand what labels and documentation are needed to legally sell your product**. Each certification will need to be documented (typically on the label, in the documentation, and on the gift box).
3. **Determine certification approval requirements**. Often, a testing lab requires the following to test and issue the certification:
   - A working final-design unit with firmware and core software.
   - Accessories for charging or replacement batteries.
   - The quick start guide and any other documentation provided to the user (this can be a close-to-final draft).
   - Some tests can be done on design verification testing (DVT) units, while others require production verification testing (PVT) units.
4. **Determine the lead time for getting your certification**. Testing and documentation can take several weeks depending on the backlog at the lab. If the product is being shipped for the Christmas season, it is likely to be in the queue with many other products that are on the same tight timeline.
5. **Decide who you are hiring to issue your certifications**. You can hire a certification lab in the country of the manufacturer or nearer to home. Often it is easier to hire a certification lab near the manufacturing location because of the logistics of getting samples and uncertified products shipped and through customs. If your product is being manufactured in China, the labs in China or Taiwan are typically less expensive than the US counterparts.
6. **Submit your samples and paperwork**. You will need to deliver all of the materials and documentation to the certification lab for them to get started.
7. **Get an initial test result**. The certification lab will often be able to get you a pass/fail report relatively quickly.
8. **Additional inspection**. Some certifications require the inspection of the production line by a certified auditor, and this may require careful scheduling to avoid delays in mass production.
9. **Waiting time to get the paperwork and certification numbers**. Getting the actual certification can take several weeks to a month or two.
10. **Update labels and documentation.** Often the actual certification numbers (you will get a unique certification number that allows anyone to look up the certification and verify that your product is legal to sell) need to be printed on the product and/or provided in the manuals. The labels and manuals can't be printed until that number is issued.

17.1.4 Avoid Letting Certification Delay Product Shipment

If the certification requires the testing of the production-intent product (typically from the DVT pilot run), it may not be possible to obtain certifications until late in the piloting process when all of the final production-intent parts are available. Also, some certifications require inspection of the production line during full production. Late failures in certification testing or not understanding the timeline can cause unexpected delays. You can avoid delays by doing the following:

- **Pre-scan or pre-certification evaluation**. Some factories can perform a pre-scan or pre-certification assessment. These unofficial tests can be done at no risk and can give the team an idea of how much margin there is in the design. For example, if the product barely passes FCC testing during this pre-scan, you can add EMF shielding by using coatings or tapes to improve the chance of passing later.
- **Understand which pilot run can be used to get the certifications**. DVT units can often be used as long as no significant design changes (in the view of the certification body) are made between DVT and PVT.
- **Build product and label later**. Depending on the certification, the production line may need to build significant inventory before the certification is issued. In some cases, it is possible to retroactively label the units that were used to obtain production approval. *It is imperative to triple-check to see if this is possible, and that you are legally labeling the product.*
- **Understand seasonal trends**. Almost any company that makes consumer products for the US or European market hopes to get their product certified in time for the Christmas shipping timeline. As a result, most certification labs are at capacity during pre-holiday months (late summer and early fall) and tend to give preferential treatment to long-term, high-value customers.

Most minor changes to the BOM and to the design during the pilot phase will not require recertification. Because of that, many organizations use DVT samples (all but the aesthetics and final fit and finish) for the certification testing. However, if there are any significant changes between the units provided and the final product, the certification may need to be re-done.

17.1.5 Cost of Certifications

If you want your product to be a huge success, you should offer it for sale everywhere around the world to maximize customers, right? Not so fast. Until your product has a track record of sales, it's probably a bad idea to go through the cost and effort required to legally sell your products in dozens of jurisdictions. Too many crowdsourced companies open sales to many countries only to discover that the certifications needed to sell in 15 countries cost more than the profits from the whole endeavor.

Getting the complete set of certifications for the US and Europe for a consumer electronic device used by adults can cost in the range of $10,000–20,000. Each additional country of certification will increase that cost by thousands of dollars. Although the standards bodies across the world continue to consolidate certification processes across countries, they aren't fully reconciled and transferable. Typically, doubling the number of countries you ship to will double the cost and the complexity.

17.2 Labeling and Documentation

If you turn over almost any electromechanical product, you will find a label of some kind. It will have a variety of information, including a bar code, a many-digit number, and a set of symbols. Information about warranties, certifications, and disclaimers will be located in many places including on the product, in the product, on the packaging, and inside manuals. Labeling needs to be thought-through carefully for many reasons, including:

- **Certification and legal requirements**. Each country or region imposes legal requirements that must be met before the product can be shipped. There are rules (sometimes obtuse and confusing) for what information is included and where these certification marks must be displayed on the product. The certification documentation can consist of symbols, marks, text, and unique certification numbers.
- **Product identification for the customer**. If the product is defective, customer service will need to know which unique product a customer has. The stock-keeping unit (SKU) is the unique identifier of the product and product version, and the serial number (SN) is the code unique to that particular device.
- **Product identification for distributors/retailers**. The master carton, inner packs and gift boxes needs to be labeled so that distributors or retailers can scan the label and manage inventory. Packing labeling can include barcodes, QR (quick response) codes, or RFID (radio frequency ID) tags.
- **Marketing**. Some labels may include marketing material that shows the brand identity of the product (e.g., the "Intel Inside" sticker on the author's laptop and the Bang & Olufsen logo near the speaker).
- **Tracking of individual subsystems**. Individual parts within your product (e.g., electronic boards, batteries, or hard drives) may have their own unique labels to enable monitoring of inventory or tracing defects. For example, if many failures are traced to one type of board, the returned failed boards can be traced back to a single production batch. In highly regulated products, there are legal requirements for batch traceability.

- **Warnings and safety**. The label, manuals, and stickers may include information about how to use the product safely (e.g., electric shock warnings on hairdryers).
- **Operating and setup instructions**. Some labels may teach the user how to set up and use the product. Many of these are stickers (e.g., arrows or "press here") that are removable after the first use.
- **Licensing**. If you have licensed software or other technology from another company that you are selling on as part of your product, you may be required to document that license information somewhere in the product, packaging, software, or manuals.

17.2.1 Where Information is Located

All of the information described above can be displayed in several locations, some visible on the outside of the product and some not. The master carton and inner packs are labeled to help identify what is inside and give guidance to the shipper on handling. The gift box will say what is in the box and may include unit-specific information to help facilitate product setup. For example, some products display the SN on the outside of the box so it can be scanned and registered to you by the retailer at the point of sale. The device itself will have a label that includes SN, electrical power characteristics, and some certification information. Manuals, quick-start guides, and online materials can contain additional certification information, safety guidelines, and warranty terms. The device ROM (read-only memory) may have the SN and other certification information in permanent memory. Internal components may be individually labeled. Finally, information about the device may be maintained in the cloud associated with user's accounts. Table 17.1 summarizes this information.

17.2.2 How to Decide What to Put Where

Figure 17.1 is an example of a label for an imaginary electronic product. It should not be used as an exact template for your product, but it shows the types of information you will typically include:

- Product logo
- Information about the power supply
- SN
- Bar code
- Manufacturing location
- Certification numbers
- Certification marks
- SKU number

You should always get advice from experts on what specifically should appear on your product label. Incorrect labeling can result in your product being rejected at customs at best, and being recalled at worst.

Packaging type	Typical information displayed	
Master carton	• Sender information • Directions for handling • Company name	• Contents • Warnings for hazardous materials such as batteries
Gift box	• SKU • Certification information such as recycling, Bluetooth • Country of production	• SN • Barcode • MAC (media access control) ID, IMEI number
Device label	• SN • SKU and model information • Barcode • Certification	• Trademark information • Country of production • Power input information • SN of key components
Device ROM	• Detailed certification information • Licensing	• Configuration
Manual, quick-start guides and other materials (either paper or electronic)	• Quick-start information • Warnings • Detailed certification	• Licenses • Warranty
Temporary labels and stickers	• Setup directions • Safety information	• Marketing information • Welcome notes
Company's websites and cloud	• Link from serial number to all associated information	
Internal labeling	• Parts SNs, barcodes and certifications	• Parts batch and mold cavity numbers

Table 17.1
Examples of information by location

Determining what information will be displayed where includes the following steps:

- **Create a list** of what information needs to be documented, displayed, or recorded.
- **Understand the legal issues around labeling**. Some rules for labeling are unambiguous, and others leave much room for interpretation. Reading the certification regulations is virtually impossible for someone not familiar with the process. You can avoid trawling through hundreds of pages of obtuse regulations by:
  - Looking at products that are in the same category as yours, see how the labeling was managed, and where they put what information.
  - Talking to a certification expert.
- **Determine where your label can be placed**. Where do you have room on your product for a label, and what is going to be visually pleasing (or just not objectionable)?
- **Assess what information your customer *needs* to access long after the product is unboxed**. You should assume that people will throw away the manuals, and most of the people who keep them stick them in a drawer

and cannot find them when needed. Anything that the customer needs for warranty, assistance, and so on, should be labeled on the product itself.

- **Plan how unit-specific information will be associated with the product.** For example, if you want the SN, MAC (media access control) ID, and other unique details programmed into ROM, teams will need to think through many questions: when to program that data, how to ensure a unique SN is assigned to each unit and stored correctly in ROM, and how the SN programming is coordinated with the external labeling.
- **Decide how you will reduce the risk of counterfeit products.** Labels can be used to ensure that counterfeit products can be easily identified (Section 17.2.4).

17.2.3 Serial Numbers

Serial numbers (SNs) are often an afterthought; however, intentionally designing your SN is critical for several reasons:

- **Warranty and product traceability.** A well-designed SN will enable teams to trace when and where a product was produced without using a lookup table. It also reduces the potential for errors in data collection on returns and complaints (Chapter 18).
- **Prevention of gray market and black market sales** (Section 17.2.4). SNs can ensure that only your legitimately sold product can be supported and that illegal products can be identified.
- **Data consistency.** If you design a SN that is too short or not expandable as your product portfolio grows, you will need to change systems; this can lead to confusion and errors.

There are two approaches to SN generation: smart SN and completely random SN. Smart SNs can be read if you know how the SN was created (the rules are often colloquially called "the Decoder ring"). Typically, a smart SN is constructed with the following information encoded into the SN:

- **Product/SKU information.** Usually, it is hard to include the complete SKU information because SKUs themselves are long, but some indication of the product type is useful.
- **Manufacturing location.** If you have multiple manufacturing facilities, you should differentiate between dual sources to prevent you from unintentionally issuing identical SNs.
- **Production date.** Depending on the batch sizes and traceability requirements, this may be a month, week, day, or shift within a given year.
- **Unique product number** (typically sequential).

For example, the SN in Figure 17.2 includes information about product SKU, production date, factory, and sequential production number.

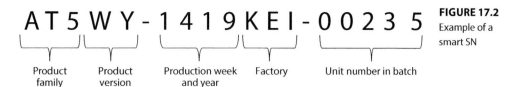

AT5WY-1419KEI-00235

Product family · Product version · Production week and year · Factory · Unit number in batch

FIGURE 17.2
Example of a smart SN

The other option is to create a completely random SN. A data file (lookup table) allows the customer support teams to look up the relevant product information when given an accurate SN. Companies often use random SNs when there is a risk of counterfeit and gray markets because it is virtually impossible to fake a SN. The SN allows customer service providers and customers to validate that they have a real product. However, it is easy to transcribe a random SN incorrectly; and tracing the product requires that the lookup tables are accurately maintained.

Other things to think about when designing an SN include:

- **Starting the SN with a letter and never with a zero**. Spreadsheets will not register an initial 0 as a digit, turning your 17-digit SN into a 16-digit one, and obviating the smart SN system. If your smart SN always starts with a letter, this will help to avoid problems introduced by data-type conversions.
- **Structuring the SN to identify a fake or incorrect SN**. A structured SN allows software to easily check if the SN is valid, which enables more accurate warranty analysis and quality traceability.
- **Mixing letters and numbers helps readability**. Long strings of numbers (especially those with lots of repeating 1s or 0s) are harder for most humans to read and transcribe correctly than mixed strings of numbers and letters.
- **Ensuring that the SNs are long enough** to accommodate high production volumes and product variety. You always want to settle on an adequate SN length and not change it over time; this makes it much easier to parse large sets of data.
- **Avoiding SNs that are too long**. The longer the number, the more real estate the number takes, and the more room for errors in transcription.

17.2.4 Gray Market and Counterfeits

Gray-market and counterfeit (black-market) products can significantly undermine a company's brand, reduce profits, and undermine pricing power. Proper labeling and traceability can help identify where bad actors are selling your product illegally or selling counterfeits. Gray markets are channels through which real products are sold at a significant discount. There are three typical ways that products get into the gray market:

- **Extra production sold illegally by the producer**. Instead of building 10,000 units, the CM might make 11,000. The factory ships ten thousand to the customer, and the additional 1,000 are sold without the knowledge of the company in a different market or through other online marketplaces at a significant discount. Some of these units may not have passed all of the quality tests and can degrade the brand.
- **Extra units sold at a lower-than-MSRP price**. Distributors may overbuy products and sell the product at a lower price than is contractually agreed. Discounts degrade the price premium for a brand and undermine the agreements that the company has with other sales channels.
- **Units purchased in one country at a lower price and exported into another country**. Units are then sold at a higher price but less than the in-country price. Selling products outside the country is a common problem with medical devices and pharmaceuticals.

Counterfeits, which are part of the black market, are products that are copies of an existing product but not manufactured at the same site, typically using less expensive and lower-quality manufacturing techniques. These can erode a brand's reputation if people are not aware that the product is not real. Also, counterfeits can create considerable risk to human life if they are safety-critical (for example, counterfeit aircraft parts).

There are several ways to reduce the risk of counterfeits and gray-market products.

- **Implement counterfeit-resistant labeling** including holograms, and educate your customer on what to look for to ensure they are getting a valid product.
- **Require registration of the product** (e.g., setting up the product requires a connection to a cloud service). The cloud service ensures that the SN is valid before registering and activating the product.
- **Create a record of all devices** when the firmware is loaded and the SN is recorded. The file of programmed devices and SNs is uploaded to you in real time from the factory, and the SNs matched with the shipments. If the factory is making product that they aren't shipping to you, the firmware upload file will show that.
- **Provide a critical and uncopiable component to the factory**. For example, modules can be pre-programmed with encrypted firmware. The CM is responsible for accounting for the inventory for these components, and that the number used matches the number of products shipped.

Organizations should continuously scan for gray-market or counterfeit products in various distribution channels. Keeping accurate records of the product in the approved distribution channels will help identify products that someone is selling illegally, are being sold at the wrong price, or are counterfeits.

Summary and Key Takeaways

❑ Certification requirements are complicated and hard to discern on your own. Teams should educate themselves on what certifications are likely to be needed.

❑ Always do your research and don't cut corners on certifications – it just isn't worth it. It may be worth your while to hire a certifications expert.

❑ Getting certified is complex and can delay a product launch. Don't wait until the last minute.

❑ Teams should identify critical information and determine where it is to be displayed and documented.

❑ Think through constructing SNs so you have both expandability and traceability.

❑ Gray market and counterfeit products pose a significant risk to product reputation and pricing.

Chapter 18
Customer Support

SELLING YOUR PRODUCT

17. CERTIFICATION/LABELING

   18. CUSTOMER SUPPORT

19 MASS PRODUCTION

Every product development company will need a customer support system to troubleshoot problems and to manage warranty returns, repairs, and reimbursements. Implementing a customer support system is more than just having a web-based FAQ page and a toll-free number. This chapter describes how warranty claims are handled, the elements of a customer support system, customer complaint data reporting, and what happens in the worst case: a recall.

Product Realization: Going from one to a million, First Edition. Anna C. Thornton.
© 2021 John Wiley & Sons, Inc. Published 2021 by John Wiley & Sons, Inc.

Think about some brands you really love, whose products you think highly of. Now think about some brands you don't like, whose products you've had bad experiences with. Chances are that the brand you have the most negative feelings about is a brand that you – or a lot of people you know – have had bad customer service experiences with.

We have all had a frustrating experience with customer service when a product fails. We spend time punching numbers getting to the right person only to be told to reset the product and call back. Helping customers after they buy a product is critical to customer satisfaction. When a product we love fails to work as expected, how the company responds can impact how we perceive (and recommend) the product. The author recently had two very different experiences dealing with customer service with two different faulty kitchen products. Both occurred around the holidays, a period when kitchen equipment reliability is critical. Her favorite (and costly) paring knife cracked and broke. Looking at the surface of the break, there was clearly a defect in the metal. The knife manufacturer's customer support responded to her email within four hours, and after sending the defective knife back, she received a brand-new knife within a week.

On the other hand, her high-end oven, which had been purchased four years earlier, overheated while baking holiday cookies. It was hot enough to catch the cookies and cookie sheet liner on fire and raise the temperature of the kitchen walls and cabinets significantly. The fire department had to be called to verify that the fire hadn't spread, and while there ate most of the previously baked cookies. The system failed in two ways. First, the auto-locks that were supposed to kick in did not. Second, the oven didn't shut itself off. It took shutting off the power at the circuit breakers to get the heating elements to stop pumping heat into the oven.

Had her daughters (who now know how to shut off the circuit breakers) been cooking alone, the outcome could have been much worse. Dealing with the manufacturer to get reimbursed for the defective product ended up being a six-month process. First, the manufacturer didn't take the safety concerns seriously. It took three escalations to talk to someone who could approve a reimbursement. Finally, after that representative finally approved a refund, the author had to call the company several more times to get them to pick up the defective unit, then several more calls to get the reimbursement check issued. Based on her experience with that customer service, the author will never again buy that brand, and has strongly discouraged friends from purchasing anything from them either.

No matter how carefully you design and build your product, some customers will have questions and problems, and some will want to return their products. Some of these services will need to be provided for free (because the product is under warranty); sometimes customers may be compensated to maintain customer loyalty; at other times you will decide

Inset 18.1: Legal Obligation to Report Critical Safety Issues

There can be considerable legal (and moral) issues if your customer contacts you with information about a potential safety hazard and you ignore the problem. For example, the following is posted on the CPSC website [139] and warrants being repeated here:

"If you are a manufacturer, importer, distributor, and/or retailer of consumer products, you have a legal obligation to immediately report the following types of information to the CPSC:

- *A defective product that could create a substantial risk of injury to consumers;*
- *A product that creates an unreasonable risk of serious injury or death;*
- *A product that fails to comply with an applicable consumer product safety rule or with any other rule, regulation, standard, or ban under the CPSA or any other statute enforced by the CPSC;*
- *An incident in which a child (regardless of age) chokes on a marble, small ball, latex balloon, or other small part contained in a toy or game and that, as a result of the incident, the child dies, suffers serious injury, ceases breathing for any length of time, or is treated by a medical professional; and*
- *Certain types of lawsuits. (This applies to manufacturers and importers only and is subject to the time periods detailed in Sec. 37 of the CPSA.)*

Failure to fully and immediately report this information may lead to substantial civil or criminal penalties. CPSC staff's advice is 'when in doubt, report.'"

Your customer support group needs to be trained on how to handle such complaints and you should have a pre-planned escalation process to immediately address any safety-related issues. The CPSC has a comprehensive Recall Handbook [50]. However, it is imperative to get legal advice on how to set up your systems to identify potential hazards and respond appropriately.

that it's appropriate to charge the customer. A poorly designed customer support process can be very costly for a company and, more importantly, lead to poor customer reviews that drive down revenue.

To avoid expensive failures and ensure a safe product, it is critical to understand your company's responsibility to provide a defect-free product (and how defect-free is defined), and your legal (and ethical) responsibilities to address customer complaints and potential safety issues (Inset 18.1). A well-designed customer support process can also provide you with invaluable data for improving the current product and designing the next-generation product.

This chapter provides a summary of several key concepts in customer service; however, it doesn't replace doing your homework and risk assessment, and getting competent legal counsel. The next four sections will discuss:

- What is a warranty? (Section 18.1).
- What is a recall? (Section 18.2).
- What is involved in setting up a customer support system? (Section 18.3).
- How should you structure customer complaint analysis to support continual improvement? (Section 18.4).

18.1 Warranty

A warranty is a legal obligation that commits a company to address any failure of a product to perform as expected during a specified period of time. Depending on the sale location (state and country), certain aspects of the warranty may vary: the coverage period, coverage scope, and even the definition of "fitness for a particular purpose." It is essential to understand your legal and financial commitment to support the product after the sale. Almost all products must be covered by a warranty which can be either implied or express. In addition, you are required to maintain a warranty accrual account. A portion of the sale of each product has to be put into an account to cover future warranty liabilities.

Express warranties are documented descriptions of what failures and defects are and are not covered under warranty. The coverage period for express warranties in the US is typically one year, but other countries may require a more extended warranty period. On a single product, you might have an express warranty of one year for most US states, a two-year warranty for others, and a six-year warranty for some. The content and coverage of express warranties are governed by law. While most warranty laws are governed by the individual states in the US, most of the laws are basically the same because all of them have adopted the Uniform Commercial Code, which governs the execution of commerce in the states. There are federal laws as well. For example, in the US, the Magnuson–Moss Warranty act of 1975 [130] was enacted to address misuse and abuse of warranties, and defines requirements when providing an express warranty.

Implied warranties occur when there is no express warranty. Implied warranties are those unwritten promises enforced by state and federal law in the US and by country-specific laws outside of the US. Implied warranties are based on the common law of "fair value for money spent" [132]. Implied warranty lengths and coverage are highly variable depending on local laws. The laws enforce the concept that when you sell a product, you are tacitly promising that the product will do what it says it will do, and there is not anything significantly wrong with it from the time you buy the product through its expected life. As with certifications, the more countries you are shipping to, the more diverse the implied warranties will be.

If you don't want to have a warranty – that is, if you intend to sell the product "as-is" – you must explicitly state that fact; but even then there are laws to protect the customer against defective products. You can disclaim the implied warranty (i.e., "sold as-is"), but most customers will ask why.

A warranty dictates that if a customer has a reasonable reason for returning your product, you are obliged to repair or replace the product or to refund the customer's money. The customer usually gets the benefit of the doubt, but some organizations are adding sensors and other devices to detect misuse of the product and thwart fraudulent warranty claims. For example, most phones now have a moisture sensor that changes color if exposed to water; unless the customer buys an extended warranty that covers water damage, the customer cannot claim a warranty replacement if that sensor indicates water damage.

The number of in-warranty returns on a given product will be a function of the reliability of the product and the warranty period over which you pay to reimburse, repair, or replace the defective product. The expected life of the product (when the product can break, but may not be in warranty) can be significantly longer. For example, your car may be covered for seven years or 70,000 miles. The automotive manufacturer will pay for defects deemed manufacturer-error rather than user-error or normal wear and tear (e.g., they would reimburse a faulty engine but not a dented fender). After seven years, you are responsible for paying for any failures unless there is a safety-related recall.

The time during which the company is responsible for addressing quality issues may be significantly longer than the express warranty because the expected life of the product may be much longer than the warranty period. When discussing the length of the warranty, you need to be explicit about what warranty promise you are referring to. These can include:

- **Express warranty promise**. This is the legal obligation of the company to repair any manufacturing defects if the product is used as expected.
- **Warranty as implemented**. Unless products are digitally registered when they are purchased (such as with a phone or other IoT [Internet of Things] devices), it can be difficult to prove when the warranty period starts and ends. As a result, customer's complaints are often treated as in-warranty if it is not possible to prove otherwise.
- **Extended warranty**. If customers purchase an optional extended warranty, the products should be designed to ensure that the revenue generated by the extended warranty is not exceeded by the cost to support the product.

- **Expected life**. Customers don't want products that fail one day after the warranty expires. When products do fail soon after the end of the warranty, the frustration of the customer can be quite high. In cases where there is a safety failure (see the example of Cuisinart in Inset 7.8), the company may be responsible for reimbursing or repairing the product even if it is well outside the warranty period.

18.2 Recall

A recall occurs when all or some of the sold product needs to be repaired, replaced, or removed from use, independent of whether or not that individual product fails. It occurs when there is a critical safety issue or failure to adhere to regulatory requirements. The company bears the entire cost of the recall. Needless to say, a recall can be very damaging to a company, both in immediate cost and in long-term reputational damage. Several well-known examples of recalls include Johnson & Johnson's Tylenol recall in 1982 [133], and Toyota's floor mat recall in 2010 [134]. Damage caused by a recall can include:

- **Customer harm**. The worst possible consequence of a faulty product is that it might injure or kill customers.
- **Brand value**. Recalls (especially those handled poorly) can have long-term impacts on the reputation of your brand, and sales of all your products might suffer.
- **Cost of recall**. The cost of a recall can run in the tens of millions to billions of dollars – for example, GM ignition or Takata airbag. You can purchase product liability insurance to avoid the risk of huge payments for injuries and lawsuits (Inset 18.2).
- **Missed deliveries**. You might miss promised delivery dates because production needs to be stopped until the recall issue is resolved, and then inventory needs to be diverted to support the recall. This again can hurt your reputation and your bottom line.
- **Delayed new product launches**. Next-generation products may be affected because critical engineering and manufacturing resources are redeployed to resolve the problem.

The decision to recall a product from the market can be voluntary or imposed by the regulatory body that oversees the product; in the US, these regulatory bodies include the Food and Drug Administration (FDA) for pharmaceuticals, NHTSA (National Highway Traffic Safety Administration) for automobiles, and CPSC (Consumer Product Safety Commission) for consumer products. In the late 1970s, the Ford Pinto was recalled; journalists reported that Ford delayed issuing this recall for financial reasons, and the delay very likely led to additional deaths and accidents [135]. Ideally, organizations should immediately address a safety risk, communicate with their customers, and proactively pull the product from the market in a way that prevents any further threat to consumers. The Johnson & Johnson Tylenol recall is often used as an example of how to address consumer safety correctly [136]. They immediately pulled product off the market, systematically improved the safety of the product going forward through adding tamper-proof packaging, and then worked to ensure consumer's confidence in the safety of the product before they re-introduced it.

The formal recall process will vary by industry, but all share a number of common processes. A formal recall typically includes notifying the right government organizations, communicating to the public through public announcements or direct contact with owners, and tracking what product is returned. The consumer is told what process to follow to bring goods in for repair, replacement, refund, or to self-repair.

If an error by one of your parts suppliers is the root cause of the recall, that supplier may be obliged to reimburse you for some or all of the costs (Section 14.5.9). However, even if the supplier reimburses all of the recall costs, the overhead and damage to the brand can be challenging to recover from. To protect against the impact of a recall, you can purchase liability insurance; many investors will require it (Inset 18.2).

The best way to avoid a recall is to pay close attention to safety, quality, product verification, and regulatory requirements. As pointed out in Section 5.3.3, reviewing similar recalled products can be used to set appropriate

---

Inset 18.2: Product Liability and Recall Insurance

No matter how careful you are in designing for safety, there is always a chance your product will result in an unexpected injury. According to one article in the Harvard Business Review [137].

"One thing is clear from case history: even if your product is as safe as anyone else's in your industry and does what customers expect it to do, if a feasible design alternative could have prevented an accident, your product is at fault."

The financial impact of a recall or lawsuit can ruin a company. According to the Insurance Information Institute, the median liability award in 2017 was $1.5 million. To protect against "unknown unknowns," companies often purchase product liability insurance and/or product recall insurance. The cost of insurance can vary widely depending on the uncertainty (higher for a new product in a market) and product complexity. Many sites quote a base rate of 0.25% of product sales; however, the price can be higher for products perceived to be riskier [138].

specifications, make design decisions, and implement testing protocols to avoid similar failures. A quick, unscientific review of the CPSC recall database resulted in the following typical causes of consumer product recalls:

- Lacerations
- Choking hazards
- Falling or tripping
- Pinch points

- Death
- Fires
- Burns
- Electrical shock
- Ingestion of magnets

- Strangulation
- Poison
- Contact with hazardous material
- Skin irritation

18.3 Customer Support

Customer support systems are designed to quickly resolve trivial problems, troubleshoot more challenging technical issues, and support warranty claims. Most of the time, customer issues can be handled over the phone or through web support pages. For example, customers can be talked through resolving software that won't load correctly, problems with learning how to use the product, or just remembering to plug in the product. However, your customer support staff needs to be able to address more challenging aspects of returning and repairing the product or, in the worst case, identifying safety issues that could trigger a recall. Figure 18.1 shows the parts of a customer support system.

A typical customer support process starts with the customer contacting the company by phone or through an online chat. Ideally, the customer service representative can resolve the issue over the phone through simple troubleshooting. If the issue can't be resolved and the product is under warranty, it can be repaired, replaced, or returned.

Some customers will go back to the place where they purchased the product. Products are returned through national retail big-box stores, or an online distribution company may be handled differently depending on the contracts you have with your distributor. In some cases, you may have to reimburse your distributor for the return they handled, and in others you may need to give the distributor a blanket percentage of their sales back to cover any returns. In some cases, the distributor will dispose of the product, while in others you will get the product back. What happens physically to the returned products depends on the agreement you negotiate with your distribution channel.

FIGURE 18.1

Customer support
paths and outcomes

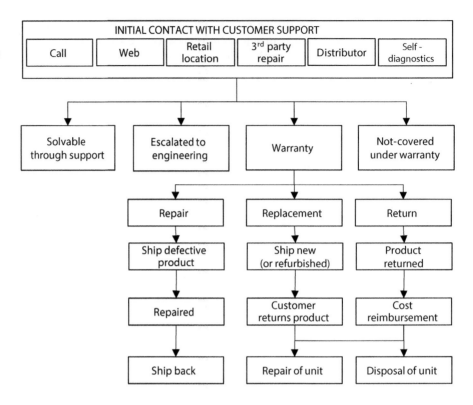

If you are handling the calls yourself, the customer support system should be designed to:

- **Manage incoming customer requests and questions**. A company will usually give customers several ways to get help, including web-based chat, call centers, in-person helpdesks (e.g., Apple's Genius Bar), third-party support (e.g., going back to Best Buy), or maintenance providers (the company you bought your appliance from may coordinate repairs).
- **Resolve the complaint**. Once a customer files a complaint, the company needs to figure out how to solve the problem. Methods to resolve issues can be as simple as a script to teach the customer to reboot the product, or as complex as referring the problem to technical experts and bringing the product back for troubleshooting and repair.
- **Manage warranty and repair claims**. If there is a warranty claim, the company needs to repair or replace the product, or reimburse the customer. Managing returned products can be handled by your company, the contract manufacturer, or third-party logistics (3PL) (Chapter 16).

18.3.1 Incoming Customer Contact

Originally, customer support was done by bringing the product back to the store or calling a repair person. In the US, the advent of toll-free calls in the late 1960s meant that people didn't have to call companies using expensive long-distance calling rates, so companies made wider use of toll-free customer support numbers. Today, companies typically provide several modes for customers to reach out for help, including:

- **Online/web-based support**. This route can reduce the cost of customer interactions and can filter out some of the more easily handled calls. For example, an internet service provider can remotely ping a modem and diagnose whether the signal is clear or non-existent. Online systems can be as simple as a FAQ (frequently asked questions) page or as sophisticated as connecting with the device remotely to run diagnostics. In addition, some high-tech companies create online communities where customers can help each other solve problems.
- **Traditional toll-free call centers**. Many customers want the reassurance of talking to a human being. Some companies manage their own call centers; others outsource to contract call centers – either in-country or offshore in places like India – who triage the call and only escalate the critical problems to experts. This allows lower-skill generalists to handle a large percentage of calls so that you don't have to pay the more highly skilled technical experts to ask the customer whether they plugged in their device. However, talking with someone who clearly doesn't understand the product and is reading from a screen can be frustrating for customers.
- **Company-owned retail locations**. Businesses who maintain their own retail locations often have support centers (e.g., Apple's Genius Bars).
- **Distributors or purchase locations**. These businesses can manage warranty and returns for products purchased there. For example, in the cycling industry, the bike shops manage the diagnosis and repair, and bill either the customer or the manufacturer (depending on whether the repair is covered under warranty) for labor and materials.
- **Third-party repair companies** can be contracted to manage the warranty complaints and repair and reimbursement processes; for example, you can call an appliance repair person who will have knowledge of a wide variety of appliances and can process the warranty for you.
- **On-board diagnostics**. Some high-tech products are designed so that predictable failures will trigger customer support calls automatically. For example, the author's car sends her a monthly email with the status of the vehicle and any required repairs. The app automatically helps her to schedule an appointment with the car dealership if any repairs are needed.

18.3.2 Solving the Customer Complaint

If customer service representatives are unable to diagnose certain issues, they escalate the complaint to a more technically savvy team. In these cases, customer support routes the challenging calls to company employees who have the technical knowledge to solve the problems. More effort will be

Vignette 9: What Happens When You Build Something that Lasts?

Danielle Applestone, Founder, Othermill

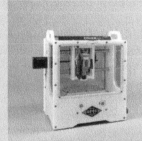

We founded Other Machine Co. on technology that was developed on a government grant. The goal was to build desktop-sized manufacturing equipment that could be used to train the next generation of the US manufacturing workforce for less money.

First, in the category of "things I wish I'd known": I wish I'd known that technology built for the government is not a product for anyone other than the government. However, there was something special about what we'd built. We'd radically reduced the barrier to entry for CNC (computer numerically controlled) machines, and any time you lower the barrier to a technology, you open up new markets.

CNC machines are typically very difficult to use and definitely not welcoming for casual hobbyists. Even though this machine was easy to use, we made the classic mistake of "this is a machine for everyone!" which is short-hand for "we have no idea what problem we're solving." So we ran a Kickstarter campaign back when it was still pretty novel and raised funding to start the company.

Just over 200 people paid about $1,500 for our product through crowdfunding. We raised $311,657 during the campaign, which ended up being about a quarter of what we needed to actually deliver those 200 machines. Even though we were late shipping the machines, the machines we sent out were essentially glorified prototypes.

The strangest and most unexpected thing was how reliable they were. About five years later, after I sold and left the company, many of our original Kickstarter Othermill owners were still using their machines. But eventually, after many years of use, the original Kickstarter Othermills started to have problems.

We had already moved onto the fourth-generation professional version of the machine (now fully designed for PCB prototyping and rebranded as the Bantam Tools Desktop PCB Milling Machine). Those original Kickstarter Othermills had been built with our best guess of how to build a machine like this, with tons of custom parts, some hobby components, and some parts that were discontinued during that four-year period. So we sent out replacement parts where we could, but when those eventually ran out, we were left with one of the biggest moral decisions that every hardware company faces: what do I owe the customer as my product ages?

It always feels like the design stage is an eternity, and getting production going is like an endless marathon of marathons through minefields beyond your control, but the real test of character is actually supporting a product that is in the hands of customers for years on end. If you're lucky, customers actually open the box and use your product at some point.

In our case, our early customers grew to rely on their machines and wanted to repair them, not replace them over time. This became a problem, since many of the parts we used in those early machines were basically custom parts. The parts were prototyped in small batches and essentially bespoke because we were still iterating on the design. Because of the way the machine was built, those parts lasted for years, but bearings wear out and motors get dust in them. Eventually, the machines needed parts that we no longer remembered how to make and maybe never properly documented in the first place. You do the best you can when you're trying to move fast, but even great teams forget how to get to the moon.

Spoiler alert: there's no good way to support old products. I never came up with the magic solution. But awareness of the entire product lifecycle is important if you're thinking of getting into hardware product development. We ended up approaching it from the following perspective:

- Without these 200 early adopters, our company never would have existed.
- The value of capital equipment like a $1,500 laptop depreciates to zero over five years.
- Customers had grown attached to their machines and wanted to keep them. (And we had unknowingly encouraged that by giving the machines fun serial numbers (SNs) like "Sassy Saxophone" and "Brave Bassoon.")
- Having outdated equipment in the field is a liability and a time pit for support and service.
- We wanted our company to survive and simply couldn't afford to replace or repair the Kickstarter Othermills.
- Without these 200 early, passionate, and patient adopters, our company would never have existed (this gets listed twice because it was important to us).

The decision of how to handle the older versions of our product is a moral decision because there is no way to make a 100% rational choice. You have to pick a path based upon what kind of company you want to be.

Ultimately, we decided that we wanted to be the kind of company that offered all the original Kickstarter Othermill owners $1,000 credit toward the purchase of a brand new $3,199 Othermill Pro, but that they had to choose to use that credit within the next year. It was a thank you, but it also acknowledged that we are a business, and told customers that if they really do need to prototype PCBs on their desktop, they'll need to spend some cash.

In hindsight, I wish we'd just used more off-the-shelf components and made our early products self-serviceable and easily upgradeable – that's what we did for all future models. But it's not always possible to build in backward compatibility and a path for replacing discontinued components. But if you care about your customers, you'll give it some thought. To me, the hardest thing about hardware is product support, not product creation.

Text source: Reproduced with permission from Danielle Applestone
Photo source: Reproduced with permission from Danielle Applestone

put into customer support for products that are very expensive to replace or repair or for issues that lower-level customer support hasn't encountered before. If a product is very expensive to replace, the company will want a second opinion from its technical experts before triggering a costly warranty response.

Rapid diagnosis and routing are critical to effective customer support. Customer support should be enabled by:

- **Databases and data entry systems** that allow customer support to accurately and quickly record the crucial information about the product, its problem, and actions taken. Technical experts who trouble-shoot the product and repair people need accurate and specific information to enable effective root cause analysis and resolution of problems.
- **Root cause analysis** data for customer support teams. The teams should have information about the common symptoms and the most likely fixes at their fingertips.
- **Engineering teams** of people who can provide technical backup and who can quickly get to the root cause of a problem for problems that are escalated to them.

18.3.3 Warranty Claim

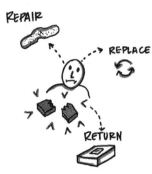

If the product failure falls under warranty, one of several actions can be taken. Some of the problems require the use of *reverse logistics* to return the defective merchandise and get a new or fixed product back to the customer. The faulty products need to be repaired, refurbished, or disposed, and your company must have an inventory of finished products and replacement parts on hand (Inset 18.3).

Reimbursement. In this case, the consumer returns the product and gets a reimbursement for the cost of the product. Even though you aren't replacing or repairing the product, you should try to get the defective product back to ensure that the customer hasn't defrauded you. The returned product can be studied to understand the cause of the failure, and in some cases it may also be refurbished and resold at a discount. If you handle the reimbursement yourself, you typically get the product back. However, if the product is sold through a distribution channel, you may or may not get the defective product back; the distributor may dispose of it themselves.

Replacement. A defective product might be replaced with a new or refurbished product. Replacing a defective unit helps to ensure that you have

---

Inset 18.3: Spares, Repairs, and Replacement Parts

To effectively support repairs both in and out of warranty, the customer support organization needs to hold inventory of replacement units and spare parts. Ideally, product designers should design the product to be easily repaired (Section 10.5). While it is not possible to predict inventory needs, estimates can be made based on initial reliability testing, reliability budgets, and history from other products. Over the life of the product, inventory levels to support repairs should be adjusted to balance the cost of holding inventory against likely repair rates and the ability to quickly deliver a replacement to a customer.

Depending on how you set up your warranty and customer expectations, you may be able to sell spares and replacement parts as additional sources of revenue. For example, a typical charging cable costs around $2–3 to purchase from an original equipment manufacturer (OEM); however, the list price for a replacement is typically several times that.

---

provided a defect-free product to the customer; however, replacing the unit usually costs more than repairing it.

Repair. For an expensive, customized product, the cost of replacing the product may be significantly higher than the cost of repairing it. In this case, the product needs to be returned and a system put in place to trace its path through the repair process. For some larger products (e.g., washing machines), the product may need on-site repair.

You have a number of options for making repairs and refurbishing the product:

- **CM**. The CM may have responsibility for repairing or refurbishing product. You have the challenge of shipping the product back to the country where it was made.
- **Third-party repair centers**. To avoid international shipping, customs, and duties, you may want to get the product fixed in the country of sale. Spares and test equipment would then need to be maintained locally to fix the product in-country.
- **Dedicated repair centers**. For many products that require routine maintenance (e.g., automobiles), customers will bring their product to an authorized center (e.g., an automotive repair center). If the product is under warranty, there is no cost to the owner other than a wasted morning in the repair center drinking lousy coffee. The repair center will send the bill to the manufacturer.
- **Repair technicians**. For products that can't easily be shipped, repair technicians generally go to the customer's site to do the repairs. In some cases, the repair people are employees of the product development company; in others, they are third-party providers.

18.3.4 Handling Returns Through Distribution Channels

A customer typically will go to one of two locations to get a complaint resolved – they can call your customer service number or they may go back to the retailer where they bought the product. Depending on how you have set up the contracts with your distributors, you may prefer to route customers back to the retailer or to handle it yourself. When the retailer handles the complaint, they take the return, and, depending on the sales channel and your agreement with them, you may or may not get the returned product back.

Some distribution channels will automatically take a fixed percentage off the sales price to cover a returned product (for any reason) when they pay your invoice. The sales channel (for example Best Buy) then owns any returned product, and can refurbish and resell the product themselves through discount channels. This kind of agreement can reduce the need for a reverse logistics process, but you lose the ability to diagnose the poorly performing products or talk to customers who are unhappy.

18.3.5 Reducing the Costs of Customer Service

Effectively managing customer support can have dramatic impacts on the cost to support the product in the field. If you don't have well-trained customer service people, customers get frustrated and go to social media with their frustrations. If you don't have a way to track warranty returns, you won't have a way to get ahead of emerging issues.

Many companies treat customer support as an afterthought and start planning for support very late in the launch process. Thinking through what will happen in the hands of the customer can reduce the number of calls, the duration of the calls, and overall dissatisfaction that can lead to poor product reviews.

Here are the most effective ways to lower the cost of customer support:

- **Design your product to reduce the number of times a customer will call for help**. Careful attention to storyboarding of the unboxing and setup process can reduce calls for help. Also, onboard diagnostics and well-defined procedures for maintenance can reduce the number of calls.
- **Pre-register products to reduce time spent gathering data**. A significant amount of time is often spent during a phone call collecting the relevant data about a product from a customer. Pre-registration of the product (with all of the customer and product information) can improve call-center efficiency. Having customers begin contact on the web where they enter the SN and other data (and get some support) can reduce the number of calls and shorten call times. Having a simple QR code on the product

that the customer can read with a smartphone can reduce SN transcription errors.

- **Design your product with built-in diagnosis tools**. The customer support teams can be provided with information on typical failures and tools to quickly diagnose a failure. If you depend on someone saying "it is not working," it will take a while to understand what "not working" means. If there are clear indicators (e.g., failure codes, flashing colored lights, or recorded messages), the customer support group can more quickly identify the type of failure.

- **Design your product for easy self-repair**. The product can be designed to allow for easy replacement of failed subsystems (e.g., replaceable batteries). Enabling the customers to repair certain aspects of the product themselves reduces the need to ship the product back for replacement and repair.

- **Plan for easy shipment of replacement parts**. FormLab's first few product lines required the entire product to be shipped back for repair in its original packaging. Most customers won't keep and store bulky boxes, and the shipping costs are very high. FormLab's new product line has an easily removable optical unit to facilitate fast repair and minimal downtime. The owner can get a replacement optical unit while theirs is repaired, thus reducing their downtime and shipping costs. Also, the maintenance costs are reduced because the whole unit doesn't need to be disassembled.

- **Avoid "entitlement creep"**. Entitlement creep happens when customers whose products are out of warranty ask to have their repairs paid for. Creep can be intentional (fixing a product to keep a customer happy) or unintentional (having no way of proving the product is out of warranty). Forcing customers to register and activate their products through cloud technology is one way to avoid unintentional creep. Finally, entitlement creep can be aggravated if you can't show that the product was damaged by the user (e.g., smartphone moisture sensors).

18.4 Customer Support Data

If you want an effective customer support system, you need a data system to record and track information about complaints and the costs to resolve them. Data is critical for managing the complaints and making sure they are resolved, for tracking the costs of warranty, and for identifying emerging quality trends.

Most small companies start with home-grown customer complaint systems kept in ever-growing and increasingly complex spreadsheets. At some point, companies grow out of homegrown systems and need to transition to formal customer-support software. Transitioning to a new software when the old one is being held together with the digital equivalent of bailing twine and duct tape can be very painful. Ideally, companies should start implementing commercial customer support systems before or soon after a product's introduction to avoid laborious crossover.

There are many customer support systems on the market, designed for all sizes of companies. To identify the best solution for your organization, start by defining what the engineering and finance teams want to do with the data, and how you want to track customer interactions. After understanding the end goals, evaluate the options to determine what system will best meet your needs. It is easy to get distracted by cool bells and whistles that don't ultimately support your core needs.

Customer support systems are challenging to design because there are so many users for the data. Often the needs of different stakeholders are in conflict. For example, the customer service team may not want to spend significant time typing up notes because note-taking increases call duration, but the repair team and quality team may need those details to facilitate repair and continual improvement. The stakeholders for the data include:

- **Consumers** who need to know whether their complaint is being addressed and when their product will be returned to them. You don't want a customer waiting months to get a resolution because their complaint got lost. Many commercial customer support systems will allow the customer to directly access the data and enable them to look up the status of the repair without calling a toll-free number.
- **The customer support representatives** who need to be able to enter the data quickly and diagnose and fix the non-warranty issues as efficiently as possible. They don't want to spend a lot of time re-entering data or clicking through multiple screens.
- **Customer support leadership** need to be able to train customer support teams on the best responses to certain failures. For example, if there is an ongoing issue with battery charging, the leadership needs to be able to roll out a recommended repair (e.g., a firmware update) and ensure that all of the customer support people know the correct response.
- **Reverse logistics team** who needs to use the data to track and route what materials are being sent where and by whom, and trace when and where shipments don't arrive correctly. This allows the teams to easily track where a customer's defective product is in the process. There is nothing worse than sending back a broken product, only to have it lost.
- **Repair center personnel** need to be able to understand what failure the customer is experiencing. In a poorly designed system, the notes taken by customer service will be cryptic and incomplete. If customer service

centers work in multiple languages, translating the notes becomes even more difficult.

- **The purchasing team** needs to track the use of spares to ensure that there is sufficient inventory to support ongoing and emerging issues.
- **The finance team** needs the data to track who owes what to whom. Also, they need the data to manage the warranty reserve accounts to understand whether the percentage held back when revenue is recognized needs to increase or decrease.
- **Quality engineering and continual improvement** teams needs to use the data to drive quality improvements and cost reductions. The data is used to identify quality issues in order to prioritize corrective action, drive continual improvement, and suggest next-generation design changes.
- **Product design teams** who need to use the data to identify and implement improvements in the next generation of products.

18.4.1 What Data is Recorded

While each company will have its own unique information requirements, Checklist 18.1 provides a list of commonly used data that should be recorded and tracked.

18.4.2 How is Data Reported

As pointed out above, the data generated by the customer support process will be used by a number of stakeholders. Tracking the quality of the product in the field is critical for many functional groups: design, supporting engineering, quality, and operations teams all need to understand the current quality failures in the field. Typically, either the customer support function or the quality team will publish the warranty and return data on a weekly or monthly basis to the whole organization. Based on the data and trends, various functional groups may take action to address the warranty rates and costs. While there are many ways to view the data, companies typically assess the trends in one of two ways. Figure 18.2 (normalized and anonymized data from a real customer complaint reporting system) gives two examples of how warranty data can be reported based on:

- **When the complaint is reported**. The upper chart shows the annualized return rate by the week the complaint was reported. This rate is calculated by dividing the number of claims by the total *installed base*. The total installed base is calculated by estimating how many products in the field are likely to be still under warranty. This chart gives an accurate indication of the volume of complaints and what you will spend next week, but doesn't give you an insight into whether the complaints are from products just purchased or products at the ends of their warranty lives.

Checklist 18.1: Customer Complaint Data Checklist

Product information

❏ **SN and product type**. The call center and website software should use verification checks and pull-down menus to reduce error entry.

❏ **Purchase date** (if possible) helps identify whether the product is in warranty and how long the product was in use before the failure occurred.

❏ **The sales channel** (e.g., "where did you purchase the product?"). Different sales channels are tracked separately. You may need to import data from retailers to get a complete picture of the return rate.

Customer information

❏ **Product use location**. Different regions will have different failure modes. For example, a product failing after being left out in a cold garage in Wisconsin in January will be evaluated very differently than one that failed after sitting in a car in Florida in August.

❏ **Shipment and contact information**. These ensure that the product goes to the right person, and that the product isn't misplaced.

Call status

❏ **Date and time of the initial call**. This helps track how long it takes to resolve a complaint.

❏ **Dates of actions taken to resolve problems**. A history of the resolution steps should be maintained. For example, a product issue may be escalated to an engineering team and then to the repair center. Tracking the responsiveness of the whole system helps identify how customers get treated.

❏ **Status of the complaint resolution**. It is necessary to track the status of an open complaint and where customer issues are getting stalled.

❏ **Date of call resolution and closeout**. This helps calculate resolution-time metrics and continually improve them.

Description of the problem

❏ **Description of the symptoms the customer is seeing**. It is very tempting to document what customer support thinks is the solution rather

than an accurate description of the symptoms. For example, "battery failure" is a diagnosis. Descriptions of symptoms might be "Customer plugged in the charger, and the light did not come on" or "When the battery was inserted into the device, it wouldn't turn on."

❏ **Documentation of multiple issues described in a single call**. Often, there isn't just one thing going wrong with a product. If only the first one is recorded, other problems may be missed.

❏ **Problem category**. Calls should be grouped to allow for Pareto charts and trend charts. However, categories should be defined in order to reduce the chance of mis-categorization. For example, there should not be broad categories for "electrical issues," "battery issues," and "charger issues" without clear criteria for what goes in which bucket. Product categories should be encoded in pull-downs to avoid spelling errors, different naming schemes, and other errors that make the data difficult to analyze.

❏ **Criticality category**. Some symptoms are more critical than others. Not being able to turn on a product is annoying, but fire and hazards need to be immediately escalated outside the standard reporting cycles.

Description of the solution

❏ **What action was taken and whether it resolved the problem**. If the symptom information from the call center is incomplete, types of failures can be grouped based on what the fixes for the problems were. Besides, it helps to prioritize which failures cost the most.

❏ **Track what parts/units were shipped**. Knowing what parts are being replaced can be used to track batches of components that are causing failures and parts that are repeatedly replaced but still fail.

❏ **Cost of resolution**. Cost information for spares, repair, and replacement is critical for accounting and calculating warranty reserve requirements.

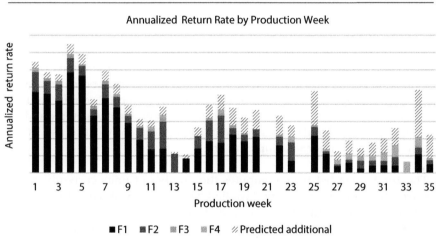

FIGURE 18.2
Customer complaint
data tracking

- **When the product was built**. The lower chart shows the annualized rate of return for all of the product produced in a given time period. F1–F4 are the failure modes the product experienced. The rate of return has two parts: products that have actually been returned to date and a prediction of what is likely to be returned over the total warranty life. The predicted return rate is calculated based on historical patterns of returns (if the product typically fails early in life or late in life). This way of charting time-based trends can give an accurate picture of the trends in quality, but it depends heavily on the accuracy of the prediction model.

18.4.3 How to Avoid Common Mistakes

Many companies make the same mistakes when setting up what data will be collected and how they will be analyzed. Here are a few of the most common mistakes.

Mistake #1: Not categorizing complaints well. As mentioned above, the calls and complaints should be grouped into categories that will allow the team to identify which problems should be focused on first (by total returns, criticality, and cost). Ensuring that the groupings are orthogonal (i.e., you can only categorize each complaint one way) is very important. Customer service might go through phases where all of the complaints will be categorized in one way, and then a week later categorized differently. Changes in categorization can skew analyses.

The number of categories is also significant. Too many, and it takes customer service too long to find the right description; too few, and the data is too aggregated. The categories need to be re-evaluated regularly, and customer service trained appropriately.

Mistake #2: Not normalizing data for sales volumes. Just counting the number of calls or failures by month fails to consider sales variability. Faults should be reported as a percentage of units produced or the number under warranty to get a normalized value for return rate. It may be necessary to normalize data for seasonal trends. For example, in the US in mid-January, there is typically a spike in calls because people finally get around to playing with their holiday gifts. Some of the options for what to use as the denominator when normalizing data are:

- **The total installed base** computes the number of units currently under warranty. Normalizing by the installed base works well when the failures are distributed evenly throughout the product's life.
- **Recent sales** uses the number of units sold X months before. If most failures are in early life, this can be tuned to capture any trends quickly.
- **Volume produced during that production period** gives a better indication of manufacturing performance over time, but depending on how slowly the data comes in, it is a lagging indicator of quality (i.e., you only know about quality issues long after the product has been in the field) unless you use a prediction model like the one in Figure 18.2.

Mistake #3: Looking at just one metric. Not all calls and complaints are the same. Addressing annoyance calls with a simple fix (software bug fix or new quick-start guide) can remove a large number of low-cost complaints. Removing these trivial issues will free up critical customer support resources and reduce wait time for calls. On the other hand, any complaint with safety issues, no matter how infrequent, should rise immediately to the top of the critical list. There are serious legal (and moral) implications of not addressing potential safety issues (Inset 18.3).

Different metrics can give different insights and drive different actions. For example,

- **Total issues per unit** measure the irritation level of the customer.
- **Total calls that don't result in service** are nuisance issues that can be designed out or supported through web-based support.
- **Critical failures that result in a total malfunction of the product** may be relatively inexpensive to fix but cause significant damage to brand and customer loyalty.
- **The cost of claims** focuses the organization on the warranty reserves to help free up cash.

Mistake #4: Reporting data inconsistently. If the organization gets a different set of charts every week, they will spend more time figuring out what the data mean than they spend figuring out what action to take. The quality organization should create a standard set of charts that are distributed weekly. Standardizing the reporting will enable the organization to identify trends more quickly because they are looking for the trends, not trying to understand a new way of reporting the data.

By avoiding these common mistakes, the data generated by customer support will proactively find problems and drive quality.

Summary and Key Takeaways

❑ Warranties are legal obligations. Companies need to design systems to support products in the field and ensure that enough money is set aside to support the warranty process.

❑ You can (and most likely should) buy insurance to protect against major warranty recalls.

❑ Designing products to be easily repaired can save calls, response time, and cost of warranty service.

❑ Recalls can have dramatic impacts on customer perception of your brand and on the financial viability of your business.

❑ Teams need to ensure that there are sufficient spares to support the warranty process.

❑ Warranty systems should be designed to enable comprehensive data collection about the causes and symptoms of failures, to support repairs and corrective actions.

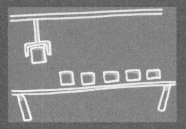

Chapter 19
Mass Production

SELLING YOUR PRODUCT

17. CERTIFICATION/LABELING

18. CUSTOMER SUPPORT

19 MASS PRODUCTION

Companies see the completion of the product realization process as the end of the marathon. However, once mass production starts, a whole new set of activities and problems arise: operational efficiency becomes critical, problems arise in the field that need to be fixed, and the whole organization needs to drive down costs and continuously improve quality. Companies will also need to start thinking about the next product in the portfolio.

Product Realization: Going from one to a million, First Edition. Anna C. Thornton.
© 2021 John Wiley & Sons, Inc. Published 2021 by John Wiley & Sons, Inc.

 Product development teams fantasize about the first time a product gets into the hands of the customers. They spend months (or years) getting from that working prototype to a production line that is churning out product. They have struggled through and overcome all of the challenges described in the preceding chapters. However, getting the first shipment into the hands of the customer is just the start of a new set of hurdles. As production volumes increase, new pressures arise: the pressure to drive your costs down, and the ensuing changes to operations, design, and the supply chain. Investors start asking about the next product. All the while, rolling forecasts need to be translated into material orders and production plans. Customers have to be supported, and quality issues addressed.

Product realization often seems like fighting a series of unexpected fires; mass production, by contrast, should be almost perfectly predictable. Mass production is about focusing on the details and keeping chaos at bay. Major changes don't happen during production, but small incremental changes do. As a company ramps up mass production, it will typically hire people with deep expertise in efficiently running manufacturing operations.

There are hundreds of books and articles on mass production; however, this chapter will focus on a selected handful of topics that teams need to be aware of and plan for during product realization. They include:

- Managing the effect of manufacturing scale on speed, cost, and quality (Section 19.1).
- Driving continual improvement (Section 19.2).
- Driving costs down systematically (Section 19.3).
- Auditing the production system to ensure continued compliance with the documented process (Section 19.4).
- Maintaining equipment to ensure quality and maximize uptime (Section 19.5).
- Thinking about the next product (Section 19.6).

19.1 Manufacturing Scalability

During piloting, the factory will start by manufacturing relatively small batches, typically at a slower speed than the final mass production rate. Even once mass production starts, rates often remain low (to accommodate continued testing and/or because consumer demand for the new product hasn't spiked yet) but quickly increase as sales drive the need for more of the product. As production rates rise, a new set of problems arise. Teams need to plan for issues that arise during scale-up, such as:

- **Changes driven by scale**. As production rates go up and batches get larger, processes that operate at a larger scale may be needed to drive down cycle times and costs. However, scaling processes isn't always easy. For example, if the process involves mixing or processing large quantities of materials (such as mixing a new biomedical substance), the methods used to make large batches are often different from those used to make small batches. Just using a larger pot may not produce the same results. New mixers and material handlers change the fluid dynamics, thermal control, and the time that materials spend outside of refrigeration.

- **Degradation of materials while they sit on a line**. Larger batches will result in more work in progress (WIP) and longer cycle times. As a result, more products will sit waiting on the production line. For example, a medical device company was using a desiccant in its packaging to keep the product dry. When production volume ramped up, the company bought a larger bag of desiccant packets and left them out for the whole shift. On humid days, the packet temperature dramatically increased as they sat and absorbed the ambient moisture. The effectiveness of the desiccant degraded and increased the risk that the humidity inside the packaging could exceed allowable limits.

- **Space**. Depending on the inventory, WIP, and line design, increasing the volume of production can have a dramatic impact on available floor space. A medical equipment company had designed a 24-hour burn-in (running the product for extended periods before shipping) for a large refrigeration system. At low volumes, running a 24-hour burn-in was not an issue, but when the production rate increased, the company did not have the space required to hold all of the product made in one day. Only when they reduced the burn-in time could they physically manage the production rates.

- **Supplier capacity**. Supplier capacity can often be a challenge when the production rates increase. Suppliers who quickly deliver ten or a hundred units may struggle to keep up as larger orders pour in. As rates go up, the quality may degrade.

- **Labor quality**. As rates increase, the factory may add new employees to the line – employees who don't have the same institutional knowledge as the initial workers. Because production is already up and running, leadership may take less care in training these new workers than they did the initial ones, who were key to getting the operation started. These new operators are therefore more likely to skip steps and make mistakes. Doing a design for assembly (DFA) improvement effort along with well-written, accurate SOPs increases the chances that new employees can quickly learn how to build quality products.

- **Parts interchangeability and time for rework**. As production rates increase, the time available for adjusting non-conforming parts decreases and the more important it is for parts to be identical and interchangeable. Parts that vary from normal will drive rework and slow down production. When rework is needed, lines back up much more quickly. An analogy is a small fender-bender on a highway. When it happens at midnight on a back road, traffic can move around it and the flow of cars isn't disturbed. When it happens at 5:00 p.m. on Massachusetts Route 128 in 70 mph traffic, the backup behind the accident can quickly grow to several miles (and probably cause more accidents upstream).
- **Product damage**. As materials move more quickly through a facility, there is more chance for damage, which also creates rework traffic jams.

19.2 Continual Improvement

Before the start of mass production, it is not possible to design out all potential product failures, to ensure that the production system is perfectly smooth, or to completely minimize rework and maximize yield. The factory, customer support, and continuing engineering groups will need to identify continual improvement activities to drive cost out, reduce scrap and returns, and reduce excess inventory. Continual improvement has the benefit of increasing profit margin on an existing product, and also puts the organization in a better competitive position if they need to drop prices. Besides, product teams can apply the lessons learned from continual improvement to the next product.

During product realization, opportunities for improvement will arise that teams can't immediately address because of limited time and resources. These opportunities should be documented and implemented once mass production begins. For example, groups may be aware they have started production using a high-cost supplier. There isn't enough time to vet a new supplier before the initial launch, but once full production starts, teams can change suppliers to significantly reduce COGS. There are several sources of ideas for continual improvement described in Checklist 19.1.

The team responsible for continual improvement – typically quality engineering or a dedicated continual improvement team – should maintain the list of improvement opportunities in a central document or system. The list (usually a spreadsheet) enables the whole organization to track the status and completion of the improvements.

Listing the potential improvements is the easy part; prioritizing and efficiently executing them is more complicated. Teams may be tempted to make long lists of all possible improvements and try to implement all of them at once. When companies work on too many projects in parallel, the

---

Checklist 19.1: Sources for Continual Improvement Opportunities

❏ **Marketing and sales**. Receive feedback on product improvements and changes from customers.
❏ **Customer service**. Analyze warranty data to identify which failures are driving returns and customer dissatisfaction.
❏ **Quality**. Analyze factory yield and rework data to identify where poor quality is driving cycle time, rework, and cost.
❏ **Factory operations**. Look for high touch assembly processes (those requiring a lot of labor to assemble) that drive up assembly costs and cycle time.
❏ **Factory operations**. Look for piles of inventory in the system. Large amounts of WIP highlight where insufficient capacity is creating an unbalanced line. Large amounts of finished goods in inventory highlight where sales forecasts aren't correctly driving production plans.
❏ **Engineering**. Identify high-cost parts, which are opportunities for reducing COGS.
❏ **Supplier management**. Weed out suppliers who are not meeting expectations (in quality, quantity, or schedule).
❏ **Purchasing**. Identify materials with long lead times or suppliers with significant delays.

---

completion of all of the projects can be significantly delayed. It is better to focus on a critical few, get them done, bank the benefits, and then move on to the next set.

Projects should be prioritized based on the cost, timeline, impact, and risk of each project. Figure 19.1 shows a graphic that can be used to prioritize projects. Obviously, safety-related product or production issues always take priority.

There will always be several "low-hanging fruit" projects that carry minimal risk and a high return on investment (ROI). These should be the focus of the first continual improvement efforts. The continual improvement team should then have a set of low-cost, low-ROI projects to knock out quickly as the opportunity arises. Also, teams should have a couple of "shoot for the moon" projects which, while high risk, also have the potential for a high return. The moon-shot projects are typically longer term, to be completed after some of the easier fixes have been implemented. The low-cost, low-ROI projects, once implemented, can provide the cash needed for the larger projects. Obviously you want to avoid projects that are high risk with little return.

Project prioritization should also ensure that critical resources aren't over-taxed. Often several essential resources become the bottlenecks if they are over-allocated too many projects. For example, if many of the projects require verification testing, the testing and quality groups may be the bottleneck.

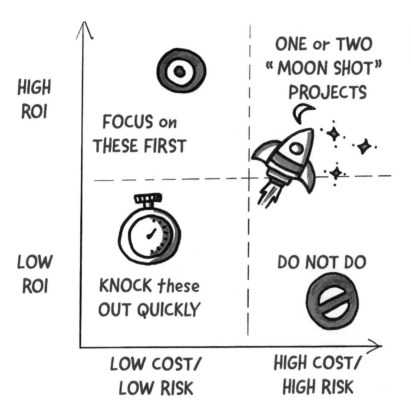

FIGURE 19.1
Project selection
matrix

The same project management discipline described in Chapter 4 should be applied to the continual improvement projects, including critical path management, risk management, and RASCI charts.

19.3 Cost Down

Cost-down efforts (as the name implies, these are any efforts to drive down costs) will likely make up a majority of the continual improvement efforts. Cost-down projects drive down COGS, improve purchasing terms with your suppliers, and reduce lead times for products. Even small reductions in cost can have a dramatic impact on the overall profitability of the organization, especially when the volumes ramp up. Saving $0.10 per unit when you are producing 1,000,000 units per year adds $100,000 to the profitability. There are a number of sources of cost reduction (summarized in Checklist 19.2) including reducing the cost of parts, of delivery (incoming and outgoing), and of production.

Checklist 19.2: Sources of Cost Reduction

Reduce parts costs
- ❑ Identifying trends in commodity pricing that can drive costs down
- ❑ Asking suppliers for volume discounts
- ❑ Consolidating orders to one supplier
- ❑ Second-sourcing or rebidding with your existing supplier
- ❑ Changing from assigned to generic parts (if assured no impactful change in quality)
- ❑ Removing non-critical parts
- ❑ Redesigning parts to take cost out (reducing undercuts, material, features)
- ❑ Checking the bill of materials (BOM) for errors in pricing

Reduce delivery costs
- ❑ Renegotiating payment terms
- ❑ Reducing lead times through alternative suppliers or negotiation
- ❑ Reducing shipping costs by bundling or shipping by sea
- ❑ Improving forecasting and production planning

Reduce production costs
- ❑ Replacing manual assembly with automation when possible
- ❑ Replacing single-cavity molds with multi-cavity tooling
- ❑ Outsourcing subsystem assembly to a less expensive facility

---

The first place to focus is the reduction of the COGS. The cost-reduction team (typically the continual engineering and supplier management teams) should evaluate every possible opportunity to reduce costs, including:

- **Using market forces and learning curves**. Most core technologies (e.g., batteries, chips/modules Wi-Fi modules, LCDs) typically continually go down in price (assuming no shortage in materials) as the technology improves, manufacturing becomes more efficient, and capacity increases. The purchasing groups should track key market trends and get requotes for critical components to take advantage of changes in the marketplace.
- **Asking for volume discounts**. Suppliers may give you a discount for ordering in larger quantities. Your suppliers may not automatically give you the price reduction; you usually need to ask.
- **Consolidating orders**. Suppliers and distributors may give price breaks if you order more types of products at the same time. For example, the distributor you order your resistors from may also be able to supply connectors and batteries, so you may be able to negotiate a lower cost on your whole order. Ordering several part types from a single distributor can also reduce the time and energy spent managing multiple parts from multiple suppliers.

- **Second-sourcing and rebidding**. Expensive parts (semi-custom and fabricated parts) may be cheaper with a new supplier. Getting quotes from other suppliers may also give you leverage in negotiations with your existing suppliers. Be advised, however, that continually threatening to change suppliers can erode the trust you have with your critical suppliers.
- **Changing from assigned to generic parts**. Often, in the first generation of a product, expensive parts are purchased from well-known suppliers. As teams better understand the performance margins on their units, generic parts may be found to be viable and less costly alternatives.
- **Removing parts**. Excess components and redundant systems are often included in a first-generation product to allow for easy design changes. For example, an electromechanical product may originally be designed with additional capacitors, diodes, and other ICs that later turn out to be redundant. The engineering team may approve the removal of these parts after the design is validated in the field, if they determine that the removal doesn't degrade performance or invalidate certifications.
- **Redesigning**. Parts may be redesigned to reduce parts count, consolidate tools, reduce material, or reduce labor costs. For example, the parts may be redesigned to combine parts or reduce the material used.
- **Checking the BOM for errors**. The BOM can have what appear to be relatively small errors. The volume of consumable materials (such as glues) may be overestimated, or the number of screws listed incorrectly. While these often account for only pennies per unit, the cumulative impact on cost can be considerable. The BOM should be regularly reviewed to identify and correct errors.
- **Outsourcing subsystem assembly**. A supplier may be able to deliver entire subassemblies at a lower cost than you can achieve in-house. For example, an injection molding supplier may be able to pre-assemble the housing for you and ship entire units. Because the parts are coming off their line and immediately assembled, the supplier's overhead and labor costs may be lower than yours. In addition, outsourcing the assemblies essentially increases your capacity; you can apply the space and labor to additional production.

In addition to addressing the COGS, your team can reduce the cost to deliver the product by:

- **Renegotiating payment terms**. As CMs or suppliers gain confidence in your business, they may give more competitive terms (such as smaller markups or lower percentage down-payments on a PO) if they cannot reduce their manufacturing costs and COGS to meet a competitor's bid.
- **Reducing lead times**. As pointed out in Chapter 15, long lead-time items make ordering, inventory, and cash flow more difficult. Finding a similar part with a slightly higher cost but a significantly shorter lead time might actually improve costs by reducing inventory carrying costs.
- **Reducing shipping costs**. If your distribution system can ship products by ocean freight rather than by air, the cost will be significantly lower.

However, to change to ocean freight involves building up a four- to six-week inventory of finished goods ahead of the demand to account for the additional time on the boat.

- **Improving forecasting and production planning**. The cost of carrying inventory can be a significant drain on available capital. Early in mass production, when the eventual sales volumes are not well understood, it may be necessary to hold excess inventory to cover uncertainty in forecasts. As the forecasts become more precise, the production rates understood, and the lead times confirmed, teams can then reduce the inventory of incoming parts and materials, WIP, and finished goods.

The capital equipment and tooling can also be a source of cost reduction, including:

- **Replacing manual assembly with automation**. At low volumes, manual assembly may be the appropriate choice; teams don't want to build automation systems until demand justifies that expense. As volumes increase, reduced labor costs will outweigh the additional cost to build automation systems, and automation will generally improve the consistency of quality.
- **Replacing single-cavity molds with multi-cavity tooling**. Early in production, when volumes are not understood and cash is in short supply, lower-cost single-cavity soft-steel tooling may be used to reduce the overall fixed cost. As volume increases, new hardened steel tools with multiple cavities may be more cost-effective.

When executing the cost-down process, teams must track when the changes are implemented and link the changes to the production batches and BOMs. The tracking is necessary to ensure accurate billing and to identify the effects of each change on product quality.

19.4 Auditing

Ensuring that workers always follow the same qualified process when making your product, day in and day out, is critical to maintaining quality. However, as rates increase and production lines age, it is easy for the actual process to deviate from the ideal. For example, as production volumes ramp up, new workers and equipment will be added to the line. Experienced workers may start to take shortcuts, change the process to make it easier, or fail to carefully inspect their own work. For example, they may not wait for the product to cool down before moving it or may eyeball how much glue is used rather than using the metered glue dispenser.

Auditing is a formal process during which an external expert evaluates what the workers are actually doing on the line and whether that matches the documented procedures (SOPs or process plans). Audits can be formal legal processes – for example, in the US, the Food and Drug Administration

(FDA) may inspect/audit a factory – or can be done by your team doing an informal (but thorough) walk around the facility. Audits can check for adherence to:

- **Manufacturing SOPs**. Are the workers following the steps?
- **Staff training**. Are the workers correctly trained and qualified for the working they are doing?
- **Material handling and control**. Are materials being subjected to temperature, humidity, or handling damage?
- **Material traceability**. Are quarantined products separated from the rest of production?
- **Calibration and preventative maintenance**. Are tools and machines routinely maintained and calibrated to ensure accuracy?
- **Parts inspections**. Are components from suppliers meeting your specifications?

Problems highlighted during auditing should be addressed through the continual improvement processes.

19.5 Equipment Maintenance

One of the significant sources of quality issues and increased cycle time is equipment failure. All production equipment *will* wear and break down eventually, and equipment performance will degrade. To minimize the impact of equipment failure, the operations group needs to create a maintenance schedule and the associated SOPs:

- **Calibration**. Most equipment will need to be regularly calibrated. For example, a gauge may read that a critical pressure is being held around 500 psi, but a degradation in the pressure transducer may mean that the actual pressure is higher or lower. The manufacturer of the equipment will typically provide guidelines on how frequently the part needs to be recalibrated.
- **Preventative maintenance**. Machines should be proactively shut down for regular maintenance. This maintenance prevents machines or tooling from failing at less convenient times. The schedule is dependent on the utilization rate, failure modes, and safety regulations. The supplier will typically provide guidance on frequency for critical systems.
- **Tool cleaning**. Tools such as molds, dies, and stencils, should be cleaned and inspected after each use to increase the tool life.
- **Tooling and equipment replacement**. Tools and equipment typically have a functional life. The operations group needs to have a plan for exactly when to replace each tool before it begins to degrade. You don't want to see a problem in a part and then have to wait for 12 weeks while a new tool is cut as the quality of parts continues to degrade.
- **Backup plans for critical equipment**. If failure of a crucial piece of equipment will shut down the whole line, the factory operations team

needs to have contingencies in place. These can include manual versions of the equipment, backup capacity, service agreements with rapid response options, or additional inventory.

19.6 Launching the Next Product

Most successful companies have more than just a single one-and-done product. If a product is successful, customers and investors will expect a newer, improved version within a few years. Companies hoping to build on initial success should introduce next-generation products, expand product families to include new products, and even add entirely new product families to your portfolio. As the product realization team starts the design of next-generation products, the lessons of the prior product realization process should be used to better execute the next product. A few things to keep in mind are:

- **Reuse spec documents, quality test plans, risk documents, and schedules**. Documents from prior products can be used as a starting point for the new product, streamlining the document creation.
- **Look at your warranty data**. The failure modes of prior products can give useful insights for your next product. Those insights should be used to plan for quality testing, design changes, and supplier requirements.
- **Evaluate the cost drivers**. Understanding the key drivers of cost from the previous product can help the engineering team identify lower-cost solutions for the next generation.
- **Assess your CM and suppliers**. Evaluate what you liked and didn't like about your supply base. Most new organizations only look at costs, but other factors such as responsiveness, trustworthiness, and quality may end up being just as critical. Identify tradeoffs between your existing trusted suppliers and new suppliers who can add to your supply base and competitively bid for the work, but have yet to prove reliability.
- **Identify where you can reuse parts and tooling**. Some parts and tooling can be shared across product lines to reduce design time, tooling costs, and complexity.
- **Determine how to better run pilots the next time**. The first-generation product may have taken five pilot runs to launch. You won't be able to launch your new product using only one pilot run, but critically evaluating prior pilots can help you to streamline the process.

19.7 Conclusions

These 19 chapters have introduced you to a dizzying array of tools and methods that all interact and whose execution is important to the success of the product. It is unlikely that you will get all of the information absorbed in the first reading, and you will need to refer back to sections to refresh your understanding. In addition, the tools, software, best practices, and methods are continuing to evolve. An additional website at productrealizationbook.com contains readings and other materials that will be added as the state of the art changes. The last piece of advice is to keep learning, ask a lot of questions, assume nothing, and most importantly, have fun building great products.

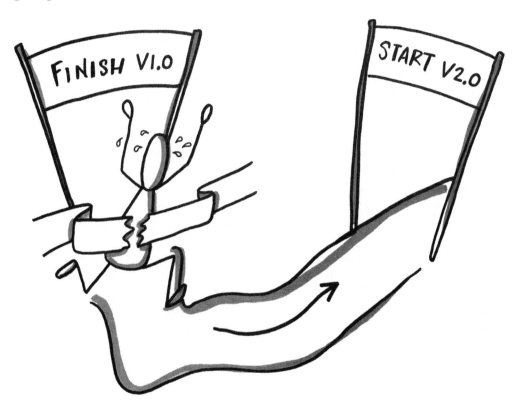

Summary and Key Takeaways

❏ You are not finished with product realization when you go into mass production.
❏ Continually improving the product, process, and quality can help drive COGS and warranty costs down.
❏ Routinely audit your facility to ensure that workers are following the documented processes.
❏ Cost reduction continues throughout the life of the product and can be achieved in several ways.
❏ Use the lessons learned from your current product to avoid similar mistakes on the next-generation product.

GLOSSARY

A

Acceptable quality limit (AQL)   The allowable number of units with defects. AQL standards determine the sampling rate and acceptable percentage of defects in batch testing.

Accessories   Products with electronic components are often packaged with additional items. These accessories include power cables, installation materials, optional parts, cleaning equipment, and spares.

Actual-to-actual (A2A)   Comparing two quotes including all BOM elements, overheads and expenses.

Aesthetics   How the product looks and feels. Aesthetics can include aspects of geometry (steps and gaps), surface finish, and visual defects.

American Society for Testing and Materials (ASTM)   The standards organization that defines testing and material standards used in the US and worldwide. Adherence to these standards may be required by law depending on the product. For example, the ASTM F963 Toy Safety Standard is federally required for all products intended for children.

Application (App)   Related software used on either a smartphone or computer to run or monitor a device. An app interacts with the device through Wi-Fi, Bluetooth, or other communication protocols.

Assigned parts   Parts for which a specific supplier and part number are required in the bill of materials (BOM). They cannot be replaced by another vendor's product without the approval of the engineering team.

Automated optical inspection (AOI)   Vision and image analysis to inspect parts and identify defects. It is most frequently used in the production of printed circuit boards.

B

Bench tests   Tests run on products using non-production testing methods and equipment. The testing equipment is typically generic (i.e., can be used on multiple products) and the tests are more comprehensive than those in production testing. Bench tests are not used in production because the equipment is too expensive or the tests too time-consuming.

Bill of materials (BOM)   The list of all of the parts, their suppliers, costs, and minimum order quantities (MOQs) to build a complete saleable unit.

Biocompatibility   The interaction between materials in a product and the human body.

Product Realization: Going from one to a million, First Edition. Anna C. Thornton.
© 2021 John Wiley & Sons, Inc. Published 2021 by John Wiley & Sons, Inc.

Bottleneck   A slowed or halted process in the production line that limits the overall throughput.

Brick   Jargon for breaking a product beyond repair (i.e., turning your electronic device into the functional equivalent of a brick).

Broker   An intermediary (middleman) who connects you to a contract manufacturer or other suppliers and manages the relationships and payments.

Build   A build is a short production run. Other names include pilot run or specific builds such as EVT, DVT, or PVT.

Build-to-order   Starting the production of a saleable unit based on a specific purchase order from a specific customer. Typically, build-to-order has a long lead time for the customer.

Build-to-stock   Starting the production of a saleable unit assuming the product will be sold but not having a PO in hand. The units are put into finished goods inventory.

Burn-in   Process that tests a product for a number of hours (or days) before shipping to the customer. It ensures that the product will perform as expected and helps find any early life failures.

Business-to-business (B2B) products   These are products sold by a company to a business customer, in contrast to business-to-customer (B2C) products that are sold directly to consumers.

C

Calibration   Ensuring that testing equipment is accurate and has the expected precision. Calibration should be traceable to NIST standards.

Capacity   The number of units that a process or facility can produce in a given amount of time.

Capital equipment   Equipment used to produce fabricated parts, usually in conjunction with a tool or CNC program. These are typically high cost and have a very long amortization period.

Cash flow   Rate at which cash comes in, minus the rate it is spent. Unless a company has a line of credit or external investments, it must maintain positive cash flow to stay in business.

Cavity   The space machined out of a mold into which material is forced to create each part. Each tool can have one or more cavities. The larger the number of cavities, the higher the production rate and the higher the tooling and equipment cost.

Certifications   Tests and documentation required to be able to legally sell a product in a given market.

Color/material/finish specification (CMF)   Specification of all of the aesthetics of the finished product including specific color (Pantone), the materials (e.g., stainless steel), and the finish (level of reflectivity, durometer, etc.).

Commercial off-the-shelf (COTS)   Parts or services that are purchased without change or customization. They can include commodity products or unique products offered without modification by a single provider.

Commodity parts   Parts that can be bought from many suppliers or distributors and are essentially interchangeable.

Conformal coating A thin film of polymer used to coat electronics to prevent water or humidity from affecting the electronics.

Conformité Européene (CE) The body that defines the EMF and safety requirements for products sold in the EU.

Consigned parts Parts that are purchased by the product teams rather than the factory. Parts may be consigned to save the material handling charge or control proprietary information. When working with a CM, a majority of the parts and materials will be purchased by the CM and not consigned.

Consumables The materials that are typically bought in bulk but are needed for assembly. They often do not show up in the CAD or assembly drawings and can include glue, tape, and wiring.

Consumer Product Safety Commission (CPSC) "The US Consumer Product Safety Commission is an independent federal regulatory agency that was created in 1972 by Congress in the Consumer Product Safety Act. In that law, Congress directed the Commission to 'protect the public against unreasonable risks of injuries and deaths associated with consumer products'." [140]

Consumer The person(s) who ultimately uses the product.

Contract manufacturer (CM) Suppliers that provide manufacturing and assembling. They may provide just manufacturing services (OEM) or can offer a range of design services as well (ODM).

Contract terms The percentages added to the material costs, minimum order quantities, and payment terms agreed to in the MSA.

Coordinate measurement machine (CMM) Equipment that accurately measures dimensions in 3D space.

Cost of goods sold (COGS) The total cost to build a product up to the point that the factory is ready to ship it.

Cost-down The process of systematically reducing product costs through design changes, sourcing changes, or volume changes.

Critical path Set of activities that defines the overall schedule of the project. A delay in any activity on the critical path will delay the overall schedule.

Current Good Manufacturing Practices (CGMP) CGMP refers to the Current Good Manufacturing Practice regulations enforced by the FDA. CGMPs provide for systems that assure proper design, monitoring, and control of manufacturing processes and facilities [141].

Customer relationship management (CRM) The system (software or other) utilized by the customer service team for managing a company's relationships with customers.

Customer The person/organization who purchases the product from your company.

Cutting tooling The process of machining the features into tools that will be used to make the product. Once you start to cut tooling, it is difficult and expensive to change the design.

Cycle time The total time from the beginning to the end of a process.

D

Datum   Theoretical plane or surface on a part from which all dimensions are measured. The datum ensures consistency in measurements and inspection.

Design for assembly (DFA)   Methods for assessing a design's fitness for assembly.

Design for manufacture (DFM)   Methods for assessing a design or part's fitness for manufacturing.

Design verification testing (DVT)   The project phase that assembles product using production-intent parts that are not built on final production assembly and test equipment. DVT units are used for the certification process as well as for sales samples.

Die   A specialized tool used to cut and form sheet metal (also can refer to extrusion tooling).

Distribution   The method for getting the product from the factory to the customer.

Distributor   A distributor (e.g., a wholesaler) purchases products from a number of suppliers and sells directly to a consumer, or to a retailer who then sells on to the consumer.

Downtime   The percentage of time a piece of equipment is not working because of failures and scheduled maintenance.

Draft angle   The degree to which parts are tapered to ensure that they can easily be removed from the tool without damaging the part or the tool.

Durability   The ability to withstand normal-use forces, loads, and exposure to environmental conditions.

Durometer   The hardness/softness of a material.

Dust cover   Packaging material that reduces scratches and contamination of the product.

E

Ejector pin   A pin that pushes a part out of a mold. After the mold opens, the pins are actuated to push the part out and then are returned back inside the mold. The pins are usually only applied to the inside surface of a part as they typically leave a small mark on the surface which would not meet aesthetic requirements.

Electromagnetic field (EMF)   A field of energy waves that emanate from all products that use electricity. Strong EMF is considered to be a health risk to people and can cause electromagnetic interference with other electronic equipment.

Electromagnetic field (EMF) shielding   Ensures that electromagnetic fields at harmful frequencies do not escape from a product and/or impact delicate electronics.

Electromagnetic interference (EMI)   Some products are sensitive to electromagnetic fields. EMI can be an issue for people with pacemakers who are often advised to avoid strong EMF fields generated by products such as airport scanners.

Electrostatic discharges (ESD)   Shocks from static or other devices that can negatively impact electronics.

End cap   The display case at the end of an aisle in a retail location. End caps are used to display product that is being promoted.

End-of-life When a product (or product line) is discontinued, companies need to ensure they have sufficient spares and parts to support the previously made products through the expected life of the product.

Engineering verification testing (EVT) The pilot phase that uses a mix of production-intent, low-volume prototype parts. Builds are based on the final product CAD and specifications and are very close to the final mass-production products, varying only in the manufacturing process used.

Enterprise data management Methods to track and manage all of the materials, documents, and data required to define and produce a product.

Enterprise resource planning (ERP) Systems used to manage resources and resources flow within a company. Many ERP systems have an MRP system as a module within them.

Ex-factory The time when product ownership is passed from the factory to the customer.

Extended warranty A warranty contract purchased by the customer that extends the product warranty beyond the standard expiration date.

F

Factoring Instead of having to wait for a customer's payment (which can be as long as 120 days), credit institutions will pay you the amount your customers owe you and take a small percentage of the invoice as payment.

Failure modes and effects analysis (FMEA) A systematic way to identify risks in design and production and prioritize them.

Federal Communications Commission (FCC) The agency in charge of regulating the electromagnetic emissions of products with electronics.

Finish Surface treatment on visible and assembly surfaces of the part. Finishes can be used to meet aesthetic requirements, cover up defects, or change material properties

Finished goods The finished products (including packaging) that have been fully assembled but not yet distributed.

Firmware Computer code that communicates between the product software and the individual component on the board. It is typically stored in ROM.

First article inspection (FAI) See First part inspection.

First part inspection (FPI) Approval of the first part (typically a fabricated part or custom part). This part is subjected to detailed dimensional measurements to ensure that it completely conforms to the drawings.

First-pass yield The percentage of parts that pass all inspections without fail. The yield is calculated by multiplying the yield rate of each step in the process. For example, if there are five steps in a process and each has a 99% yield rate, the first-pass yield would be 95%.

First shots First parts off a new tooling. New tooling is rarely ready for mass production, and first shots are used to fine-tune the tooling.

Fixtures Pieces of equipment that hold parts to enable and facilitate assembly.

Flash Excess material that squeezes out between the two halves of a tool or mold.

Flex circuits   Electronics assembled on a flexible substrate. This method allows for much thinner geometries and for electronics to be routed through complex spaces.

Flexible fixtures   Fixtures that can be used on multiple projects or parts, or that can be easily reconfigured for different usages.

Forecasting   Predicting the number of units that will be purchased or need to be produced to meet demand.

Foreign objects and debris (FOD)   Unwanted material on and in a product. It can include everything from dirt to loose bolts to insects.

Free on board (FOB)   FOB means that the product is owned by the organization purchasing the materials and that they are liable for any damage after the material is designated FOB.

Freight forwarders   Intermediaries who coordinate shipment from the factory to its destination. They may coordinate between multiple modes of transport; for example, moving a product from the airport onto a truck for shipment.

G

Gate   An opening in the mold that allows molten material to flow between the injection point and the cavity that forms the part. The gates need to be placed in areas on the part where they can't be seen as the gates are typically cut or broken off and leave a defect on the surface.

Gauge   A tool used to measure a part.

Gauge repeatability and reliability (Gauge R&R)   Standard method that quantifies the inherent variation in a measurement system.

Generic parts   Commodities or custom parts that are described generally, but to which a specific manufacturer part number is not given. This gives the procurement team the ability to order material from alternative suppliers to reduce costs.

Geometric dimensioning and tolerancing (GD&T)   Systematic method for defining engineering tolerances.

Gerbers   CAD files used to define the printed circuit board assemblies.

Gift box   This is the packaging that the customer purchases, including the product, accessories, and manuals.

Glass-filled   Glass fibers are mixed into polymers to increase strength and stiffness of parts such as injection molded parts.

Golden samples   Samples that are signed off as the standards for approving product.

H

Hard tooling   Tooling that is produced with harder and more durable materials (typically hardened steel instead of aluminum). These tools take longer to produce and modify than do soft tooling, but they are more durable and last longer.

Highly accelerated life testing/highly accelerated stress testing (HALT/HASS testing)   Accelerated testing of products designed to rapidly identify weaknesses and quality problems.

I

IMEI number A unique number (globally) that identifies a cell phone.

Industrial design (ID) National Association of Schools of Art and Design describes industrial design as the process to "create and develop concepts and specifications that optimize the function, value, and aesthetics of products, environments, systems, and services for the benefit of user, industry, and society. Industrial design involves combinations of the visual arts disciplines, sciences, and technology, and requires problem-solving and communication skills" [142].

Ingress protection (IP) Standard for defining the resistance of a product to dust and water.

Inner packs Smaller (typically cardboard) boxes packaged within the master cartons in which a small number of gift boxes of the same SKU are packaged. These are used to protect gift boxes and can also be used as shipping packaging for order fulfillment when the master carton is opened.

Inserts (packaging) Includes foam, thermoformed trays, cardboard inserts, platforms, and dividers. The term "inserts" can also be used to describe additional materials that are included in packaging to provide the customer with additional information or value.

Integrated circuit testing (ICT) Also called a "bed of nails" test, used to test the connections in and function of a PCBA.

International Traffic in Arms Regulations (ITAR) Export regulations imposed in the US on some technologies that may have military applications.

ISTA-2A A typical standard for shipment testing used in many companies and governed by the International Safe Transit Association. The standard is used for products weighing 150 lb (68 kg) or less.

J

Jig Structure used to hold parts while guiding tools to create features.

Job shop Facility that contains a collection of various manufacturing processes that are used to make a wide variety of products in low volumes.

L

Landed cost The total cost to get an individual product to the customer. Includes COGS as well as customs and shipping costs.

Last-mile The distribution process required to get the product from the in-country distribution center into the hands of the customer.

Launch Entry of a product into the market. It involves ensuring that marketing and distribution are ready at the same time as a product becomes available for sale.

Lead time Time between putting an order in for certain parts and the time those parts arrive. Lead time can range from days to months depending on the part, the supplier, and the demand for the product being purchased.

Less than truckload (LTL)   A shipment that does not completely fill a single truck or a container used for sea, road, or rail transport. LTL shipments might be combined with other customers' products to fill a shipping container.

Life testing   Testing that evaluates whether the product can perform correctly over its expected life. The product is tested using repeated loads and stresses and analyzed for degradation in performance.

Life-cycle cost   See: Total cost of ownership.

M

MAC ID   The unique identifier given to each communication device (e.g., Wi-Fi, Bluetooth, etc.).

Make vs. buy   Decisions made by an organization to produce in-house or to outsource. Can be applied to parts or services.

Manufacturer's suggested retail price (MSRP)   The cost the company recommends that the product be sold for by its distributors. Depending on the contract with the distributor, the actual cost to the customer may be different.

Manufacturing execution system (MES)   Tracks where the parts are on the factory floor in real time. MESs are typically used for more sophisticated or complicated processes where it is critical to track the real-time status of important materials and partially assembled product.

Manufacturing readiness level (MRL)   Used to assess the maturity of manufacturing processes and how close the process is to being commercially viable.

Markup   The percentage a supplier adds to the cost of purchased material to cover overhead, fixed costs, and profit.

Mass production (MP)   Stage at which the product is being produced at full rates.

Master carton   Usually a cardboard box used to pack multiple inner packs and/or gift boxes. It is the primary packaging used to stack on pallets and ship over long distances. A master carton can contain multiple SKUs.

Master Service Agreement (MSA)   The contract reached between a supplier (typically a CM) and its customer. The MSA defines all of the terms and conditions that govern all future transactions.

Material authorization (MA)   Pre-purchases materials ahead of a PO. MAs are usually made for parts/materials with long lead times.

Material resource planning (MRP)   Controls when orders are released to the factory to ensure that products are produced to meet purchase orders from customers.

Mechanicals   Mechanical parts in a product, as opposed to the electronics or the software.

Metrology   Set of methods and tools used to measure dimensions, weights, forces, etc. It is the "science of measurement."

Minimum order quantity (MOQ)   The minimum number of units that have to be ordered to get a quoted cost. The per-unit cost of the parts typically goes up with smaller MOQs.

Mixed-model production   Production lines that run multiple SKUs intermixed on the same production line.

Model-based definition   All design and build information attached to the 3D CAD model to minimize the need for 2D drawings.

N

Net payments   The agreed-upon time-frame for how much time you are given to pay the supplier from the time the part/product is delivered. For example, net 30 days means you need to pay the invoice within 30 days from delivery.

Non-disclosure agreement (NDA)   A legal document that identifies what information, data, CAD, etc., can and cannot be released by both parties signing the document.

Non-recurring engineering (NRE) expenses   Expenses that are incurred during design and product realization. NRE expenses typically include the engineering- and manufacturing-related costs (as opposed to marketing).

O

Ongoing production testing (OPT)   Life testing and durability testing that is performed throughout the production life of the product. It is used to ensure that no change in the processes will result in quality failures.

Open book quote   A quote from a supplier that provides a detailed breakdown on part costs and overhead costs, as opposed to a closed-book quote which only gives you the total price and no details.

Opening a tool   Re-cutting or altering a tool to make changes to the geometry of the part.

Order fulfillment   The process of going from receiving a customer order to getting the product into the hands of the customer. In logistics and distribution it is the "last-mile" of the process.

Original design manufacturer (ODM)   A contract manufacturer who takes responsibility for both the design and manufacturing of a product.

Original equipment manufacturer (OEM)   A contract manufacturer who produces what the product design team designs.

Outsource   To hire an outside firm to perform a service or to purchase a part/product from another firm.

P

Packaging   All of the materials required to store, ship, and sell a product.

Pallet   A standard-sized structure used to group multiple master cartons. It is typically designed to be lifted with a forklift.

Pantone   The company that sets the standards for uniquely defining colors [56].

Part envelope   Overall size and geometry of a part or system. Typically given as an overall height, width, and length of the part.

Pattern   An object that is replicated by casting.

Persona   Description of a typical consumer/customer as if they were a real person. It includes demographics, motivations, and behavior.

Pick-and-pack   A typically manual (more pick-and-pack facilities are being automated) process that collects multiple components into a single shipment or package.

Pilot lines   Assembly or production lines used exclusively for pilot runs to facilitate learning without disturbing existing production.

Pilot process   Short production run used to find problems in the product design and its manufacturing process.

Point of purchase (PoP) displays   Packaging that is used to display a product in a retail location, often placed at the end of an aisle.

Poke-a-yoke   The Japanese term for error-proofing an assembly or manufacturing process.

Post-processing   Processes applied to a part after a primary process (e.g., injection molding). It can include heat treatments, finishing, and machining.

Potting   Manufacturing process that fills an electronic assembly with a solid or gel. Potting is used to reduce the negative effects of vibration and humidity.

Preventative maintenance (PM)   Maintenance done ahead of equipment failures, such as regular cleaning or changing out critical components that are likely to fail.

Printed circuit board, and printed circuit board assembly (PCB and PCBA)   A PCB is a board that supports the integrated circuits and electronic modules. It both holds the components as well as connects them electronically. A PCBA is the combination of the board and the components, which have been assembled by soldering the components onto the PCB.

Procurement team   Group tasked with identifying suppliers and purchasing parts and subsystems.

Product data management (PDM)   Systems used to manage critical documents, their versions, and the process for updating them. They are essentially glorified filing systems.

Product lifecycle management (PLM)   Software systems that manage documents related to the product development processes and the processes around them. PLM systems have embedded product data management systems within them. They ensure that the right people are informed and that the documents are signed off in the correct way.

Production-intent parts   Parts that are built on equipment and tooling that are representative of the final production equipment.

Production verification testing (PVT)   The product realization pilot phase in which bugs in production are worked out. The final production line equipment and workers are used, but production is at a low enough volume to catch problems and implement corrective action.

Project manager   Person responsible for the planning and execution of a particular project and the management of the day-to-day tasks required to complete it.

Prototype process   A process used to simulate the final production process, but usually used to make small numbers of product.

Prototype product A sample product that represents the function (works-like) and/or aesthetics and user interface (looks-like) of the final product. It is typically made using processes that are not used in mass production.

Purchase order (PO) Document used to purchase a product or service or to place an order for materials to be delivered later. A PO is considered a binding contract.

Q

Qualification Ensures a manufacturing process can conform consistently to documented specifications. In some cases (e.g., the FDA), process qualifications are legally required.

Quick-start guide Shortened version of a product manual. It is required when filing for FCC certifications in the US.

R

Radio frequency identification (RFID) tags Tags that can be read by scanners using radio frequency scanning. These allow for tracking of packages without line-of-sight to the label as well as preventing shrink (theft).

Ramp The process of increasing production volumes and reducing cycle times to bring the initial production rate up to mass production rates.

Rapid prototyping Range of production methods that can be used to make sample products quickly. The methods include techniques like urethane casting, additive manufacturing, and CNC machining.

RASCI (responsible, accountable, supporting, consulted, and informed) A method for documenting roles and responsibilities.

Rate The number of units that can be produced in a set time.

Reliability How well and consistently a product performs over time.

Request for quote (RFQ) A formal invitation for a supplier to provide a quote and payment terms for a product or service. The RFQ typically provides significant details about the product so an accurate quote can be done.

Re-spin Jargon used to describe redesigning a board to accommodate a component change. Re-spin requires changes to both the board design and component selection.

Return material authorization (RMA) Approval to return material to a supplier because of a quality issue.

Reverse logistics The distribution methods used to get failed or returned products back to the manufacturer and then get repaired or replacement products back to the customer.

Rib A thin wall added to parts to improve stiffness and prevent warping.

Robustness How well a product performs when manufacturing and operation variations are present.

RoHS European regulation that restricts the use of hazardous materials in products (e.g., lead).

Roles and responsibilities Responsibilities are the tasks and deliverables that a person or group is assigned. Individuals and groups are assigned to roles that are responsible for various tasks, deliverables, and contributions to the teams. In small teams a single person may have multiple roles, whereas, in larger companies, entire groups are assigned to roles.

Root cause analysis (RCA)   Systematic method for identifying the underlying cause of a given problem.

S

Safety stock   An amount of inventory held to account for variability in demand, lead time, and scrap.

Samples   Products built for non-commercial use including marketing, testing, and certification.

Scrap   Material that has to be disposed of (and hopefully recycled). This term can apply to offcuts (material that is cut off the stock when making a part) as well as complete products.

Second-sourcing   Identifying a second source in addition to your primary supplier for components. This can be done to reduce risk, create competition, or grow capacity ahead of additional product launches.

Semi-custom parts   Parts that are designed specifically for a product but are versions or modifications of a standard design.

Serial number (SN)   A unique number assigned to each product sold.

Setup   The process of getting a production batch run started. This can include tool installation, material loading, and warm-up cycles.

Shenzhen electronics markets   Approximately 12 separate markets in the Shenzhen region of China that sell a wide variety of electronic systems and components.

Shipment audits   An inspection and quality check done after final packaging but before the product is designated FOB or ex-works. The products are inspected for functional and aesthetic failures. If the rate of failure is higher than the negotiated AQL rate, the shipment is rejected and needs to be 100% inspected.

Shot   In injection molding, a single cycle of the injection molding tool.

Sink marks   Surface defects in injection-molded parts that develop where there is insufficient support and the material deforms while cooling.

Sleeve   The wrap that goes around a generic gift box. It is used to inhibit theft and allows for easy changes to the packaging information.

Soft tooling   Tooling that mimics final production tooling but is made with less expensive materials using faster machining processes. These tools do not last as long as hard tools.

Soft-touch paint   Durometer coating than makes a surface feel like rubber, suede, or leather. Typically has a matte finish.

Software as a service (SaaS)   A method for delivering software on a license or subscription basis. The software is hosted in the cloud (i.e., no local copies).

Source (v.)   Find an acceptable vendor or supplier for a part.

Spot market   A market for components that can be purchased directly with no lead time.

Sprue A channel that is used to bring molten material to the cavity. It is removed after molding using a hot knife or it is mechanically broken off.

Standard operating procedures (SOPs) Documented processes followed by the manufacturer to ensure that every product is made correctly using the same manufacturing processes.

Statistical process control (SPC) Statistical methods to track the quality of a product or its process characteristics.

Steel-safe Mode of cutting a tool. The cavity is cut such that the part will be undersized. As the team learns where added material on the part is needed, material is removed from the tool. This reduces the risk of over-cutting the tool and having to add material (which always results in a poor surface finish).

Stencil A mask used in the fabrication of a PCBA. It is used to apply the solder paste to the PCB and is unique to each part number.

Stock-keeping unit (SKU) Unique product identifier for a specific product configuration. A single product may have multiple SKUs that denote different colors, versions, configurations, and countries of sale.

Storyboarding Documenting the story of how your product will be used in various scenarios.

Supply chain tier Refers to the place of a given supplier in the supply chain. Tier 1 suppliers include contract manufacturers. Tier 2 typically supplies tier 1, and so on.

Supporting engineering The team of engineers whose responsibility it is to make design changes in existing products.

Surface mount technology (SMT) Method for soldering electronic components directly onto a surface of a PCB (as opposed to through-hole).

T

Target cost The cost that the design team is aiming for to ensure sufficient profit margins. The total cost target for the product can be rolled down to create target costs for individual subsystems and components.

Technology Readiness Level Standard framework used to qualitatively assess how close a technology is to commercialization.

Third-party logistics (3PL) An outside company hired to manage logistics, supply chain, and order fulfillment.

Through-hole Older method (than SMT) for assembling components on a PCB. The leads are inserted through holes and soldered on the back.

Tier 1/2/3 contract manufacturer/suppliers Tier 1 CMs refer to the larger, well-known CMs such as Foxconn. Tier 3s are small local businesses.

Tooling Specialized pieces of hardware – such as dies, fixtures, molds, patterns, and stencils – that are designed and built to be used for a specific product.

Total cost of ownership (also called life-cycle costs) This is the cost to your customer to buy the product, consumables, training, maintenance, repair, etc. Depending on the design and manufacturing decisions, the design team can dramatically reduce the life-cycle cost and increase the appeal of the product.

Tote   A container used to safely move material between stations in a factory.

Trace width   Distance between traces in a PCBA. The smaller the trace width, the higher the component density, but the harder the board is to manufacture.

Traveler (also called a work-order traveler)   Paper or electronic record that tracks the status of a part, subsystem, or batch.

U

Ultraviolet resistance   UV rays from the sun or other light sources can change the chemical properties, color, or material properties of a product. Some materials are more sensitive to UV than others. UV resistance is the ability for materials to not be chemically altered in a significant way.

Unboxing experience   The experience of the customer from purchase through unpacking and then through first use of the product.

Underwriters Laboratory (UL)   Testing body that certifies products for electrical safety.

Uptime   The percentage of time manufacturing equipment is available for production.

Urethane casting   A method for making a mold of silicone from a pattern and then using a thermoset material to fill the mold. This can be used in low-volume production as a stand-in for injection molding.

V

Validation   The process of checking whether the product specifications are aligned with customer needs.

Verification   The process of checking whether a product meets the specifications defined in the specification document.

Via   Connection between layers in a printed circuit board (PCB).

W

Wall wart   The electrical plug that goes into the wall. Typically, wall warts are rated based on amperage and voltage output and can be used in both 110 and 220V wall sockets.

Warranty   The legal guarantee of product's performance and safety. A typical warranty expires after one year of usage (in the US), unless an extended warranty is purchased.

Warranty accrual rate   Amount of cash set aside against warranty claims divided by the total revenue for the same period.

Warranty liability account   Account for which a portion of revenue is held to cover potential warranty liability costs.

White goods   Large electrical goods such as washers or dryers.

Wholly Foreign-Owned Enterprise (WFOE)   A financial setup that allows US-based companies to pay salaries in a foreign country. WFOE is typically used in China.

Wire bonding   A method of attaching a chip or module to a PCB using fine wires.

Wire harness   Cables and wires that are pre-bundled and assembled with connectors prior to installation into a product.

Work in progress (WIP)   The inventory of not-yet-finished products on the factory floor.

Y

Yield   The percentage of products or parts that successfully pass all inspections.

ACRONYMS

3D	three-dimensional	COTS	commercial off-the-shelf
3PL	third-party logistics	CPA	certified public accountant
A2A	actual-to-actual	CPO	Chief Product Officer
Ah	Amp–hours	CPSC	Consumer Product Safety Commission
AIAG	Automotive Industry Action Group	CRM	customer relationship management
ANSI	American Nation Standards Institute	CTO	Chief Technology Officer
AOI	automated optical inspection	DDP	delivered duty paid
API	application programming interface	DFA	design for assembly
APQP	Advanced Product Quality Planning	DFM	design for manufacturing
AQL	acceptable quality limit	DFX	design for X
AR	augmented reality	DOA	dead on arrival
ASME	American Society of Mechanical Engineering	DoD	Department of Defense
AST	accelerated stress testing	DP	design pilot
ASTM	American Society for Testing and Materials	DRAM	dynamic random-access memory
B2B	business-to-business	DVT	design verification testing
B2C	business-to-customer	EAR	Export Administration Regulations
BOM	bill of materials	E-BOM	engineering bill of materials
BPA	bisphenol A	EDM	enterprise data management
CAD	computer-aided design	EMF	electromagnetic field
CAM	Computer Aided Manufacturing	EMI	electromagnetic interference
CCL	Commerce Control List	EOL	end-of-life
CE	Confomité Européene	EP	engineering pilot
CEO	Chief Executive Officer	ERP	enterprise resource planning
CGMPs	Current Good Manufacturing Practices	ESD	electrostatic discharge
CM	contract manufacturer	EU	European Union
CMF	color, material, finish	EVT	engineering verification testing
CMM	coordinate measurement machine	EXW	ex-works
CNC	computer numerically controlled	FAA	Federal Aviation Administration
CNY	Chinese New Year	FAI	first article inspection
COGS	cost of goods sold	FAQ	frequently asked questions
COO	Chief Operating Officer	FCC	Federal Communications Commission
		FDA	Food and Drug Administration
		FMEA	failure modes and effects analysis

Product Realization: Going from one to a million, First Edition. Anna C. Thornton.
© 2021 John Wiley & Sons, Inc. Published 2021 by John Wiley & Sons, Inc.

FMVT	failure mode validation test	MOQ	minimum order quantity
FOB	free on board	MP	mass production
FOD	foreign objects and debris	MPS	master production schedule
FPI	first part inspection	MRL	manufacturing readiness level
FT	functional test	MRP	material/manufacturing
GD&T	geometric dimensioning and		resource planning
	tolerancing	MSA	master services agreement
GST	goods and services tax	MSRP	manufacturer's suggested
HALT	highly accelerated life testing		retail price.
HR	human resources	MSST	multi-step stress testing
HTB	higher the better	MVP	minimum viable product
HTHH	high temperature, high humidity	NDA	non-disclosure agreement
HTLH	high temperature, low humidity	NHTSA	National Highway Traffic Safety
HTS	Harmonized Tariff Schedule		Administration
IC	integrated circuit	NIST	National Institute of Standards
ICC	International Chamber		and Technology
	of Commerce	NRE	non-recurring engineering
ICT	in-circuit testing	NTSB	National Transportation
ICT	integrated circuit testing		Safety Board
ID	industrial design	NVOCC	non-vessel operating
IEC	International Electrotechnical		common carrier
	Commission	ODM	original design manufacturer
IoT	Internet of Things	OEM	original equipment manufacturer
IP	ingress protection	OPT	ongoing production testing
IP	intellectual property	PAC	purchasing and production
IQC	incoming quality control		activity control
ISO	International Standards	PCB	printed circuit board
	Organization	PCBA	printed circuit board assembly
ISTA	International Safe Transit	PDM	product data management
	Association	PIA	Plastics Industry Association
IT	information technology	PLM	product lifecycle management
ITAR	International Traffic in Arms	PM	preventative maintenance
	Regulations	PM	project management
JSF	Joint Strike Fighter	PO	purchase order
LCD	liquid crystal display	PoP	point of purchase
LED	light-emitting diode	PP	production pilot
LiPo	lithium-ion polymer	PPAP	production part
LTB	lower the better		approval process
LTL	less than truckload	PPE	personal protective equipment
LTLH	low temperature, low humidity	PRD	product requirements document
MAC	media access control	PSD	product specifications document
MA	material authorization	PVT	production verification testing
MBD	model-based definition	QA	quality assurance
M-BOM	manufacturing bill of materials	QMS	Quality Management System
MEOST	multiple environment overload	QR	quick response
	stress test	R&D	research and development
MES	manufacturing execution system	R&R	repeatability and reproducibility

RASCI	responsible, accountable, support, consulted and informed	SN	serial number
		SOP	standard operating procedure
RCA	root cause analysis	SPC	statistical process control
RFI	request for information	SS	stainless steel
RFID	radio frequency ID	TPL	third-party logistics
RFP	request for proposal	TRL	Technology Readiness Level
RFQ	request for quote	UI	user interface
RMA	return material authorization	UL	Underwriters Laboratory
RoHS	Restriction of Hazardous Substances	US-CBP	US Customs and Border Protection
ROI	return on investment	USD	US dollars
ROM	read-only memory	USDOT	US Department of Transportation
RPN	risk priority number	USML	United States Munitions List
S&OP	sales and operations plan	UV	ultraviolet
SaaS	software as a service	VAT	value added tax
SAE	Society of Automotive Engineers	VOC	voice of the customer
SEZ	special economic zone	VR	virtual reality
SKU	stock-keeping unit	WFOE	Wholly Foreign-Owned Enterprise
SMT	surface mount technology	WIP	work in progress

REFERENCES

1. Isidore, C. (2018). Tesla will start working 24/7 to crank out Model 3s. CNN Money. https://money.cnn.com/2018/04/18/news/companies/elon-musk-tesla-model-3-production/index.html (accessed 28 July 2020).

2. Randall, T. and Halford, D. (2020). Tesla Model 3 Tracker. *Bloomberg.com*. https://www.bloomberg.com/graphics/tesla-model-3-vin-tracker/ (accessed 28 July 2020).

3. Insinna, V. (2019). Inside America's dysfunctional trillion-dollar fighter-jet program. *The New York Times Magazine* (21 August).

4. Gates, D. (2011). Boeing's tab for the 787: $32 billion and counting. *The Seattle Times* (25 September), p. A14.

5. Wakabayashi, D. (2014). Inside Apple's broken Sapphire factory. *Wall Street Journal* (20 November), p. B1.

6. Jensen, L.S. and Özkil, A.G. (2018). Identifying challenges in crowdfunded product development: a review of Kickstarter projects. *Design Science* 4: e18.

7. Carpenter, N. (2017). The 5 biggest crowdfunding failures of all time. https://www.digitaltrends.com/cool-tech/biggest-kickstarter-and-indiegogo-scams/ (accessed 28 July 2020).

8. Rinpoche, Y.M. and Tworkov, H. (2014). *Turning Confusion into Clarity: A Guide to the Foundation Practices of Tibetan Buddhism*. Snow Lion.

9. Urban, T. (2016). Inside the mind of a master procrastinator. https://www.ted.com/talks/tim_urban_inside_the_mind_of_a_master_procrastinator (accessed 28 July 2020).

10. Carreyrou, J. (2018). *Bad Blood: Secrets and Lies in a Silicon Valley Start-Up*. New York: Alfred A. Knopf.

11. Zaleski, O. and Huet, E. (2017). Silicon Valley's $400 juicer may be feeling the squeeze. https://www.bloomberg.com/news/features/2017-04-19/silicon-valley-s-400-juicer-may-be-feeling-the-squeeze (accessed 28 July 2020).

12. Levin, S. (2017). Squeezed out: Widely mocked start-up Juicero is shutting down. *The Guardian* (1 September).

13. Einstein, B. (2017). Here's why Juicero's press is so expensive. https://blog.bolt.io/heres-why-juicero-s-press-is-so-expensive-6add74594e50 (accessed 28 July 2020).

14. Bodley, M. (2016). Brewing a new strategy: Keurig looks to learn from the failure of its Kold system. *Boston Globe* (1 June), p. A10.

15. Nieto-Rodriguez, A. (2017). Notorious project failures – Google Glass. https://www.cio.com/article/3201886/notorious-project-failures-google-glass.html (accessed 28 July 2020).

16. Edwards, O. (2006). The Death of the EV-1. *Smithsonian Magazine* (June).

17. Ewing, J. (2017). Engineering a deception: what led to Volkswagen's diesel scandal. *New York Times* (16 March).

18. Kitroeff, N. (2019). Boeing 737 Max safety system was vetoed, engineer says. *New York Times* (29 October).

19. GAO (2016). *Technology Readiness Assessment Guide: Best Practices for Evaluating the Readiness of Technology for Use in Acquisition Programs and Projects (GAO-16-410G)*. US Government Accountability Office.

20. Banke, J. (2010). *Technology Readiness Levels Demystified*. https://www.nasa.gov/topics/aeronautics/features/trl_demystified.html (accessed 28 July 2020).

21. Department of Defense (2018). *Manufacturing Readiness Level (MRL) Deskbook*. http://www.dodmrl.com/MRL_Deskbook_2018.pdf (accessed 28 July 2020).

22. Carman, A. (2019). Crowdfunding disaster Coolest Cooler is shutting down and blaming tariffs for its downfall. https://www.theverge.com/2019/12/9/21003445/coolest-cooler-update-business-tariffs-kickstarter (accessed 28 July 2020).

23. Cooler, C. (2019). Coolest Cooler: 21st Century Cooler that's Actually Cooler. https://www.kickstarter.com/projects/ryangrepper/coolest-cooler-21st-century-cooler-thats-actually (accessed 28 July 2020).

24. Soper, T. (2017). Glowforge delays shipment of 3D printers again as buyers express frustration with Seattle start-up. *GeekWire* (13 October).

25. Ulrich, K.T. and Eppinger, S.D. (2016). *Product Design and Development*. New York, NY: McGraw-Hill.

26. Aulet, B. (2013). *Disciplined Entrepreneurship: 24 Steps to a Successful Start-Up*. Hoboken, NJ: Wiley.

27. Ries, E. (2011). *Lean Start-Up: How Today's Entrepreneurs Use Continuous Innovation to Create Radically Successful Businesses*. New York: Random House.

28. AIAG (2008). *Advanced Product Quality Planning and Control Plan*, 2e. Automotive Industry Action Group.

29. Quality System Regulation (2015). 21 CFR § 820.

30. ASTM F963 (2017). *Standard Consumer Safety Specification for Toy Safety*. American Society for Testing and Materials.

31. CPSC, *CPSC Recall List*. https://www.cpsc.gov/Recalls (accessed 28 July 2020).

32. International Traffic in Arms Regulations, 22 CFR §120–130.

33. Export Administration Regulations, 15 CFR § 730–774.

34. ISO 9001:2015 (2015). *Quality Management Systems*. International Standards Organization.

35. European Parliament and The Council on Medical Devices, Regulation (EU) 2017/745.

36. ISO 13485:2016 (2016). *Medical Devices – Quality Management Systems Requirements for Regulatory Purposes*. International Standards Organization.

37. ISO 14971:2012 (2012). *Medical Devices – Application of Risk Management to Medical Devices*. International Standards Organization.

38. ISO/TS 16949. (2016). *Automotive Quality Management Standard*. International Automotive Task Force (IATF).

39. AS9100D (2016). *Quality Systems – Aerospace Model for Quality Assurance in Design, Development, Production, Installation, and Servicing*. Society of Automotive Engineering.

40. TL9000 (2017). *Quality Management System R6.1*. Telecommunications Industry Association.

41. Declaration of Daniel W. Squiller in Support of Debtors' Motion (2014). 14-11916-HJB. US Bankruptcy Court, District of NH.

42. PMI (2019). What is project management? http://www.pmi.org/about/learn-about-pmi/what-is-project-management (accessed 28 July 2020).

43. PricewaterhouseCoopers (2012). *Insights and Trends: Current Portfolio, Programme, and Project Management Practices: The Third Global Survey on the Current State of Project Management*. PricewaterhouseCoopers.

44. Shepardson, D. (2015). GM compensation fund completes review with 124 deaths. *Detroit News* (24 August).

45. Reynolds, T., Gutierrez, G., and Gutierrez, G. (2014). Document shows GM Engineer approved ignition switch change. *NBC News*. https://www.nbcnews.com/storyline/gm-recall/document-shows-gm-engineer-approved-ignition-switch-change-n68371 (accessed 28 July 2020).

46. Christensen, C.M. (2016). *Competing Against Luck: The Story of Innovation and Customer Choice*. New York, NY: Harper Business.

47. Stringham, G. (2010). *Hardware/Firmware Interface Design Best Practices for Improving Embedded Systems Development*. Elsevier.

48. ISTA-2A (2011). *Partial Simulation Test Procedure.* International Safe Transit Association.

49. IBM (2014). IBM commits $100M to globally expand unique consulting model that fuses strategy, data and design [press release]. http://www-03.ibm.com/press/us/en/pressrelease/43523.wss (accessed 28 July 2020).

50. Consumer Product Safety Commission (2012). *Recall Handbook.* https://www.cpsc.gov/s3fs-public/8002.pdf (accessed 28 July 2020).

51. CPSC (2015). Laceration injuries prompt SharkNinja to recall Ninja BL660 blenders to provide new warnings and instructions. http://www.cpsc.gov/Recalls/2016/Laceration-Injuries-Prompt-SharkNinja-to-Recall-Ninja-BL660-Blenders (accessed 28 July 2020).

52. CPSC (2013). Calphalon recalls blenders due to injury hazard. http://www.cpsc.gov/Recalls/2014/Calphalon-Recalls-Blenders (accessed 28 July 2020).

53. CPSC (2016). Denon recalls rechargeable battery packs due to fire and burn hazards. http://www.cpsc.gov/Recalls/2016/Denon-Recalls-Rechargeable-Battery-Packs (accessed 28 July 2020).

54. CPSC (2008). Atico International USA recalls personal blenders due to laceration hazard. http://www.cpsc.gov/Recalls/2008/Atico-International-USA-Recalls-Personal-Blenders-Due-to-Laceration-Hazard (accessed 28 July 2020).

55. Thompson, R. (2007). *Manufacturing Processes for Design Professionals.* New York: Thames & Hudson.

56. Pantone (2019). What is the pantone color system? https://www.pantone.com/color-systems/pantone-color-systems-explained (accessed 28 July 2020).

57. ISO/CIE 11664-4:2019(E) (2019). *Colorimetry – Part 4: CIE 1976 L*A*B* Colour Space.* International Commission on Illumination.

58. Standex Engraving. Mold-tech. https://www.mold-tech.com (accessed 28 July 2020).

59. Plastics Industry Association. www.plasticsindustry.org (accessed 28 July 2020).

60. SPI (1994). *Cosmetic Specifications of Injection Molded Parts* AQ 103. Society of the Plastic Industry.

61. ASME Y14.100 – 2017 (2017). *Engineering Drawing Practices.* American Society of Mechanical Engineering.

62. ASME Y14.5 – 2018 (2018). *Dimensioning and Tolerancing.* American Society of Mechanical Engineering.

63. Dotcom Distribution (2018). *Great(er) Expectations: The Rapid Evolution of Consumer Demands in eCommerce.* Dotcom Distribution 2018 eCommerce Study. https://dotcomdist.com/wp-content/uploads/2019/06/The_Rapid_Evolution_of_Consumer_Demands_in_eCommerce_eGuide_v3_8.3-1.pdf (accessed 28 July 2020).

64. Hubert, M., Hubert, M., Florack, A. et al. (2013). Neural correlates of impulsive buying tendencies during perception of product packaging. *Psychology & Marketing* 30 (10): 861–873.

65. FEFCO Code (2019). *International Fibreboard Case Code.* FEFCO.

66. Oxford English Dictionary (2007). *Quality.* Oxford University Press.

67. Taguchi, G. (1993). *Taguchi Methods: Design of Experiments.* Dearborn, MI: ASI Press.

68. Deming, W.E. (1986). *Out of the Crisis.* Cambridge, MA: Massachusetts Institute of Technology, Center for Advanced Engineering Study.

69. Juran, J.M. and Gryna, F.M. (1988). *Juran's Quality Control Handbook.* New York: McGraw-Hill.

70. Consumer Reports (2019). Takata airbag recall: Everything you need to know. https://www.consumerreports.org/car-recalls-defects/takata-airbag-recall-everything-you-need-to-know (accessed 28 July 2020).

71. IEC 61000-4-2. *ESD Immunity and Transient Current Testing.* IEC. https://www.iec.ch/emc/basic_emc/basic_emc_immunity.htm (accessed 28 July 2020).

72. IEC 60529:1989+AMD1:1999+AMD2:2013 CSV (2013). *Degrees of Protection Provided by Enclosures (IP Code).* International Electrotechnical Commission.

73. Krystal, B. (2016). 8 million Cuisinart food processor blades have been recalled. Yours may be one of them. *The Washington Post* (14 December).

74. Pushkar, R. (2002). Comet's Tale – Making its inaugural flight 50 years ago, the world's first passenger jetliner, the British-built Comet 1, was sleek, fast, and, it would turn out, fatally flawed. *Smithsonian* 33 (3): 59–62.

75. The numbers used in the reliability budgets and the lists for the MSA are based on discussions with Bill Drislane.

76. CPSC Recalls (2014). Fitbit recalls Force activity-tracking wristband due to risk of skin irritation. https://www.cpsc.gov/Recalls/2014/fitbit-recalls-force-activity-tracking-wristband (accessed 28 July 2020).

77. Park, J. (2014). A letter from the CEO. https://www.fitbit.com/forcesupport (accessed 28 July 2020).

78. CPSC Recalls (2016). McDonald's recalls "Step-iT" activity wristbands due to risk of skin irritation or burns. https://www.cpsc.gov/Recalls/2016/mcdonalds-recalls-step-it-activity-wristbands (accessed 28 July 2020).

79. Baker, J. (2016). *Wearable Products: Biocompatibility Insights*. CDP Blogs. https://www.cambridge-design.com/news-and-articles/blog/biocompatibility (accessed 28 July 2020).

80. Niazi, A., Dai, J.S., Balabani, S., and Seneviratne, L. (2006). Product cost estimation: technique classification and methodology review. *Journal of Manufacturing Science and Engineering* 128 (2): 563–575.

81. NASA (2015). *NASA Cost Estimating Handbook, Version 4.0*. NASA.

82. Astor, M. (2017). Your Roomba may be mapping your home, collecting data that could be shared. *New York Times* (25 July).

83. Rosenberg, R.P. (2018). Strava fitness app can reveal military sites, analysts say. *New York Times* (29 January).

84. Harwell, D. (2019). Doorbell-camera firm Ring has partnered with 400 police forces, extending surveillance concerns. *The Washington Post* (28 August).

85. Strom, S. (2015). Big companies pay later, squeezing their suppliers. *The New York Times* (6 April).

86. Segan, S. (2017). Inside Samsung's Galaxy S8 testing facility. *PC Magazine* (29 March). https://uk.pcmag.com/smartphones/88614/inside-samsungs-galaxy-s8-testing-facility (accessed 28 July 2020).

87. Sang-Hun, C. and Chen, B. (2016). Why Samsung abandoned its Galaxy Note 7 flagship phone. *New York Times* (11 October).

88. Tibken, S. (2017). CPSC urges better battery safety after Samsung's Note 7 fiasco. https://www.cnet.com/news/us-safety-agency-cpsc-battery-safety-samsung-galaxy-note-7 (accessed 28 July 2020).

89. Womack, J.P. (2007). *The Machine that Changed the World: The Story of Lean Production – Toyota's Secret Weapon in the Global Car Wars that is Revolutionizing World Industry*. New York: Free Press.

90. Birch, S. (2012). How activism forced Nike to change its ethical game. *The Guardian* (6 July).

91. Byrne, E.S. (2017). A clear correlation: Ethical companies outperform. Ethisphere (9 June), pp. 40–41.

92. Boothroyd, G., Dewhurst, P., and Knight, W.A. (2010). *Product Design for Manufacture and Assembly*. Boca Raton, FL: CRC Press.

93. Mizell, D.W. (1997). Virtual reality and augmented reality in aircraft design and manufacturing. In: *Frontiers of Engineering: Reports on Leading Edge Engineering from the 1996 NAE Symposium on Frontiers of Engineering*, National Academy of Engineering. The National Academies Press.

94. Nielsen. (2014). *Doing Well by Doing Good. Nielsen Annual Global Survey on Corporate Social Responsibility*. https://www.nielsen.com/wp-content/uploads/sites/3/2019/04/global-corporate-social-responsibility-report-june-2014.pdf (accessed 28 July 2020).

95. *Restriction of the Use of Certain Hazardous Substances in Electrical and Electronic equipment (RoHS)*, Directive 2011/65/EU of the European Parliament and of the Council of 8 June 2011.

96. *Fair Trade Certified*. https://www.fairtradecertified.org (accessed 28 July 2020).

97. Musk, E. (2018). Yes, excessive automation at Tesla was a mistake. To be precise, my mistake. Humans are underrated. https://twitter.com/elonmusk/status/984882630947753984?lang=en (accessed 28 July 2020).

98. SAE AS9102B. (2014). *Aerospace First Article Inspection Requirement*. SAE.

99. Munk, C.L., Nelson, P.E., and Strand, D.E. (2004). Assignee: Boeing. *Determinant wing assembly*. US patent. US20050116105A1.

100. Barboza, D. (2000). Firestone workers cite lax quality control. *New York Times* (15 September).

101. NIST Physical Measurement Laboratory (2019). *Calibration Procedures*. https://www.nist.gov/pml/weights-and-measures/laboratory-metrology/calibration-procedures (accessed 28 July 2020).

102. Mil-Std-105E *Military Standard: Sampling procedures and tables for inspection by attributes*. US Department of Defense.

103. ANSI/ASQ Z1.4-2003 (R2018) (2018). *Sampling Procedures and Tables for Inspection by Attributes*. American Society for Quality.

104. Montgomery, D.C. (1985). *Introduction to Statistical Quality Control*. New York: Wiley.

105. *Massachusetts Trusted Manufacturing Advisor*. https://massmep.org (accessed 28 July 2020).

106. NYSERDA – New York State Energy Research & Development Authority. https://www.nyserda.ny.gov (accessed 28 July 2020).

107. ISO (n.d.). *Certification*. https://www.iso.org/conformity-assessment.html (accessed 28 July 2020).

108. ISO/IEC 90003:2014 (2014). *Guidelines for the application of ISO 9001:2008 to computer software*. International Standards Organization.

109. ISO 14000:2015 (2015). *Environmental Management*. International Standards Organization.

110. ISO 45001:2018 (2018). *Occupational Health and Safety*. International Standards Organization.

111. ISO/IEC 27001:2013 (2013). *Information Technology – Security Techniques– Information Security Management Systems – Requirements*. International Standards Organization.

112. ISO/IEC 17025:2017 (2017). *General Requirements for the Competence of Testing and Calibration Laboratories*. International Standards Organization.

113. SA8000 (2014). *SA8000® Standard*. Social Accountability International.

114. Federico-O'Murchu, L. (2014). Why can't Apple meet demand for the iPhone 6? https://www.cnbc.com/2014/12/05/why-cant-apple-meet-demand-for-the-iphone-6.html (accessed 28 July 2020).

115. Statt, N. (2017). Snap lost nearly $40 million on unsold Spectacles. https://www.theverge.com/2017/11/7/16620718/snapchat-spectacles-40-million-lost-failure-unsold-inventory (accessed 28 July 2020).

116. Kubota, Y., Mochizuki, T., and Mickle, T. (2018). Apple suppliers suffer with uncertainty around iPhone demand. *Wall Street Journal* (19 November). https://www.wsj.com/articles/apple-suppliers-suffer-as-it-struggles-to-forecast-iphone-demand-1542618587 (accessed 28 July 2020).

117. Magretta, J. (1998). The power of virtual integration: an interview with Dell Computer's Michael Dell. *Harvard Business Review* 76 (2): 72–84.

118. Shah, A. (2017). PC prices will continue to go up due to SSD, DRAM, LCD shortages, Lenovo says. *PCWorld*. https://www.pcworld.com/article/3171366/pc-prices-will-continue-to-go-up-due-to-shortage-of-components.html (accessed 28 July 2020).

119. Wolfe, D. (2019). The global shortage of capacitors impacts all consumer electronics. https://qz.com/1575735/a-mlcc-shortage-is-stifling-electronics-hardware-auto-makers (accessed 28 July 2020).

120. US Customs and Border Protection (2018). Becoming a customs broker. https://www.cbp.gov/trade/programs-administration/customs-brokers/becoming-customs-broker (accessed 28 July 2020).

121. United States International Trade Commission. *Harmonized Tariff Schedule of the United States Revision 7*.

122. International Chamber of Commerce (2010). *Incoterms® 2010*. International Chamber of Commerce. https://icc-wbo.org/resources-for-business/incoterms-rules/incoterms-rules-2010/ (accessed 28 July 2020).

123. California Proposition 65 (1986). *Safe Drinking Water and Toxic Enforcement Act of 1986*. California Office of Environmental Health Hazard Analysis.

124. Underwriters Laboratory *UL Certification*. http://www.ul.com/certification (accessed 28 July 2020).

125. Equipment Authorization Procedures, 47 CFR §§ 2.901–2.1093 (2017).

126. Hazardous material regulations, 49 CFR Subchapter C. https://www.govinfo.gov/content/pkg/CFR-2012-title49-vol2/xml/CFR-2012-title49-vol2-subtitleB-chapI-subchapC.xml (accessed 28 July 2020).

127. IATA (2020). *61st Dangerous Goods Regulations*. International Air Transport Association.

128. Regulation No 1275/2008 (2008). *Directive 2005/32/EC of the European Parliament and of the Council with regard to ecodesign requirements for standby and off mode electric power consumption of electrical and electronic household and office equipment*. European Commission.

129. Directive 2012/19/EU (2012). *Waste Electrical and Electronic Equipment (WEEE)*. European Parliament and of the Council.

130. Bluetooth (2020). *Qualify your product*. https://www.bluetooth.com/develop-with-bluetooth/qualification-listing (accessed 28 July 2020).

131. Zigbee Alliance. *Get Certified*. https://zigbeealliance.org/certification/get-certified (accessed 28 July 2020).

132. Magnuson–Moss Warranty Act (P.L. 93-637), 15 U.S.C. § 2301 et seq. (1975).

133. O'Rourke, M. (2010). Tylenol's headache. *Risk Management* 57 (5): 8–9.

134. O'Rourke, M. (2010). Toyota's total recall. *Risk Management* 57 (3): 8, 10–11.

135. Birsch, D. (1994). *The Ford Pinto Case : A Study in Applied Ethics, Business, and Technology, SUNY Series, Case Studies in Applied Ethics, Technology, and Society*. Albany, NY: State University of New York Press.

136. Greyser, S.A. (1992). *Johnson & Johnson: The Tylenol Tragedy*. Harvard Business School Case 583-043.

137. Manley, M. (1987). Product liability: you're more exposed than you think. *Harvard Business Review* 65 (Sept.–Oct.): 28–40.

138. Insurance Information Institute (2019). *Facts + Statistics: Product liability*. https://www.iii.org/fact-statistic/facts-statistics-product-liability (accessed 28 July 2020).

139. Consumer Product Safety Commission (2016). Duty to report to CPSC: rights and responsibilities of businesses. https://www.cpsc.gov/Business--Manufacturing/Recall-Guidance/Duty-to-Report-to-the-CPSC-Your-Rights-and-Responsibilities (accessed 28 July 2020).

140. Consumer Product Safety Commission. Who we are - what we do for you. https://www.cpsc.gov/Safety-Education/Safety-Guides/General-Information/Who-We-Are---What-We-Do-for-You (accessed 28 July 2020).

141. US Food and Drug Administration (2018). Facts about the current good manufacturing practices (CGMPs). http://www.fda.gov/drugs/pharmaceutical-quality-resources/facts-about-current-good-manufacturing-practices-cgmps (accessed 28 July 2020).

142. National Association of Schools of Art and Design (2019). National Association of Schools of Art and Design: Handbook 2019-2020. https://nasad.arts-accredit.org/accreditation/standards-guidelines/handbook/ (accessed 28 July 2020).

INDEX